Methods in Cell Biology

Biomolecular Interactions Part A

Volume 166

Series Editors

Lorenzo Galluzzi
Weill Cornell Medical College,
New York, NY, United States

Methods in Cell Biology

Biomolecular Interactions Part A

Volume 166

Edited by

Arun K. Shukla
Biological Sciences and Bioengineering,
Indian Institute of Technology,
Kanpur, India

ELSEVIER

ACADEMIC PRESS
An imprint of Elsevier

Academic Press is an imprint of Elsevier
50 Hampshire Street, 5th Floor, Cambridge, MA 02139, United States
525 B Street, Suite 1650, San Diego, CA 92101, United States
The Boulevard, Langford Lane, Kidlington, Oxford OX5 1GB, United Kingdom
125 London Wall, London, EC2Y 5AS, United Kingdom

First edition 2021

ISBN: 978-0-12-823351-1
ISSN: 0091-679X

For information on all Academic Press publications
visit our website at https://www.elsevier.com/books-and-journals

Publisher: Zoe Kruze
Developmental Editor: Tara Nadera
Production Project Manager: Denny Mansingh
Cover Designer: Christian J. Bilbow

Typeset by STRAIVE, India

Contents

Contributors

Mariangela Agamennone
Department of Pharmacy, University "G. d'Annunzio" of Chieti-Pescara, Chieti, Italy

Tatsuya Aonuma
Department of Anatomy, Cell Biology and Physiology, Indiana University School of Medicine, Indianapolis, IN, United States

Wesley B. Asher
Department of Psychiatry, Columbia University; Division of Molecular Therapeutics, New York Psychiatric Institute; Department of Molecular Pharmacology and Therapeutics, Columbia University, Vagelos College of Physicians and Surgeons, New York, NY, United States

Nipuni P. Barupala
Department of Anatomy, Cell Biology and Physiology, Indiana University School of Medicine, Indianapolis, IN, United States

Sebastian Bayer
Leibniz-Institute for Food Systems Biology at the Technical University of Munich, Freising, Germany

Luca Borro
Department of Imaging, Advanced Cardiovascular Imaging Unit, Bambino Gesù Children's Hospital, IRCCS, Rome, Italy

Michel Bouvier
Department of Biochemistry and Molecular Medicine, Institute for Research in Immunology and Cancer, University of Montreal, Montreal, QC, Canada

Chino C. Cabalteja
National Institutes of Health, National Institute of Diabetes, Digestive, and Kidney Diseases (NIDDK), Laboratory of Bioorganic Chemistry, Bethesda, MD, United States

Tao Che
Department of Anesthesiology; Center for Clinical Pharmacology, University of Health Sciences and Pharmacy in St. Louis and Washington University School of Medicine, St. Louis, MO, United States

James W. Checco
Department of Chemistry; The Nebraska Center for Integrated Biomolecular Communication (NCIBC), University of Nebraska-Lincoln, Lincoln, NE, United States

Ross W. Cheloha
National Institutes of Health, National Institute of Diabetes, Digestive, and Kidney Diseases (NIDDK), Laboratory of Bioorganic Chemistry, Bethesda, MD, United States

Priya Chouhan
Division of Cell Biology and Immunology, CSIR-Institute of Microbial Technology (IMTECH), Chandigarh, India

Antonella Di Pizio
Leibniz-Institute for Food Systems Biology at the Technical University of Munich, Freising, Germany

Louis Dwomoh
The Centre for Translational Pharmacology, Institute of Molecular, Cell and Systems Biology, College of Medical, Veterinary and Life Sciences, University of Glasgow, Glasgow, United Kingdom

Amal El Daibani
Department of Anesthesiology, Washington University School of Medicine, St. Louis, MO, United States

Marialuigia Fantacuzzi
Department of Pharmacy, University "G. d'Annunzio" of Chieti-Pescara, Chieti, Italy

Celia R.S. Garcia
Department of Clinical and Toxicological Analysis, School of Pharmaceutical Sciences, University of São Paulo, São Paulo, Brazil

Natasha L. Grimsey
Department of Pharmacology and Clinical Pharmacology, School of Medical Sciences; Centre for Brain Research, Faculty of Medical and Health Sciences, University of Auckland; Maurice Wilkins Centre for Molecular Biodiscovery, Auckland, New Zealand

Shailendra Gupta
Department of Systems Biology and Bioinformatics, University of Rostock, Rostock, Germany

Matthew Harris
Department of Pharmacology, University of Cambridge, Cambridge, United Kingdom

Maria Hauge Pedersen
Department of Psychiatry, Columbia University, Vagelos College of Physicians and Surgeons; Division of Molecular Therapeutics, New York Psychiatric Institute, New York, NY, United States; NNF Center for Basic Metabolic Research, Section for Metabolic Receptology, Faculty of Health and Medical Sciences, University of Copenhagen, Copenhagen, Denmark

Kenneth A. Jacobson
Molecular Recognition Section, Laboratory of Bioorganic Chemistry, National Institute of Diabetes and Digestive and Kidney Diseases, National Institutes of Health, Bethesda, MD, United States

Shanu Jain
Molecular Recognition Section, Laboratory of Bioorganic Chemistry, National Institute of Diabetes and Digestive and Kidney Diseases, National Institutes of Health, Bethesda, MD, United States

Jonathan A. Javitch
Department of Psychiatry, Columbia University; Division of Molecular Therapeutics, New York Psychiatric Institute; Department of Molecular Pharmacology and Therapeutics, Columbia University, Vagelos College of Physicians and Surgeons, New York, NY, United States

Frédéric Gaëtan Jean-Alphonse
CNRS, IFCE, INRAE, Université de Tours, PRC, Nouzilly; Université Paris-Saclay, Inria, Inria Saclay-Île-de-France, Palaiseau, France

Satoshi Kawaguchi
Department of Anatomy, Cell Biology and Physiology, Indiana University School of Medicine, Indianapolis, IN, United States

Il-man Kim
Department of Anatomy, Cell Biology and Physiology; Krannert Institute of Cardiology; Wells Center for Pediatric Research, Indiana University School of Medicine, Indianapolis, IN, United States

Gaurav Kumar
Division of Cell Biology and Immunology, CSIR-Institute of Microbial Technology (IMTECH), Chandigarh, India

Ei Thet Htar Kyaw
Pharmacology and Biomolecular Science Graduate Program, Faculty of Pharmacy, Mahidol University, Bangkok, Thailand

Graham Ladds
Department of Pharmacology, University of Cambridge, Cambridge, United Kingdom

Corina T. Madreiter-Sokolowski
Gottfried Schatz Research Center, Molecular Biology and Biochemistry, Medical University of Graz, Graz, Austria

Bidisha Mandal
Division of Cell Biology and Immunology, CSIR-Institute of Microbial Technology (IMTECH), Chandigarh, India

Supachoke Mangmool
Department of Pharmacology, Faculty of Science, Mahidol University, Bangkok, Thailand

Bruno Moukette
Department of Anatomy, Cell Biology and Physiology, Indiana University School of Medicine, Indianapolis, IN, United States

Alessandro Nicoli
Leibniz-Institute for Food Systems Biology at the Technical University of Munich, Freising, Germany

Caitlin R.M. Oyagawa
Department of Pharmacology and Clinical Pharmacology, School of Medical Sciences; Centre for Brain Research, Faculty of Medical and Health Sciences, University of Auckland; Maurice Wilkins Centre for Molecular Biodiscovery, Auckland, New Zealand

Warisara Parichatikanond
Department of Pharmacology, Faculty of Pharmacy, Mahidol University, Bangkok, Thailand

Pedro H.S. Pereira
Department of Clinical and Toxicological Analysis, School of Pharmaceutical Sciences, University of São Paulo, São Paulo, Brazil; Department of Biochemistry and Molecular Medicine, Institute for Research in Immunology and Cancer, University of Montreal, Montreal, QC, Canada

Nicole A. Perry-Hauser
Department of Psychiatry, Columbia University, Vagelos College of Physicians and Surgeons; Division of Molecular Therapeutics, New York Psychiatric Institute, New York, NY, United States

Shivani Sachdev
National Institutes of Health, National Institute of Diabetes, Digestive, and Kidney Diseases (NIDDK), Laboratory of Bioorganic Chemistry, Bethesda, MD, United States

Veronica Salmaso
Molecular Recognition Section, Laboratory of Bioorganic Chemistry, National Institute of Diabetes and Digestive and Kidney Diseases, National Institutes of Health, Bethesda, MD, United States

Marisa Sepulveda
Department of Anatomy, Cell Biology and Physiology, Indiana University School of Medicine, Indianapolis, IN, United States

Abhishek Sharma
Division of Cell Biology and Immunology, CSIR-Institute of Microbial Technology (IMTECH), Chandigarh, India

Sheetal Sharma
Division of Cell Biology and Immunology, CSIR-Institute of Microbial Technology (IMTECH), Chandigarh, India

Sheryl Sharma
Department of Chemistry, University of Nebraska-Lincoln, Lincoln, NE, United States

Silvia Sposini
Department of Metabolism, Digestion and Reproduction, Institute of Reproductive and Developmental Biology, Imperial College London, London, United Kingdom; University of Bordeaux, CNRS, Interdisciplinary Institute for Neuroscience, Bordeaux, France

Anna Suchankova
Department of Pharmacology, University of Cambridge, Cambridge, United Kingdom

Amit Tuli
Division of Cell Biology and Immunology, CSIR-Institute of Microbial Technology (IMTECH), Chandigarh, India

Kshitiz Walia
Division of Cell Biology and Immunology, CSIR-Institute of Microbial Technology (IMTECH), Chandigarh, India

Verena Weber
Leibniz-Institute for Food Systems Biology at the Technical University of Munich, Freising, Germany

Preface

The interaction of biomolecules in cellular systems is arguably the most central aspect of cellular physiology and regulatory mechanisms. Every biological process involves integrated biomolecular interactions at some level leading to their precise activation, localization, and regulation. For example, DNA-protein interaction is the key in fundamental processes of DNA replication, transcription, and translation. Similarly, protein-protein interactions and protein-carbohydrate interactions are critical for the functional responses encoded in the cellular proteome. Probing biomolecular interactions is not only crucial for developing a better understanding of cellular processes, but it also constitutes an important target for designing better therapeutics for human diseases. Volumes 166 and 169 of *Methods in Cell Biology* include chapters on a diverse spectrum of topics in the area of biomolecular interactions encompassing various aspects of intra- and intermolecular interactions pertaining to nucleic acids, lipids, carbohydrates, and proteins. The chapters cover a broad range of methods including microscopy, coimmunoprecipitation, BRET, ligand binding assays, sensor design, protein engineering, reconstitution assays, and computational approaches.

I take this opportunity to thank all the authors for contributing important and interesting chapters to volumes 166 and 169 despite their busy schedule. I also express my sincere gratitude to the editorial staff and production team of *Methods in Cell Biology* for organizing the two volumes most efficiently and in a timely manner. I hope the readers find volumes 166 and 169 of *Methods in Cell Biology* focused on biomolecular interactions useful and timely, and I look forward to receiving comments and feedback.

Best wishes

Arun K. Shukla, PhD
Biological Sciences and Bioengineering
Indian Institute of Technology, Kanpur 208016, India
Email: arshukla@iitk.ac.in

CHAPTER

Measuring the rapid kinetics of receptor-ligand interactions in live cells using NanoBRET

1

Anna Suchankova, Matthew Harris, and Graham Ladds*

Department of Pharmacology, University of Cambridge, Cambridge, United Kingdom

**Corresponding author: e-mail address: grl30@cam.ac.uk*

Chapter outline

Abstract

The importance of receptor-ligand binding kinetics has often been overlooked during drug development, however, over the past decade it has become increasingly clear that a better understanding of the kinetic parameters is crucial for fully evaluating pharmacological effects of a drug. One technique enabling us to measure the real-time kinetics of receptor-ligand interactions in live cells is NanoBRET, which is a bioluminescence resonance energy transfer (BRET)-based assay that uses Nano luciferase. The assay described here allows the

Methods in Cell Biology, Volume 166, ISSN 0091-679X, https://doi.org/10.1016/bs.mcb.2021.06.013

measurement of kinetic parameters of a fluorescent ligand and an unlabeled ligand binding to the same place at the receptor, as well as monitoring the effects of another compound like an allosteric modulator on the ligand binding.

1 Introduction

Receptor-ligand interactions play a fundamental role in many biological processes in living cells, including drugs' interactions with their targets, and are a crucial factor to consider during drug development. In the past decades, drugs have been failing clinical trials due to a lack of sufficient in vivo efficacy (Waring et al., 2015) and the optimization has been focused on developing candidates with high affinity and selectivity for its target, with these characteristics often being measured at equilibrium in cell systems over-expressing the target receptor. As multiple approved drugs currently on the market show non-equilibrium binding characteristics (Schuetz et al., 2017), it is now becoming clear that measuring the binding properties at equilibrium might not be optimal for estimating in vivo efficacy and more detailed comprehension of the kinetics of association and dissociation of a receptor-ligand complex is needed to evaluate the full pharmacological effect of a drug and its mode of action. This deeper understanding will then help with selecting better compounds for further profiling.

There is a variety of different ways to detect receptor-ligand interactions and indirect second messenger assays like cAMP accumulation or calcium mobilization at G protein-coupled receptors (GPCRs) are widely used. However, to be able to measure detailed kinetic parameters, techniques looking more directly at the receptor-ligand complex need to be used. Historically, radioligand binding assays were employed for this purpose, but these carry multiple drawbacks including impractical and hazardous use of radioactivity, high cost, and often measurement at a non-physiological temperature of 4 °C (Zwier et al., 2010). Alternative ways of measuring receptor-ligand interactions would then include, for example, atomic force microscopy used to carry out force-clamp measurements on receptor-ligand bonds (Rico, Chu, & Moy, 2011) or nuclear magnetic resonance (NMR) that relies on a comparison of NMR parameters of the free and bound states of the molecules (Cala, Guillière, & Krimm, 2014). Another rather popular method is surface plasmon resonance (SPR) spectroscopy, which is a rapidly developing technique for the study of ligand binding interactions with membrane receptors (Patching, 2014). Its advantage lies in not requiring any labels and being capable of measuring real-time binding and kinetic information. However, the requirement of purified protein can make its use on membrane receptors like GPCRs challenging. Techniques that do not require receptor purification include those using radiolabeled ligands, mentioned above, or fluorescently labeled ligands. In contrast to radiolabeled ligands, fluorescent labels are more stable and safer to use and can be utilized in Förster resonance energy transfer (FRET) assays, which are an extremely powerful technique that can determine if two fluorophore-tagged molecules are within a certain distance of each other and therefore identify molecular interactions (Ergin, Dogan, Parmaksiz, Elçin, & Elçin, 2016). Being the first, FRET has set a precedent for the applicability of resonance energy transfer approaches for the measurement of molecule interactions, until it was in many ways surpassed by bioluminescence resonance energy transfer (BRET) assays.

BRET has been used extensively in the past years for monitoring protein-protein interactions. For example, Hager, Johnson, Wootten, Sexton, and Gellman (2016) used Rluc8 luciferase-tagged glucagon-like peptide 1 receptor (GLP-1R) and green fluorescent protein (GFP) tagged β-arrestin 1/2 to assess recruitment of β-arrestin 1 and β-arrestin 2 to the GLP-1R in response to GLP-1 analogs, whereas Namkung et al. (2016) has used BRET sensors for monitoring of GPCR and β-arrestin trafficking in live cells. BRET technology has also been extensively used for quantitative analysis of hetero- and homo-dimerization of GPCRs like the exploration of constitutive dimerization of human serotonin 5-HT4 receptors in living cells by Berthouze et al. (2005). Because of the advantage of not requiring a light source and thereby eliminating the problems of photobleaching and autofluorescence in FRET assays, BRET has become the proximity assay of choice for many researchers (Koterba & Rowan, 2006; Stoddart et al., 2015). In 2015, Machleidt et al. (2015) introduced a new BRET methodology called NanoBRET utilizing a Nanoluciferase (Nluc), which was in the same year adapted by Stoddart et al. (2015) for monitoring ligand binding to GPCRs using fluorescently-labeled ligand and Nanoluciferase-tagged receptor. The adaption of the technique for measuring receptor-ligand interactions has been made possible by the development of many different fluorescent agonists and antagonists and can be adapted to both GPCR and non-GPCR receptors. NanoBRET has been successfully utilized, for example, by Bouzo-Lorenzo et al. (2019) to measure the kinetic parameters of adenosine A3 receptor (A_3R) antagonists binding at physiological temperatures, by Deganutti et al. (2020) to decipher the agonist binding mechanism to the adenosine A1 receptor (A_1R) or by Stoddart et al. (2018) to study ligand-binding kinetics of novel fluorescent histamine H1 receptor antagonists in live cells. Outside of the GPCR receptor family, Peach et al. (2018) used BRET to evaluate Fluorescent VEGF-A isoforms for their ability to discriminate between VEGFR2 and NRP1 in real-time ligand binding studies in live cells.

There are three major requirements for successful implementation of NanoBRET for measuring receptor-ligand interactions: (1) receptors need to be tagged with a Nanoluciferase in a way that does not compromise their function, (2) availability of an appropriate fluorescent ligand, and (3) access to an appropriate instrument to monitor energy transfer, most often a microplate reader that can measure in real-time using live cells (Pfleger & Eidne, 2006; Stoddart et al., 2015). NanoLuc is a small 19 kDa luciferase subunit derived from a larger multi-component luciferase found in deep sea shrimp, Oplophorus, that uses an imidazopyrazinone substrate furimazine (Hall, Kozakov, & Vajda, 2012). It is its small size and superior luminescence profile that has led to its preferential use over some other luciferases like

Rluc (Dale, Johnstone, White, & Pfleger, 2019). In the protocol below, NanoLuc fused to the N-termini of gastric inhibitory polypeptide receptor (Nluc-GIPR) is used as an example, accompanied by a Tag-lite® GIPR red agonist, which is a GIP derivative labeled with a red emitting HTRF (Homogeneous Time Resolved Fluorescence) fluorescent probe sold by Cisbio (PerkinElmer Inc.). BRET-ligand binding assays work by exploiting the naturally occurring phenomenon of dipole-dipole energy transfer from a bioluminescent donor (NanoLuc-tagged receptor) to an acceptor (fluorescent ligand) following enzyme-mediated oxidation of furimazine (Fig. 1). Since this energy transfer only happens when the receptor and ligand are in close proximity (<10nM), BRET can be used to measure the kinetics of receptor-ligand association or dissociation (Dale et al., 2019; Koterba & Rowan, 2006; Pfleger & Eidne, 2006; Stoddart et al., 2015). It also means that separation of free and bound fluorescent ligand is not required due to the exceptional distance-dependence of BRET resulting in no wash steps, which makes BRET assays very user friendly (Stoddart et al., 2015).

Probably the biggest challenge of using NanoBRET for measuring receptor-ligand interactions is that BRET can only directly measure the kinetics of a fluorescent ligand. To overcome this, NanoBRET can be used to measure the association kinetics of a fluorescent ligand in the presence of non-fluorescent ligand and through the knowledge of fluorescent ligand kinetics, both the association and dissociation rate constants of the unlabeled ligand can be calculated (Motulsky & Mahan, 1984; Stoddart et al., 2015; Sykes, Stoddart, Kilpatrick, & Hill, 2019). This, coupled with the ease of use and homogeneity of the assay, makes NanoBRET a very desirable technique for monitoring interactions between receptors and orthosteric ligands. However, BRET might not be as suitable for allosteric ligands as a fluorescent variant is often not available. Here, we describe the steps required to measure kinetic parameters of receptor-ligand interactions using GIPR-GIP as an example.

2 Materials

1. HEK 293T cells (gifted by Prof. David Poyner).
2. Dulbecco's modified Eagle's medium (DMEM)/F12 (Life Technologies, 31331028) supplemented with 10% heat-inactivated FBS (Sigma-Aldrich, F9665) and 1% antibiotic antimycotic (AA) solution (Scientific Laboratory Supplies, A5955). This is referred to as "complete medium" in the protocol below.
3. Phosphate buffered saline (PBS) was made using tablets (Sigma-Aldrich, P4417) dissolved in distilled H_2O and autoclaved. One tablet was dissolved in 200mL to produce a solution consisting of 2.7mM KCl, 0.137M NaCl and 0.1M phosphate buffer at a pH of 7.4.
4. Trypsin-EDTA (0.05%) (Life Technologies, 25300054).
5. Dimethylsulfoxide (DMSO) (Chem Cruz, sc-358801).

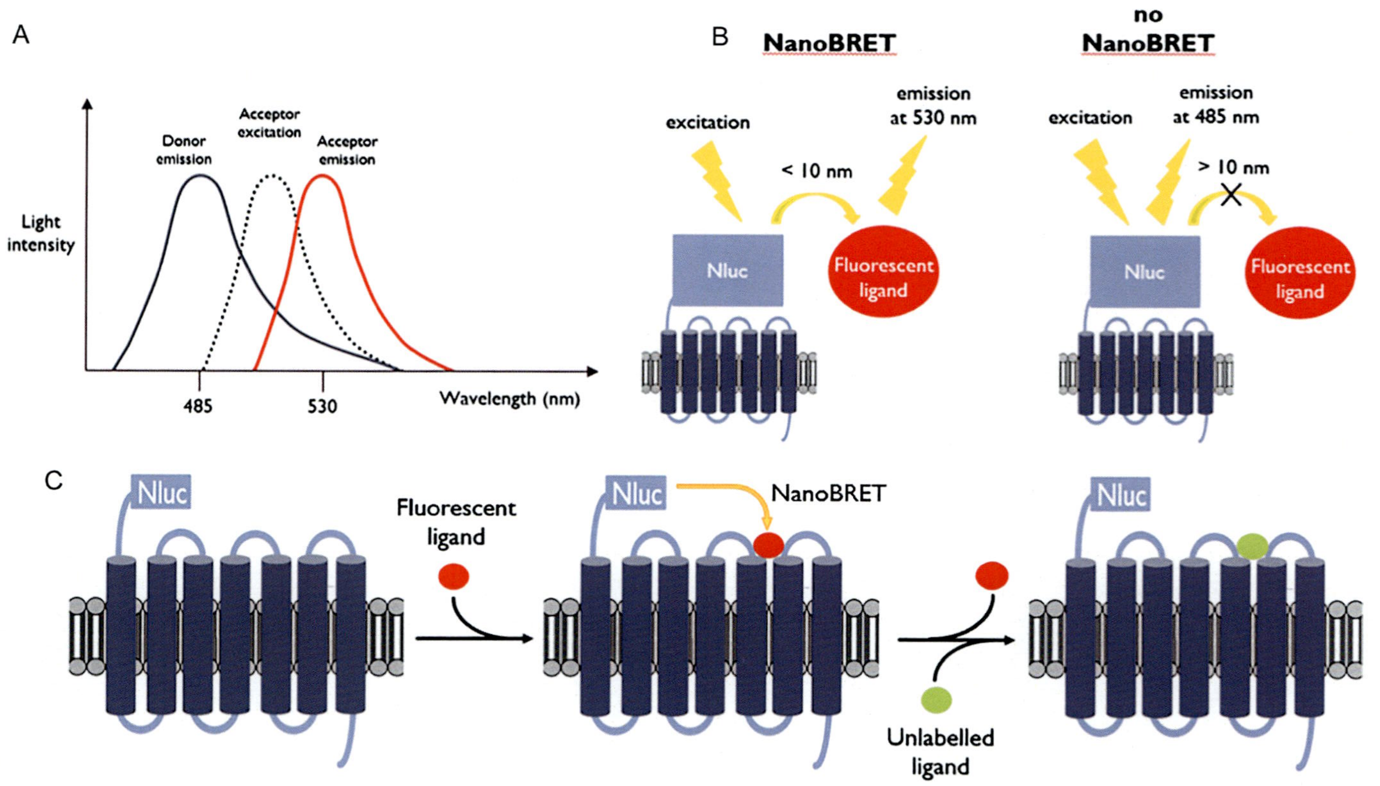

FIG. 1

The principle of the NanoBRET assay for fluorescent ligand binding. (A) A simplified diagram of the overlapping light spectra of donor (Nluc) emission and acceptor (Tag-lite® GIPR Red Agonist) excitation that enables a successful resonance energy transfer. (B) NanoBRET relies on the energy transfer from an excited luciferase-tagged receptor donor (Nluc) to a suitable fluorescent ligand acceptor (Tag-lite® GIPR Red Agonist). If the donor and acceptor are in close proximity (less than 10 nM), NanoBRET takes place and emission at 530 nm occurs. On the other hand, if no suitable acceptor is in close proximity, the emission at 485 nm occurs. (C) A diagram of the association and dissociation experiment. Addition of the fluorescent ligand causes it to bind the luciferase-tagged receptor and NanoBRET to occur. When an unlabeled ligand is added, it outcompetes the fluorescent ligand from its binding place at the receptor, initiating the dissociation phase of the receptor-ligand binding kinetics.

6. 25 cm^2 Rectangular Canted Neck Cell Culture Flask with Vent Cap (Corning, 430639). This is referred to as "T25 flask" in the protocol below.
7. 6-well culture plates (Appleton Woods, CC010).
8. White 96-well plates (Perkin Elmer, 6005688).
9. 2 mL tube (VWR international, 211-2165).
10. 15 mL tube (Sarsedt, 62.554.502).
11. Poly-L-lysine (0.01%) (Merck, p4707).
12. Polyethylenimine (PEI, Polyscience Inc., 23966).
13. Nano-Glo luciferase assay substrate (Promega, N1110). Store at −20 °C.
14. Tag-lite® GIPR Red Agonist (Cisbio, 40244). Store at −20 °C.
15. GIP (1-42) (Bachem, 4030658) made to 1 mM stocks in water containing 0.1% bovine serum albumin (BSA) (Sigma-Aldrich).
16. Compound X (Enamine) made to 10 mM stocks in DMSO.
17. Mithras LB 940 multimode microplate reader (Berthold Technologies).
18. pcDNA3.1-Nluc-GIPR (made by S. Carvalho, University of Cambridge).

3 Methods

The following protocol describes how to measure the kinetic parameters of receptor-ligand interactions using NanoBRET. As an example, a luciferase tagged GPCR receptor GIPR (Nluc-GIPR) and a fluorescent derivative of its natural agonist GIP (Tag-lite® GIPR Red Agonist) are used, but the protocol can be adapted for non-GPCR use. The method consists of five parts described in detail below: (1) cell culture, (2) transfections, (3) harvesting cells, (4) NanoBRET measurement, and (5) data analysis.

3.1 Cell culture

1. The HEK 293T cells are routinely grown in complete media in T25 flasks at 37 °C with 5% CO_2 (see Note 1).
2. At full cell confluency, aspirate the media from the flask, add 2 mL trypsin-EDTA (0.05%) to the flask well and incubate at 37 °C with 5% CO_2 (see Note 2).
3. After 5 min, add 3 mL complete media and pipet up and down to dislodge the cells from the flask and ensure homogeneous mixture of the media with cells (see Note 3).
4. Collect the media with cells into a sterile 15 mL tube and centrifuge the cells at 1400 rpm for 4 min.
5. Discard the supernatant and resuspend the cells in 5 mL complete media.
6. Count the cells using a hemocytometer.
7. Seed 1 000 000 HEK 293T cells per well of a 6-well culture plate in a final volume of 2 mL per well and incubate the cells at 37 °C with 5% CO_2 for approximately 24 h.

3.2 Transfection of cells

1. When the cells are at approximately 80% confluency, aspirate the medium from the well and replace it with 2 mL of fresh complete media. Return the cells to 37 °C with 5% CO_2, while the next steps to prepare the transfection reagents are performed.
2. In a first 2 mL tube mix 1.5 µg NLuc-GIPR and 150 mM NaCl to make up total volume of 100 µL (see Note 4).
3. In a second 2 mL tube mix 9 µL PEI and 528 µL 150 mM NaCl, again making a total volume of 100 µL.
4. Mix the contents of each tube by pipetting and incubate them for 5 min at room temperature.
5. Combine the contents of both tubes, mix the transfection mixture by pipetting and incubate for 10 min at room temperature.
6. Add the transfection mixture dropwise into a single well. Mix by gently rocking the wells and incubate the cells at 37 °C with 5% CO_2.

3.3 Harvesting cells

1. Approximately 24 h after transfections, aspirate the medium from each well, add 400 µL trypsin-EDTA (0.05%) to each well and incubate at 37 °C with 5% CO_2 for 5 min.
2. Add 1.6 mL complete media per well and pipet up and down to dislodge the cells from the plate and ensure homogeneous mixture of media with cells (see Note 5).
3. Collect the media with cells from all the 6 wells into a sterile 15 mL falcon and centrifuge the cells at 1400 rpm for 4 min.
4. Discard the supernatant and resuspend the cells in 12 mL complete media.
5. Count the cells using a hemocytometer.
6. Seed 50 000 HEK 293T cells per well of poly-L-lysine coated white 96-well plates in a final volume of 100 µL per well and incubate the cells at 37 °C with 5% CO_2 for 24 h (see Note 6).

3.4 NanoBRET measurement

The NanoBRET can be applied to receptor-ligand interactions in three different assay settings. Firstly, it can be used to measure the association kinetic parameters of the fluorescent ligand (e.g., Tag-lite® GIPR Red Agonist) itself (Section 3.4.1), which is also necessary for the determination of a suitable concentration of the fluorescent ligand for the following two applications. Secondly, NanoBRET can be utilized to measure both the association and dissociation kinetics of a fluorescent ligand (Section 3.4.2). Thirdly, it can be used to measure the effect of another compound (e.g., an allosteric modulator Compound X) on the fluorescent ligand kinetics or determine the K_i of an unlabeled ligand (Section 3.4.3).

3.4.1 Association studies

1. Aspirate the media from the cells, wash the cells with 100 µL PBS and add 80 µL PBS, containing 0.49 mM $MgCl_2 \cdot 6H_2O$, 0.9 mM $CaCl_2 \cdot 2H_2O$ and 0.1% BSA to each well.
2. Prepare the working substrate solution by diluting the Nano-Glo® Substrate 1:4000 in PBS, containing 0.49 mM $MgCl_2 \cdot 6H_2O$, 0.9 mM $CaCl_2 \cdot 2H_2O$ and 0.1% BSA.
3. Add 10 µL of the Nano-Glo® Substrate solution to each well (see Note 7) and incubate in the dark for 5 min at room temperature.
4. Add 10 µL H_2O or 10 × concentrated Tag-lite® GIPR Red Agonist diluted in PBS, containing 0.49 mM $MgCl_2 \cdot 6H_2O$, 0.9 mM $CaCl_2 \cdot 2H_2O$ and 0.1% BSA to each well to create final concentrations of 0, 0.1, 0.5, 1, 2, 5 10 and 20 nM (see Note 8).
5. Measure luminescence at both 460 m and 530 nm every 30 s for 7 min in a plate reader capable of detecting dual emissions (see Note 9).

3.4.2 Association and dissociation studies

1. Rinse the tubing and injector first with excess sterile deionized H_2O, unload the injector and then prime it by pumping through 1 µM unlabeled GIP (1-42) solution through the tubing. Set up the plate reader software to inject 10 µL of this agonist solution per well.
2. Aspirate the media from the cells, wash the cells with 100 µL PBS and add 80 µL PBS, containing 0.49 mM $MgCl_2 \cdot 6H_2O$, 0.9 mM $CaCl_2 \cdot 2H_2O$ and 0.1% BSA to each well.
3. Prepare the working substrate solution by diluting the Nano-Glo® Substrate at 1:4000 in PBS, containing 0.49 mM $MgCl_2 \cdot 6H_2O$, 0.9 mM $CaCl_2 \cdot 2H_2O$ and 0.1% BSA.
4. Add 10 µL of the Nano-Glo® Substrate solution to each well and incubate in the dark for 5 min at room temperature.
5. Add 10 µL 100 nM Tag-lite® GIPR Red Agonist to each well to create final concentrations of 10 nM.
6. Measure luminescence at both 460 nm and 530 nm every 30 s in a plate reader capable of detecting dual emissions (see Note 9). At 6.5 min a saturating concentration of unlabeled GIP (1-42) (1 µM) is injected and luminescence is measured every 30 s for a further 10.5 min. Repeat the NanoBRET kinetics measurement individually for each condition.

3.4.3 Compound studies

1. Rinse the tubing and injector first with excess sterile deionized H_2O, unload the injector and then prime it by pumping through 1 µM unlabeled GIP (1-42) solution through the tubing. Set up the plate reader software to inject 10 µL of this agonist solution per well.

2. Aspirate the media from the cells, wash the cells with 100 μL PBS and add 100 μL of 100 μM Compound X diluted in PBS, containing 0.49 mM $MgCl_2 \cdot 6H_2O$, 0.9 mM $CaCl_2 \cdot 2H_2O$ and 0.1%. BSA (see Note 10).
3. After 15 min, aspirate the media from the cells, wash the cells with 100 μL PBS and add 80 μL PBS, containing 0.49 mM $MgCl_2 \cdot 6H_2O$, 0.9 mM $CaCl_2 \cdot 2H_2O$ and 0.1% BSA to each well.
4. Prepare the working substrate solution by diluting the Nano-Glo® Substrate at 1:4000 in PBS, containing 0.49 mM $MgCl_2 \cdot 6H_2O$, 0.9 mM $CaCl_2 \cdot 2H_2O$ and 0.1% BSA.
5. Add 10 μL of the Nano-Glo® Substrate solution to each well and incubate in the dark for 5 min at room temperature.
6. Add 10 μL 50 nM concentrated Tag-lite® GIPR Red Agonist to each well to create final concentrations of 5 nM (see Note 11).
7. Measure luminescence at both 460 nm and 530 nm every 30 s in a plate reader capable of detecting dual emissions. At 6.5 min unlabeled 1 μM GIP (1-42) is injected and luminescence is measured for every 30 s for a further 10.5 min (see Note 12). Repeat the BRET kinetics measurement individually for each condition (Note 15).

3.5 Data analysis

1. To determine the BRET ratio for each specific time point, divide the 530 nm emission value of the acceptor (Tag-lite® GIPR Red Agonist) by the 460 nm emission value of the donor (Nanoluciferase). The BRET signal is then calculated by subtracting the BRET ratio of H_2O treated cells from Tag-lite® GIPR Red Agonist treated cells (see Note 13).
2. For the association ligand binding experiments, a "One-phase association" equation from GraphPad Prism 8.4.3 was used to determine the association rate constant of the fluorescent ligand (Eq. 1).

$$y = y_0 + (plateau - y_0)\left(1 - e^{-k_{on}x}\right) \tag{1}$$

where y is the BRET signal, y_0 is the initial BRET signal, plateau is the BRET signal at infinite times, k_{on} is the association rate constant and x is time (see Note 14). The total binding (maximum values for each concentration of the fluorescent ligand from Eq. 1) was then analyzed using "One site—Total binding" equation from GraphPad Prism 8.4.3 to determine the equilibrium dissociation constant of the fluorescent ligand (Eq. 2).

$$y = \frac{B_{max}x}{K_d + x} + NSx \tag{2}$$

where y is the BRET signal, B_{max} is the maximum specific binding, x is fluorescent ligand concentration, K_d is the equilibrium dissociation constant and NS is the slope of nonspecific binding (see Note 14). Finally, using the Eq. 3 below, the dissociation rate constant of the fluorescent ligand can be calculated.

$$K_d = \frac{k_{off}}{k_{on}} \tag{3}$$

where K_d is the equilibrium dissociation constant, k_{on} is the association rate constant and k_{off} is the dissociation rate constant.

3. For the association and dissociation experiments, the non-specific binding was determined by the Tag-lite® GIPR Red Agonist displacement with 1 μM GIP (1-42) and the data was fitted to the "Association the dissociation" equation from GraphPad Prism 8.4.3 to determine the k_{on} and k_{off} for the fluorescent ligand (Eq. 4).

$$\begin{aligned} &\text{If } x \le \text{Time}_0 \quad y = \frac{B_{max}[L]}{[L] + \frac{k_{off}}{k_{on}}}\left(1 - e^{-[L]k_{on}k_{off}x}\right) + NS \\ &\text{If } x \ge \text{Time}_0 \quad y = \frac{B_{max}[L]}{[L] + \frac{k_{off}}{k_{on}}}\left(1 - e^{-[L]k_{on}k_{off}x}\right)e^{-k_{off}(x - Time_0)} + NS \end{aligned} \tag{4}$$

where y is the BRET signal, B_{max} is the maximum specific binding, [L] is the concentration of the fluorescent ligand, k_{on} is the association rate constant of the fluorescent ligand and k_{off} is the dissociation rate constant of the fluorescent ligand, x is time, NS is the nonspecific binding of the fluorescent ligand and Time_0 is the time at which unlabeled ligand is injected to initiate dissociation (see Note 14). K_d for the fluorescent ligand can then be calculated using Eq. 3.

4. For the compound studies, the analysis is the same as for the association and dissociation experiments above.

4 Notes

1. When growing the cells and during cell culture, transfection of cells and harvesting cells steps, work under sterile conditions in a laminar flow hood.
2. When aspirating the media, do it from the corner of the T25 flask and be careful not to disturb the cells.
3. The incubation time can vary from 5 to 10 min, so check under the microscope that the cells have detached. Alternatively, you can also tap against the flask with the palm of your hand to help dislodge the cells.
4. Transfections are performed with PEI at a 6:1 ratio of reagent to DNA (v/w) and a total amount of 1.5 μg DNA per well of a 6-well plate is used.
5. Avoid frothing the media when pipetting up and down.
6. To coat the white 96-well plates with poly-L-lysine, add ~40 μL poly-L-lysine to each well and leave the plate for 1 h at room temperature. Then take off the

poly-L-lysine and leave the plate to air-dry overnight at room temperature under sterile conditions.

7. The final Nano-Glo® Substrate concentration in each well should be 0.1 μM.
8. It is necessary to first establish a saturation isotherm of the fluorescent ligand. Due to Tag-lite® GIPR Red Agonist being supplied as 100 μL at a concentration of 7.2 μM, we could not test a higher concentration than 20 nM. As the Tag-lite® GIPR Red Agonist in the range of 0.1–20 nM resulted in NanoBRET between the Nluc-GIPR and the Tag-lite® GIPR Red Agonist approaching saturation, we selected the concentration of 10 nM for further association and dissociation experiments.
9. We use Mithras LB 940 multimode microplate reader, but a similar plate reader capable of reading the NanoBRET assay at desired wavelengths can be used. The assays are measured at a constant temperature of 30 °C to minimize possible variations from day to day. It is possible to set a different temperature (e.g., 37 °C), if desired.
10. Compound X is a GIPR positive allosteric modulator that binds to an allosteric site distinct from the orthosteric site, where GIP (1-42) binds.
11. To maximize the window to detect any effect of Compound X on the Tag-lite® GIPR Red Agonist kinetic parameters, a lower concentration of 5 nM Tag-lite® GIPR Red Agonist was used (compared to 10 nM above) to slow the association phase so that increases can more easily be detected.
12. To ensure that the NanoBRET signal of Tag-lite® GIPR Red Agonist had plateaued before the injection of the unlabeled GIP (1-42), a 6.5 min time gap was used.
13. H_2O treated cells have no Tag-lite® GIPR Red Agonist present and therefore any BRET signal measured there is a background signal.
14. In Fig. 2 below, an example of the GraphPad Prism analysis is shown using the Tag-lite® GIPR Red Agonist binding at GIPR example. The K_d of the Tag-lite® GIPR Red Agonist was measured to be 4.77 ± 0.57 nM.
15. Alternatively, different concentrations of the unlabeled ligand can be tested to determine affinity (pK_i) values for the unlabeled agonist using the maximum total binding for ligand-induced NanoBRET and fitting it with the "one-site—Fit Ki" equation in GraphPad Prism 8.4.3 (Eq. 5) taking into account the K_d of the fluorescent ligand as was done, for example, by Deganutti et al. (2020).

$$y = bottom + \frac{(top - bottom)}{1 + 10^{([X] - \log EC_{50})}}, \log EC_{50} = \log\left(10^{\log K_i}\left(1 + \frac{[L]}{K_d}\right)\right) \quad (5)$$

where EC_{50} is the half maximal effective concentration, K_i is the equilibrium dissociation constant of unlabeled ligand, [L] is the concentration of the fluorescent ligand, K_d is the equilibrium dissociation constant of the fluorescent ligand, top and bottom are plateaus of the curve and [X] is the concentration of the unlabeled ligand.

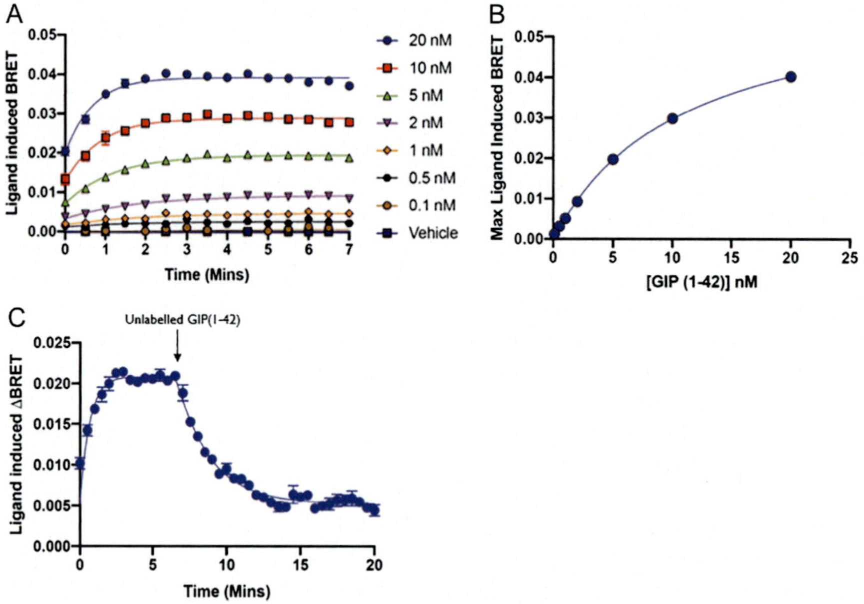

FIG. 2

Tag-lite® GIPR Red Agonist binding at GIPR has K_d of 4.77±0.57nM. (A) HEK 293T cells transfected with Nluc-GIPR were stimulated by increasing concentrations of Tag-lite® GIPR Red Agonist. Ligand binding was measured by a Tag-lite® GIPR Red Agonist-induced increase in NanoBRET. (B) Maximum total binding for ligand-induced NanoBRET for each concentration of Tag-lite® GIPR Red Agonist from A. (C) Association of 5nM Tag-lite® GIPR Red Agonist was measured in HEK 293T cells expressing Nluc-GIPR, followed by displacement with 1mM GIP (1-42) after 6.5min. Data were fitted using the association followed by dissociation model in GraphPad Prism 8.4.3 and are expressed with non-specific binding subtracted. Data are expressed as the mean±SEM of four individual experiments performed in triplicate.

Acknowledgments

This work was supported by Cambridge Trust European Scholarship (A.S). M.H. and G.L. were supported by an MRC confidence in concept award (MC_PC_17156).

References

Berthouze, M., Ayoub, M., Russo, O., Rivail, L., Sicsic, S., Fischmeister, R., et al. (2005). Constitutive dimerization of human serotonin 5-HT4 receptors in living cells. *FEBS Letters*, *579*, 2973–2980.

Bouzo-Lorenzo, M., Stoddart, L. A., Xia, L., IJzerman, A. P., Heitman, L. H., Briddon, S. J., et al. (2019). A live cell NanoBRET binding assay allows the study of ligand-binding kinetics to the adenosine A3 receptor. *Purinergic Signal*, *15*, 139–153.

Cala, O., Guillière, F., & Krimm, I. (2014). NMR-based analysis of protein-ligand interactions. *Analytical and Bioanalytical Chemistry*, *406*, 943–956.

Dale, N. C., Johnstone, E. K. M., White, C. W., & Pfleger, K. D. G. (2019). NanoBRET: The bright future of proximity-based assays. *Frontiers in Bioengineering and Biotechnology*, *7*, 56.

Deganutti, G., Barkan, K., Preti, B., Leuenberger, M., Wall, M., Frenguelli, B., et al. (2020). *Deciphering the agonist binding mechanism to the adenosine A1 receptor*. BioRxiv. 2020.10.22.350827.

Ergin, E., Dogan, A., Parmaksiz, M., Elçin, A. E., & Elçin, Y. M. (2016). Time-resolved fluorescence resonance energy transfer [TR-FRET] assays for biochemical processes. *Current Pharmaceutical Biotechnology*, *17*, 1222–1230.

Hager, M. V., Johnson, L. M., Wootten, D., Sexton, P. M., & Gellman, S. H. (2016). β-Arrestin-biased agonists of the GLP-1 receptor from β-amino acid residue incorporation into GLP-1 analogues. *Journal of the American Chemical Society*, *138*, 14970–14979.

Hall, D. R., Kozakov, D., & Vajda, S. (2012). Analysis of protein binding sites by computational solvent mapping. *Methods in Molecular Biology*, *819*, 13–27.

Koterba, K. L., & Rowan, B. G. (2006). Measuring ligand-dependent and ligand-independent interactions between nuclear receptors and associated proteins using Bioluminescence Resonance Energy Transfer (BRET). *Nuclear Receptor Signaling*, *4*, e021.

Machleidt, T., Woodroofe, C. C., Schwinn, M. K., Méndez, J., Robers, M. B., Zimmerman, K., et al. (2015). NanoBRET—A novel BRET platform for the analysis of protein-protein interactions. *ACS Chemical Biology*, *10*, 1797–1804.

Motulsky, H. J., & Mahan, L. C. (1984). The kinetics of competitive radioligand binding predicted by the law of mass action. *Molecular Pharmacology*, *25*, 1–9.

Namkung, Y., Le Gouill, C., Lukashova, V., Kobayashi, H., Hogue, M., Khoury, E., et al. (2016). Monitoring G protein-coupled receptor and β-arrestin trafficking in live cells using enhanced bystander BRET. *Nature Communications*, *7*, 12178.

Patching, S. G. (2014). Surface plasmon resonance spectroscopy for characterisation of membrane protein–ligand interactions and its potential for drug discovery. *Biochimica et Biophysica Acta (BBA)—Biomembranes*, *1838*, 43–55.

Peach, C. J., Kilpatrick, L. E., Friedman-Ohana, R., Zimmerman, K., Robers, M. B., Wood, K. V., et al. (2018). Real-time ligand binding of fluorescent VEGF-A isoforms that discriminate between VEGFR2 and NRP1 in living cells. *Cell Chemical Biology*, *25*, 1208–1218.e5.

Pfleger, K. D. G., & Eidne, K. A. (2006). Illuminating insights into protein-protein interactions using bioluminescence resonance energy transfer (BRET). *Nature Methods*, *3*, 165–174.

Rico, F., Chu, C., & Moy, V. T. (2011). Force-clamp measurements of receptor-ligand interactions. *Methods in Molecular Biology*, *736*, 331–353.

Schuetz, D. A., de Witte, W. E. A., Wong, Y. C., Knasmueller, B., Richter, L., Kokh, D. B., et al. (2017). Kinetics for drug discovery: An industry-driven effort to target drug residence time. *Drug Discovery Today*, *22*, 896–911.

Stoddart, L. A., Johnstone, E. K. M., Wheal, A. J., Goulding, J., Robers, M. B., Machleidt, T., et al. (2015). Application of BRET to monitor ligand binding to GPCRs. *Nature Methods*, *12*, 661–663.

Stoddart, L. A., Vernall, A. J., Bouzo-Lorenzo, M., Bosma, R., Kooistra, A. J., de Graaf, C., et al. (2018). Development of novel fluorescent histamine H1-receptor antagonists to study ligand-binding kinetics in living cells. *Scientific Reports*, *8*, 1572.

Sykes, D. A., Stoddart, L. A., Kilpatrick, L. E., & Hill, S. J. (2019). Binding kinetics of ligands acting at GPCRs. *Molecular and Cellular Endocrinology*, *485*, 9–19.

Waring, M. J., Arrowsmith, J., Leach, A. R., Leeson, P. D., Mandrell, S., Owen, R. M., et al. (2015). An analysis of the attrition of drug candidates from four major pharmaceutical companies. *Nature Reviews Drug Discovery*, *14*, 475–486.

Zwier, J. M., Roux, T., Cottet, M., Durroux, T., Douzon, S., Bdioui, S., et al. (2010). A fluorescent ligand-binding alternative using Tag-lite® technology. *Journal of Biomolecular Screening*, *15*, 1248–1259.

CHAPTER

Evaluating functional ligand-GPCR interactions in cell-based assays

2

Sheryl Sharma[a] and James W. Checco[a,b,*]

[a]*Department of Chemistry, University of Nebraska-Lincoln, Lincoln, NE, United States*

[b]*The Nebraska Center for Integrated Biomolecular Communication (NCIBC), University of Nebraska-Lincoln, Lincoln, NE, United States*

**Corresponding author: e-mail address: checco@unl.edu*

Chapter outline

Methods in Cell Biology, Volume 166, ISSN 0091-679X, https://doi.org/10.1016/bs.mcb.2021.06.006

Abstract

G protein-coupled receptors (GPCRs) are a family of transmembrane proteins that act as major mediators of cellular signaling, and are the primary targets for a large portion of clinical therapeutics. Despite their critical role in biology and medicine, a large number of GPCRs are poorly understood, lacking validated ligands or potent synthetic modulators. Ligand-induced GPCR activation can be measured in cell-based assays to test hypotheses about ligand-receptor interactions or to evaluate efficacy of synthetic agonists or antagonists. However, the techniques necessary to develop and implement a cell-based assay to study a given receptor of interest are not commonplace in all laboratories. This chapter outlines methods to develop a cell-based assay to evaluate agonist-induced activation for a GPCR of interest, which can be useful to evaluate the effectiveness of predicted ligands. Examples of sample preparation protocols and data analysis are provided to help researchers from interdisciplinary fields, especially those in fields with relatively little molecular biology or cell culture experience.

1 Introduction

G protein-coupled receptors (GPCRs) are membrane-bound signaling proteins that play critical roles in a wide array of biological processes. At present more than 30% of approved drugs in the United States utilize GPCRs as their primary targets, and many additional receptors represent promising targets for future drug discovery efforts (Hauser, Attwood, Rask-Andersen, Schiöth, & Gloriam, 2017). Over 90 Class A GPCRs have been designated "orphan" receptors, a term used to define receptors without known or rigorously validated ligands (Alexander et al., 2020; Laschet, Dupuis, & Hanson, 2018). In addition, there are a large number of biologically active compounds (e.g., newly discovered natural products, cell-cell signaling peptides, etc.) without known receptors (Fricker & Devi, 2018; Muratspahić, Freissmuth, & Gruber, 2019). Identifying and validating functional ligand-receptor pairings remain important directions both for basic science and for medicine (Bauknecht & Jekely, 2015; Foster et al., 2019; Gomes et al., 2013, 2016; Quiroga Artigas et al., 2020; Shiraishi et al., 2019). Even for well-established receptors, developing new synthetic modulators and evaluating the effectiveness of these compounds remain critical research efforts.

Once putative ligand-receptor interactions are proposed or new modulators for a target GPCR are developed, a powerful step toward evaluating this prediction is to monitor ligand-induced receptor activation in a recombinant system. The purpose of this chapter is to guide the researcher from an initial hypothesis regarding a functional ligand-GPCR interaction to directly testing the proposed GPCR for ligand-induced activation in cell-based assays. We outline methods to isolate the transcript of interest from tissue, clone this gene into a plasmid for expression in mammalian cells, and evaluate receptor activation in living cells in response to exogenous ligands. With the increasingly interdisciplinary nature of research, many laboratories without strong molecular biology or cell culture backgrounds may be interested in developing and testing ligand analogs for GPCR activation. Inspired by our own such experiences, our goal is to provide guidance to help researchers from a variety of fields to develop the techniques required for running these experiments in their own labs.

2 Overview of the protocol

The goal of this protocol is to develop a cell-based assay to evaluate ligand-induced activation of a GPCR of interest. GPCR activation can be monitored by measuring the intracellular accumulation of second messengers in response to stimulus. Upon activation, most GPCRs signal through a heterotrimeric G protein, which leads to a variety of downstream effects depending on the coupled Gα subunit. This protocol utilizes transient co-expression of the GPCR with a promiscuous $G\alpha_q$-family protein that couples to most GPCRs to activate phospholipase C (PLC) (Conklin, Farfel, Lustig, Julius, & Bourne, 1993; Heydorn et al., 2004; Kostenis, Martini, et al., 2005; Kostenis, Waelbroeck, & Milligan, 2005; Offermanns & Simon, 1995; Wedegaertner, Chu, Wilson, Levis, & Bourne, 1993). Activation of PLC leads to intracellular accumulation of D-*myo*-inositol 1-monophosphate (IP1) in the presence of lithium chloride (LiCl). Intracellular IP1 concentration (which is proportional to receptor activation) is measured via commercially available homogenous time-resolved fluorescence (HTRF) assay (Fig. 1) (Trinquet et al., 2006). Dose-response experiments are then used to evaluate and compare ligand potencies.

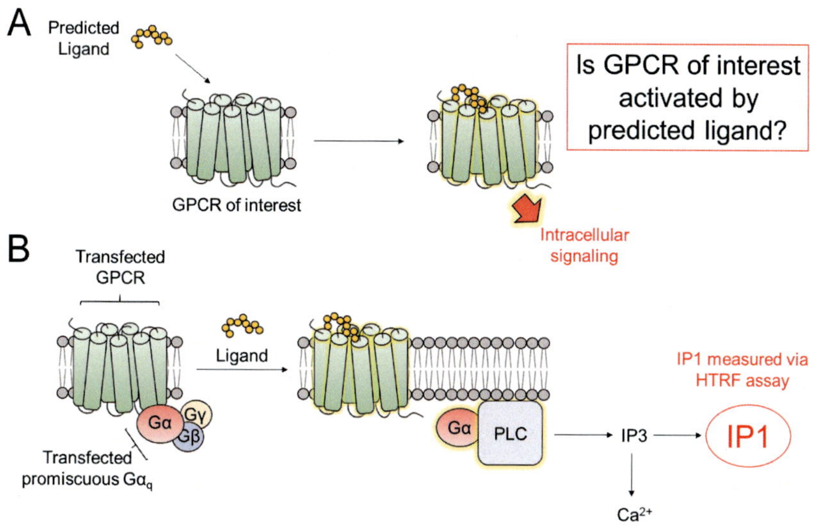

FIG. 1

Graphical overview of the method. (A) The central question answered by the method: for a proposed ligand-GPCR interaction, does the ligand functionally activate the receptor? (B) The present method describes transient co-transfection of the GPCR of interest with a promiscuous $G\alpha_q$-family protein in mammalian cells. Upon ligand stimulation, coupling to the $G\alpha_q$ pathway activates phospholipase C (PLC), which leads to an accumulation of IP1. IP1 is then measured using an HTRF assay.

There are several limitations to the described protocol. The most prominent limitation is that receptor expression is evaluated and optimized via functional activation. For understudied GPCRs, appropriate positive control ligands to verify expression in this manner may not be available. Thus, while positive results can be used to confirm a ligand-receptor interaction, negative results in receptor stimulation experiments could be due to several factors, such as non-optimal recombinant expression in cell lines. As a result, false negatives are possible without additional experiments to verify receptor expression (which are beyond the scope of this protocol). This limitation, and several others, are addressed in more detail in Section 4.

3 Step-by-step methods

3.1 Preparing cDNA library to isolate gene of interest directly from cells or tissue

If the researcher is already confident in the receptor sequence they wish to express, this step may be replaced with purchasing the gene of interest as a synthetic oligonucleotide. However, for some systems (especially non-model organisms without well validated genomes or transcriptomes), it may be beneficial to amplify the gene of interest directly from the tissue itself. Amplifying the transcript directly from tissue may reveal mutations or errors in the database-deposited sequence and inform the researcher of the true sequence of the receptor. This section describes a method to isolate mRNA transcripts from cells/tissue and convert this mRNA to a library of cDNA that may be used for subsequent amplification of a desired receptor sequence.

3.1.1 Materials and reagents

All glassware and plasticware used in this section should either be autoclaved or purchased sterile. Note that all aqueous solutions that come in contact with RNA are prepared using commercially available nuclease-free water or treated with diethylpyrocarbonate (DEPC). To inactivate nucleases with DEPC, dilute DEPC to a final concentration of 0.1% in the solution of interest, stir vigorously at room temperature overnight, and autoclave.

- Nuclease-free water or DEPC-treated water.
- TRIzol RNA isolation reagent (ThermoFisher, 15596026).
- Chloroform.
- Aqueous solution containing 0.8 M sodium citrate, 1.2 M NaCl.
- Isopropanol.
- 5 mg/mL glycogen solution (ThermoFisher, AM9510).
- 70% ethanol/water solution.
- RQ1 DNase, RNAse-free (Promega, M6101) or equivalent.

- Zymo RNA clean and concentrator kit (Zymo Research, R1013) or equivalent.
- iScript cDNA synthesis kit (Bio-Rad, 1708890).
- Plastic tissue homogenization pestle (SP Bel-Art, F19923-0000) or equivalent.
- Microcentrifuge tubes.
- Micropipettes and tips capable of accurately dispensing 1–1000 μL.
- Temperature-controlled microcentrifuge capable of 15,000 × g, 4 °C.
- Microvolume UV–Vis Spectrophotometer capable of measuring absorbance from ≤3 μL of sample (e.g., ThermoFisher NanoDrop instrument or Biotek plate reader with Take3 plate).

3.1.2 Before you begin

Regardless of if a researcher purchases a gene or decides to amplify from cDNA, a few steps can be taken before beginning the cloning procedure to maximize chances of success.

1. Ensure gene sequence of interest contains a full open reading frame (ORF), beginning with an ATG start codon and ending with a stop codon.
2. This protocol is designed to test exclusively for GPCR activation. Thus, the protein of interest must be a GPCR to allow monitoring of receptor activation. GPCRs are characterized by seven transmembrane helices, and these regions can often be predicted based on the receptor's primary amino acid sequence using computational tools. Freely available online tools such as GPCRHMM (https://gpcrhmm.sbc.su.se/) (Wistrand, Kall, & Sonnhammer, 2006) and the NCBI Conserved Domains Database (https://www.ncbi.nlm.nih.gov/cdd/) (Marchler-Bauer et al., 2011, 2015, 2017) can be used to aid in protein family prediction based on amino acid sequence.

3.1.3 Protocol

1. Suspend cell pellet or tissue in 500 μL TRIzol. Samples can be homogenized immediately or stored at −80 °C until homogenization.
2. Homogenize tissues with plastic pestle.
3. Add an additional 500 μL TRIzol and incubate at room temperature for 5 min.
4. Centrifuge suspension at 12,000 × *g*, 4 °C, 8 min. Transfer the supernatant to clean microcentrifuge tubes.
5. To the supernatant from the previous step, add 250 μL chloroform and shake vigorously by hand for several seconds to mix.
6. Centrifuge at 12,000 × *g*, 4 °C, 15 min.
7. Remove the tube from the centrifuge, ensuring not to disrupt the resulting phase separation. Carefully transfer the top layer to clean microcentrifuge tubes for use in the next steps. Discard the remaining solution.

8. To the top layer from the previous step, add 0.5 volume equivalents of the 0.8 M sodium citrate + 1.2 M NaCl solution (e.g., for 500 μL of sample, add 250 μL of the 0.8 M sodium citrate + 1.2 M NaCl solution), followed by the same volume of isopropanol (e.g., 250 μL in previous example).
9. Add 4 μL of 5 mg/mL glycogen solution and incubate at room temperature for 10 min.
10. Centrifuge at 15,000 × *g*, 4 °C, 12 min to pellet RNA.
11. Remove supernatant, being careful not to disrupt RNA pellet at the bottom of the tube.
12. Resuspend pellet in 1 mL of 70% ethanol/water.
13. Centrifuge at 15,000 × *g*, 4 °C, 12 min.
14. Remove supernatant, being careful not to disrupt RNA pellet at the bottom of the tube.
15. Allow RNA pellet to dry open to the air for 5–30 min.
16. Dissolve pellet in nuclease-free water and treat with RQ1 DNase, according to manufacturer's instructions.
17. Purify RNA using RNA clean and concentrator kit, according to manufacturer's instructions. Elute RNA with 16 μL of nuclease-free water.
18. Determine concentration and quality of eluted RNA using Nanodrop instrument or equivalent.

Note: A_{260}/A_{280} ratio should be ≥1.8. If A_{260}/A_{280} ratio is <1.8, repurify RNA with RNA clean and concentrator kit.

19. Synthesize cDNA using iScript cDNA Synthesis kit, according to manufacturer's instructions. We have had good success using 2 μg of RNA in a 40 μL reaction. After reverse transcriptase reaction, cDNA is diluted to a total of 100 μL using nuclease-free water and stored at −20 °C until use.

3.2 Cloning receptor sequence into expression plasmid

This section describes one of several possible methods to clone the receptor gene of interest into pcDNA3.1(+) vector for mammalian expression. This method starts with either a synthetic oligonucleotide corresponding to the ORF or a cDNA library containing the gene of interest (as generated in the previous section). All glassware and plasticware used in this section should either be autoclaved or purchased sterile.

3.2.1 Materials and reagents

- Forward and reverse primers designed to amplify gene open reading frame (ORF) and to install restriction enzyme sites on the resulting PCR product (see primer design step below).
- Phusion high-fidelity DNA polymerase (New England Biolabs, E0553S), or equivalent.

- Restriction enzymes corresponding to restriction sites in designed primers. In the example given below, we used BamHI-HF (New England Biolabs, R3136) and NotI-HF (New England Biolabs, R3189S). Other restriction enzymes can be used, depending on their compatibly with the receptor sequence.
- pcDNA3.1(+) mammalian expression vector (ThermoFisher, V79020).
- Shrimp alkaline phosphatase (SAP) (ThermoFisher, 783901000UN).
- Zymo DNA clean and concentrator kit (Zymo Research, D4013).
- Agarose dissolving buffer (Zymo Research, D4001-1-100).
- T4 DNA ligase (New England Biolabs, M0202L).
- Plasmid mini-prep kit (e.g., GeneJET plasmid miniprep kit, ThermoFisher, K0502).
- DH5α competent *E. coli* cells (ThermoFisher, 18265017).
- LB Broth base (ThermoFisher, 12780052).
- LB agar, powder (ThermoFisher, 22700025).
- Ampicillin sodium salt (Fisher Scientific, BP1760-5).
- Razor blades, or alternative disposable utensil appropriate for excising bands from agarose gel.
- Supplies and equipment to run a 1% agarose gel, and appropriate imager to visualize the results.
- Microcentrifuge tubes.
- Micropipettes and tips capable of accurately dispensing 1–1000 μL.
- Microcentrifuge capable of $\geq$12,000 × *g* (or speeds required for chosen DNA clean and concentrator and plasmid mini-prep kits).
- Thermal cycler for PCR amplification.
- Microvolume UV–Vis Spectrophotometer capable of measuring absorbance from $\leq$3 μL of sample (e.g., ThermoFisher NanoDrop instrument or Biotek plate reader with Take3 plate).
- Variable temperature water bath or heat block.
- Incubator/shaker set to 37 °C.
- *Optional*: S.O.C. medium (ThermoFisher, 15544034).
- *Optional*: PCR cloning kit (New England Biolabs, E1202S).
- *Optional*: Plasmid DNA Maxi-prep kit (e.g., E.Z.N.A. Plasmid Maxi kit, Omega Biotek, D6922-02).

3.2.2 Before you begin

1. Prepare 100 mg/mL ampicillin solution (1000 ×) by dissolving 1 g ampicillin in 10 mL ultrapure water. Sterile filter with 0.22 μm syringe filters. Aliquot and store at −20 °C until use.
2. Prepare LB-agar plates with 100 μg/mL ampicillin.
3. Design and order DNA primers for PCR amplification of receptor ORF and installation of appropriate restriction enzyme sites.
   a. Examine the DNA sequence for gene of interest and identify regions outside of ORF (5′ of the ATG start codon and 3′ of the stop codon) for primer annealing.

b. Choose a stretch of ~20 nucleotides in each of your chosen regions as targets for designed primers. There are many tools available to help with appropriate primer design (e.g., Primer3), and many different methods to design effective primers. We aim for a T_m of ~60 °C, ~50% GC content, and 1–2 G or C nucleotides on the 3′ end (a "CG clamp") of each primer.
c. Choose appropriate restriction enzymes for insertion of gene into pcDNA3.1 (+), paying attention to orientation of restriction enzymes relative to CMV promotor (refer to vector map in the pcDNA3.1(+) user manual to determine relative order of restriction enzyme sites). After ligation, the 5′ end of the ORF should be oriented closer to the CMV promotor than the 3′ end. For example, we may choose BamHI to incorporate at the 5′ end via the forward primer and NotI to incorporate at the 3′ end via the reverse primer.
d. Ensure chosen restriction enzyme sites are not naturally present in ORF, as these will lead to undesired cleavage events during restriction enzyme digestion. If restriction enzyme sites are found within ORF, choose alternative restriction enzymes.
e. To design final primers, synthesize 5′-CTC-[X]-[Y]-3′ where [X] is the chosen restriction enzyme sequence and [Y] is the chosen sequence from gene of interest. See Fig. 2 for an example. Custom synthetic primers can be purchased from a number of commercial sources.

3.2.3 Protocol

1. PCR amplify the receptor ORF using Phusion high-fidelity DNA polymerase and the designed primers, following instructions provided for the polymerase. This step will amplify your ORF while simultaneously incorporating appropriate restriction enzyme sites into the amplicon.

Note: If starting from synthetic oligonucleotide as template, start with 2 ng template in 50 μL reaction. Analyze product by 1% agarose gel to ensure amplification. PCR conditions may need to be optimized.
Note: If starting from cDNA library as template, amount of cDNA mixture used in reaction will vary based on gene expression level and polymerase. It may be useful to screen multiple conditions initially to identify those that amplify gene of interest. Analyze product by 1% agarose gel to evaluate amplification. PCR conditions may need to be optimized.

2. Clean-up amplified PCR product with DNA clean and concentrator kit according to manufacturer's instructions, eluting with 10 μL of nuclease-free water.
3. Determine concentration and quality of eluted PCR product using Nanodrop instrument or equivalent.

Note: A_{260}/A_{280} ratio should be ≥1.8. If A_{260}/A_{280} ratio is <1.8, repurify with DNA clean and concentrator kit.
Pause point: Purified PCR products can be stored at −20 °C until further use.

4. Perform a digest of PCR product using both selected restriction enzymes simultaneously, following manufacturer's protocol for selected enzymes.

A

...TCATCTTGTGTAATTGTCCTAGATAAGTTAAATAGTTAAGGAAACAAATTACTAACAATTTCAACATCAGTGTT
CCCTTCCCAACAACCAAAACAA*AACATGGGGTCGAACGATAC*ATTCTTCAACTTGACAATGCTCAATGTAACTGACG
GCAACGGAACATCCAGTAATGGTTCTGAAAACGGGACGGCCTGTTGGAATGAGTACTGCTGGGACGAGGAATACTAC
AACGACCTTCTAGAGGACCACGTTTTCCCAAAAGCTTACGAATGGGTCTTCATCCTTCTGTATTTCTTGACGTTCAC
CGTAGGGCTGGTGGGGAACGCTTTAGTGTGCTGGGCAGTCTGGAGAAACCACAACATGCGGACGGTGACCAATGTGT
TCATCGTCAATCTGGCTGTGGGAGATTTCCTGGTGATTCTCGCCTGCCTCCCGCCCACACTGGTGCAGGACGTCACG
GAGTCCTGGTTCTTGGGGACCCGGGTGTGCAAGGCTGTTCTCTATCTGCAGTCGACATCGGTGTCCGTATCAGTGCT
AACGCTGAGTGCCATTGCAGTAGAGCGCTGGTGGGCCATCTGCTACCCCCTCAAGTTCAAATCCACCATGTCCAGAG
CCAGAAAAATCATCTTTCTCATTTGGGTAGTTTCTTTCCTGTCGGCGCTACCGGAAGTCATTGTTGCCACGACCCAC
CCTTATCCTTTCCCTGGACATTTCACATCCATATATTTGACCAGATGCAGCCCGTCATGGAATTCCCTGAATCAGGG
CATCTACCAGATCCTGGTGGCTCTGTTTTTCTACCTTCTGCCCATGGTGATGATGGGAGCCACGTACACACACATCG
CGACAGTGCTGTGGAAGCACGAGATCCCGGGCGCCGTTGTCGGGGGTCGTAAGCCGATGCTGGACGGAAACCGGAAC
TCCCAGGACGAGCAAATCATGTCAAGACGGAAAGCTGCGCGCATGCTCATCGCTATCGTCATTGTGTTTGGGATCTG
CTACTTGCCAGTTCACATCATGAATCTTCTACGATACTTTGAAGCTGTGGACTACGGAACAGACATGCATGCCCCGA
TTGTCCAGTCCCTGGTCAGTCACTGGCTACCTTACTTCAACTCAGCGTTAAACCCTGTCATATACAACTTCATGAGT
GCCAAGTTCCGGAAAGAGTTCAAGTCCGCGTGTTTTTGCTGTTTTTATGGGATGCGGACGCGGCCCTTTCGAGGCCG
TAGAGATCATACCTTTACGATGACCTTCTCCAACAGCAACTACAGCAACTGCCACACAGAAGAGGTCACTCTCGCCA
GCATCTGAGTTACCAGGACTTCGTGAGAAGGAAGA*CTTCAGCCGGTCGTCATAAC*GATAAGCGGCGGAAAATAAAAA
GAAACAGACCTCTCTCTGTACCCTCAAATCAAACTTTAAGAATAAATTACTCATATTGATTTGAGTTCAC...

Bold, underlined = Predicted receptor ORF
Blue italics = Chosen site for forward primer
Red italics = Chosen site for reverse primer

B

Forward: 5'-CTC**GGATCC***AACATGGGGTCGAACGATAC*-3'
Reverse: 5'-CTC**GCGGCCGC***GTTATGACGACCGGCTGAAG*-3'

Bold, underlined = Restriction enzyme sites
Blue italics = Forward primer receptor-specific sequence
Red italics = Reverse primer receptor-specific sequence

FIG. 2

Example primer design for amplification of receptor open reading frame. (A) Excerpt of the deposited mRNA sequence encoding for the *Aplysia californica* allatotropin-related peptide receptor (ATRPR, XM_005106157.3) (Checco, Zhang, Yuan, Le, et al., 2018). The predicted ORF is bold and underlined, while the chosen sites for forward and reverse primers are shown italicized and in color. (B) Selected forward and reverse primers for cloning the ATRPR ORF into pcDNA3.1(+) vector. Both primers incorporate a restriction enzyme site (BamHI for the forward primer, and NotI for the reverse primer), depicted in bold and underlined.

We have had success digesting ~300–400 ng PCR product in 60 μL reaction with two restriction enzymes.

5. Perform a digest of pcDNA3.1(+) using both selected restriction enzymes simultaneously, following manufacturer's protocol for selected enzymes. We have had success digesting ~3 μg plasmid in 60 μL reaction with two restriction enzymes.

Note: Steps 4 and 5 may be performed simultaneously if desired.

6. Treat double-digested pcDNA3.1(+) vector with shrimp alkaline phosphatase (SAP), according to manufacturer's instructions.

7. Purify double-digested PCR product (from Step 4) and double-digested, SAP-treated vector (from Step 6) using 1% agarose gel in the following manner:

(a) Pour 1% agarose gel in wells large enough to hold entire digestion reaction.

(b) Add loading dye to each digest and load into separate wells on gel. If space permits, it is helpful to leave empty lanes between different samples to avoid the possibility of cross-contamination when excising bands.
(c) Run the gel long enough to ensure adequate separation of bands (e.g., 120 V for 50 min for a 7×10 cm gel).
(d) Visualize products under UV light and excise desired bands with razor or similar tool for the vector backbone and the digested PCR product.

Note: When excising bands from gel, cut as close to desired band as feasible. Use a clean blade for excising each band to minimize chances of cross contamination.

8. Dissolve excised bands in 3–5 volumes of agarose dissolving buffer, incubating at 37–55 °C until fully dissolved.
9. Clean-up both the digested vector and the digested PCR product with DNA clean and concentrator kit according to manufacturer's instructions, eluting with $\geq 6\,\mu L$ of nuclease-free water.
10. Determine concentration and quality of eluted PCR product using Nanodrop instrument or equivalent.

Note: A_{260}/A_{280} ratio should be ≥ 1.8. If A_{260}/A_{280} ratio is <1.8, repurify with DNA clean and concentrator kit.

Pause point: Purified DNA products can be stored at −20 °C until further use.

11. Ligate digested insert and vector using T4 DNA ligase, following manufacturer's instructions. We have had success using a 1:5 ratio of vector to insert (e.g., 0.02 pmol vector and 0.1 pmol insert), incubating for 10 min–2 h at room temperature, for a 20 μL-scale reaction.

Note: In addition to "Vector+Insert" reaction, it is often helpful to include a "Vector only" control during this and subsequent steps.

12. Thaw competent DH5α cells on ice. Proceed with next step immediately after cells are thawed.
13. Add 1–5 μL of above ligation reaction mixture to 50 μL aliquot of DH5α cells. Gently tap sides of tube to mix.
14. Incubate cell/ligation reaction mixture on ice for 15 min.
15. Heat shock DH5α cells by rapidly transferring tube from ice to 42 °C for 30–90 s (e.g., using water bath or heat block), followed by ice for 2 min.
16. Add 200 μL LB (without antibiotics) or S.O.C. medium.
17. Incubate at 37 °C for 30–60 min with shaking.
18. Plate onto pre-warmed LB-agar plates supplemented with 100 μg/mL ampicillin.

Note: It is often helpful to plate several plates for each ligation reaction (e.g., one plate using 200 μL of the reaction mixture and another plate using 20 μL of the reaction mixture). This maximizes chances of obtaining a plate with appropriate density of colonies to facilitate isolation of a single colony.

19. Incubate plates at 37 °C overnight.

Note 01. As noted above, it is helpful to perform this step with a "Vector only" control. This control is expected to show no (or very few) colonies. In contrast, successful ligation reactions should show a number of colonies after incubation. Colony growth in the "Vector only" control could indicate insufficient restriction enzyme digestion or unintended religation. If a large amount of colony growth is observed on "Vector only" control, optimization of restriction enzyme digestion or SAP treatment step is likely needed. In the event of no growth on "Vector+Insert" reactions, optimization of PCR amplification or ligation step is likely necessary.

Note 02. If amplifying reaction from cDNA, a low amount of PCR product may make it difficult to successfully ligate into the pcDNA3.1(+) vector using this method. If repeated attempts of restriction enzyme cloning into pcDNA3.1(+) vector are unsuccessful, an alternative approach is to directly clone PCR amplicons via blunt- or TA-end cloning into pMiniT 2.0 vector using a PCR cloning kit (New England Biolabs, E1202S), following manufacturer's instructions. Direct cloning of PCR product into this vector will allow for the amplification and purification of large amounts of plasmid containing the gene of interest. The desired ORF can then be subcloned into pcDNA3.1(+).

Pause point: LB-agar plates with DH5α colonies can be stored at 4°C for a least 1 week.

20. Using a sterile pipette tip, transfer an individual colony from LB-agar plate to 5mL LB supplemented with 100μg/mL ampicillin. Repeat for three total individual colonies.
21. Incubate 5mL cultures at 37°C with shaking overnight.
22. After overnight incubation, isolate plasmid using plasmid Miniprep kit, following manufacturer's instructions.
23. After plasmid isolation, determine concentration and quality of eluted plasmid product using Nanodrop instrument or equivalent.

Note: A_{260}/A_{280} ratio should be ≥ 1.8. If A_{260}/A_{280} ratio is <1.8, repurify with DNA clean and concentrator kit.

24. Submit isolated plasmids for DNA sequencing. For pcDNA3.1(+), sequencing should be done in both forward and reverse directions using T7-promotor and BGH-reverse primers, respectively. Confirm that full insert is positioned at correct location in the vector and that no mutations have been introduced during the cloning process. Ensure complete coverage of the ORF by sequencing reactions. If portions of the gene are not confidently sequenced, it may be necessary to design custom primers to sequence these regions.

Note: Mini-prep kits can provide up to ~20μg of plasmid DNA, which is likely enough material for DNA sequencing and for initial activity screens. However, once successful cloning has been confirmed by DNA sequencing, it is recommended to re-transform DH5α cells with the validated plasmid and purify from a large-scale culture (e.g., 200 mL) using a DNA Maxi kit. Performing such a large-scale purification will provide ample material for the experiments described below.

3.3 Testing receptor activation and determining optimal expression conditions

This section will describe how to perform initial tests for ligand-induced receptor activation using cells transiently transfected with both the receptor of interest and a promiscuous $G\alpha_q$ subunit. Because optimal transfection conditions may differ for each receptor chosen, this protocol aids in optimizing transfection conditions to generate the maximum IP1 response. Optimal transfection conditions determined here will then be used to generate dose-response curves for a given ligand in Section 3.4. Steps 1–21 of this protocol are performed in a biosafety hood and sterile conditions are maintained throughout the experiment. Steps 21–29 of the protocol can be performed on benchtop and under non-sterile conditions. Fig. 3 gives a broad timeline of transient transfection, ligand stimulation, and IP1 accumulation measurement steps.

3.3.1 Materials and reagents

- CHO-K1 cell line (ATCC, CCL-61), or similar easily maintained and transfectable cell line.
- Plasmid containing the receptor of interest in pcDNA3.1(+) vector (generated in Section 3.2)
- pcDNA3.1(+) vector with gene encoding for a promiscuous $G\alpha_q$-family protein. (Refer to discussion in Section 3.3.2, as well as additional discussion in Section 4.)
- F-12K nutrient medium (ATCC, 30-2004).
- TurboFect Transfection reagent (ThermoFisher, R0531).
- Fetal bovine serum (FBS) (VWR, 97068-085).
- HyClone Penicillin Streptomycin 100 × Solution (10,000 U/mL penicillin, 10,000 μg/mL streptomycin) (Fisher Scientific, SV30010).
- Opti-MEM reduced serum medium (ThermoFisher, 31985070).
- 0.05% Trypsin solution, with 0.53 mM EDTA in HBSS without calcium, magnesium, or sodium bicarbonate (Fisher Scientific, MT25052CI).
- IP-One Gq kit (Cisbio, 62IPAPEC).

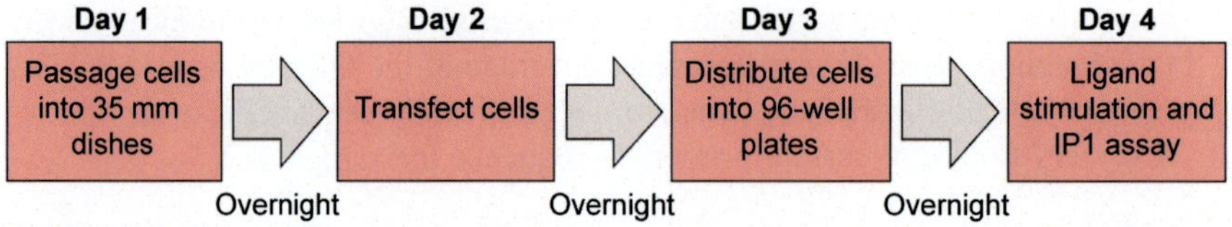

FIG. 3

Timeline for IP1 accumulation assays.

- 75 cm^2 tissue culture flasks (T75) (Fisher Scientific, FB012937).
- 35 mm diameter tissue culture-treated dishes (Fisher Scientific, FB012920).
- 96-well, cell culture-treated, flat-bottom, half-area, white microplates (Fisher Scientific, 07-200-309).
- Sterile, single-use 0.2 μm vacuum filter units with media bottles (Fisher Scientific, 09-741-02).
- Disposable sterile serological pipettes (for cell culture).
- Micropipettes and tips capable of accurately dispensing 1–1000 μL.
- Hemocytometer (Fisher Scientific, 02-671-6), or alternative method for cell counting.
- HTRF-compatible microplate reader (e.g., Biotek Synergy Neo2). For more information on HTRF-compatible plate readers, see assay guidance for Cisbio's IP-One Gq kit.
- *Optional*: Plasmid encoding for expression of green fluorescent protein (GFP) or similar protein in mammalian cells, as a positive control for transfections.

3.3.2 Before you begin

1. Prepare CHO-K1 cell culture medium with antibiotics: F-12K nutrient medium supplemented with 10% FBS, 100 U/mL penicillin, and 100 μg/mL streptomycin (F-12K + 10% FBS + P/S). Combine solutions under sterile conditions and filter through 0.2 μm vacuum filtration units into media bottles. Store this medium at 4 °C when not in use. Medium should be used within 1 month of preparation.
2. Prepare CHO-K1 cell culture medium without antibiotics: F-12K nutrient medium supplemented with 10% FBS (F-12K + 10% FBS). Combine solutions under sterile conditions and filter through 0.2 μm vacuum filtration units into media bottles. Store this medium at 4 °C when not in use. Medium should be used within 1 month of preparation.
3. Reconstitute and aliquot components from the IP-One Gq kit according to manufacturer's instructions: IP1-d2, anti-IP1-cryptate, and IP1 standard solution. Store these solutions at −20 °C.
4. Choose and purchase/clone a pcDNA3.1(+) vector for expression of a promiscuous $G\alpha_q$-family subunit. These Gα proteins associate with a wide variety of GPCRs to facilitate signaling through the PLC pathway. This means that even GPCRs who natively signal through other downstream signaling pathways can be studied by monitoring PLC activation (e.g., through IP1 accumulation, as performed in this protocol). There are several Gα proteins reported that may be useful, including Gα-15/16, $G\alpha_q$, and chimeric $G\alpha_q$-family proteins with mutations designed to enhance promiscuity (Kostenis, Waelbroeck, & Milligan, 2005; Offermanns & Simon, 1995). Our lab uses a $G\alpha_q$-derived chimeric protein bearing a G66D mutation (Heydorn et al., 2004; Kostenis, Martini, et al., 2005), N- and C-termini from $G\alpha_i$ proteins (Conklin et al., 1993; Kostenis, Martini, et al., 2005), and an internal HA tag (Wedegaertner et al., 1993) (Fig. 4).

Promiscuous $G\alpha_q$-family protein

MGCTLSEEAKEARRINDEIERQLRRDKRDARRELKLLLLGTGESG
KSTFIKQMRIIHGSDYSDEDKRGFTKLVYQNIFTAMQAMIRAMDT
LKIPYKYEHNKAHAQLVREVDVEKVSAFDVPDYAAIKSLWNDPGI
QECYDRRREYQLSDSTKYYLNDLDRVADPSYLPTQQDVLRVRVPT
TGIIEYPFDLQSVIFRMVDVGGQRSERRKWIHCFENVTSIMFLVA
LSEYDQVLVESDNENRMEESKALFRTIITYPWFQNSSVILFLNKK
DLLEEKIMYSHLVDYFPEYDGPQRDAQAAREFILKMFVDLNPDSD
KIIYSHFTCATDTENIRFVFAAVKDTILQLNLKECGLF

FIG. 4

Primary amino acid sequence of one of several promiscuous $G\alpha_q$-family proteins that can be used to couple GPCRs to PLC pathway. Substitutions from natural $G\alpha_q$ sequence are shown in red, most of which are designed to increase promiscuity (Conklin et al., 1993; Heydorn et al., 2004; Kostenis, Martini, et al., 2005; Kostenis, Waelbroeck, & Milligan, 2005; Offermanns & Simon, 1995; Wedegaertner et al., 1993).

5. Maintain cultures of CHO-K1 cells in F-12K + 10% FBS + P/S in T75 flasks, incubating at 37 °C, 5% CO_2. Passage cells regularly (every 2–3 days) to avoid reaching overconfluency.
6. Note that the volumes and concentrations of reagents used in the protocol are provided as a starting guide for the researcher. The reader can use different dilutions based upon the need of their experiment and the availability of materials and reagents.

3.3.3 Protocol

1. On the day prior to transfection, passage CHO-K1 cells into 35 mm diameter tissue culture-treated dishes with 2 mL F-12 K + 10% FBS + P/S. Incubate at 37 °C in 5% CO_2 overnight to allow cells to attach and recover.

Note 01. Optimal confluence for transfection with TurboFect is 70–90%. Thus, it may be helpful to practice seeding several dishes at different densities to optimize cell density at the time of transfection.

Note 02. The number of dishes prepared will depend on how many different transfection conditions need to be tested (one dish for each transfection condition). For the example shown in Figs. 5–7, we prepared six dishes.

2. The next day, when cells are 70–90% confluent, exchange the medium for 2 mL of F-12K + 10% FBS (no antibiotics).
3. Prepare mixtures of plasmid DNA and Turbofect in 200 μL Opti-MEM. We recommend testing different amounts of total plasmid DNA, as well as different DNA/TurboFect ratios.

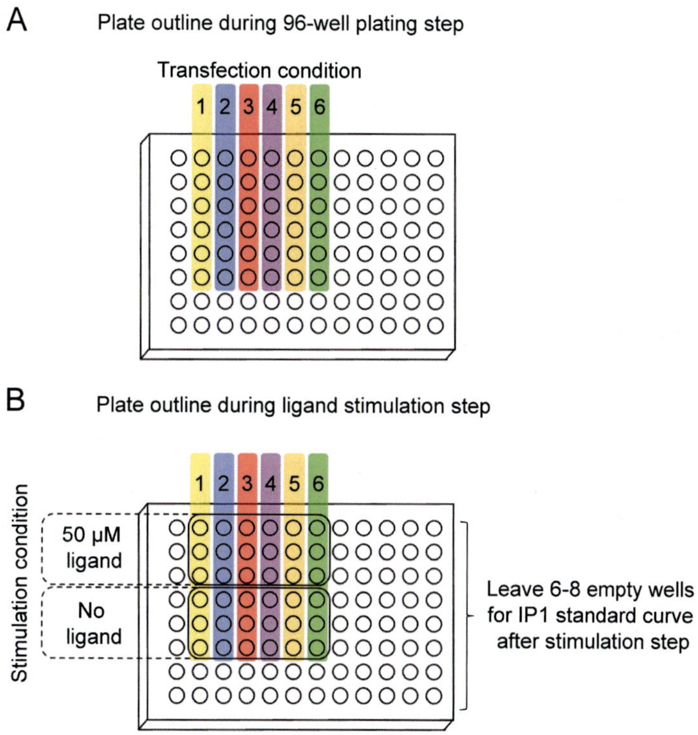

FIG. 5

Example 96-well plate map for testing transfection conditions. (A) After transfecting six different dishes of cells with varying conditions, each condition is seeded into a separate column. (B) During ligand stimulation, three wells for each transfection condition are stimulated with high ligand concentration (e.g., 50 μM) in stimulation buffer, while three different wells are treated with stimulation buffer alone ("No ligand"). Empty wells are left on one side of the plate to place the IP1 standards to generate the standard curve.

Note: We generally vary total DNA plasmid from 2 to 4 μg of total DNA (with 1:1 ratio of receptor plasmid to promiscuous Gα protein) and 4–8 μL of TurboFect. For the example shown in this section (Figs. 5–7), six different DNA/TurboFect mixtures were prepared covering these ranges. It is also a good idea to include a "No transfection" control to account for endogenous signaling by ligand on CHO-K1 cells.

4. Carefully mix each DNA + Turbofect mixture by slowly pipetting up and down several times.
5. Allow the mixtures to sit undisturbed at room temperature for 15 min.
6. Evenly distribute each DNA + Turbofect mixture in a dropwise manner to one dish of CHO-K1 cells prepared above.

A

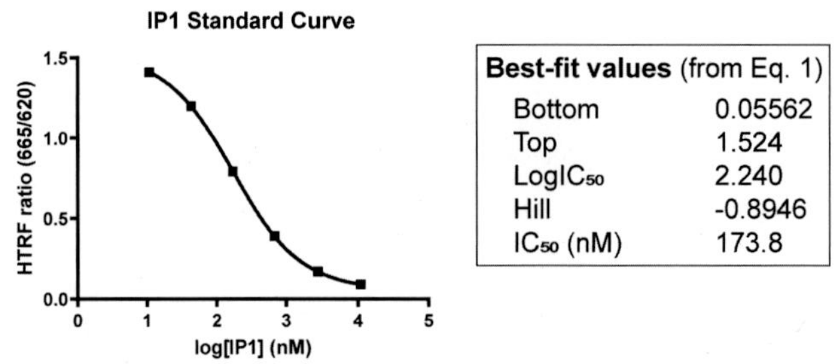

B **log(inhibitor) vs. response -- variable slope (four parameters)**

$$\text{Ratio} = \text{Bottom} + \frac{\text{Top} - \text{Bottom}}{1 + 10\text{^}((\text{LogIC}_{50} - \text{Log[IP1]})\text{*Hill})} \quad \text{(Eq. 1)}$$

C **Conversion of response ratio to [IP1]**

$$[\text{IP1}] = \frac{\text{IC}_{50}}{[(\text{Top} - \text{Ratio})/(\text{Ratio} - \text{Bottom})]\text{^}(1/\text{Hill})} \quad \text{(Eq. 2)}$$

FIG. 6

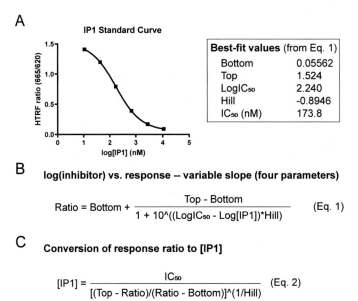

An IP1 standard curve is used to relate IP1 concentration to HTRF ratio (665 nm/620 nm fluorescence intensities). (A) An example of an IP1 standard curve, along with associated values after fitting to a four-parameter logistical model. (B) The equation for the four-parameter logistical model (Eq. 1). (C) Using an IP1 standard curve together with Eq. (2), the concentration of IP1 in a given experimental well can be calculated (Burford, Watson, & Alt, 2017).

7. Incubate the dishes at 37 °C in 5% CO_2 overnight.
8. The next day, rinse cells with 2 mL of sterile PBS and detach the cells from the dish using 1 mL of 0.05% trypsin solution.
9. Quench the trypsin reaction by adding 2 mL of F-12K + 10% FBS.
10. Centrifuge cells at 800 × *g*, 3 min.
11. Aspirate media and resuspend cells in 1 mL of F-12K + 10% FBS.
12. Count cell density using hemocytometer.
13. Dilute cells to a concentration of 200,000 cells/mL in F-12K + 10% FBS. Ensure thorough mixing of cells by gently pipetting up and down, or with gentle inversion.
14. Distribute 100 μL (20,000 cells) into individual wells of an opaque white 96-well half-area tissue culture-treated plates.

Note: An example of a 96-well plate layout is shown in Fig. 5.

15. Incubate the 96-well plate at 37 °C in 5% CO_2 overnight.
16. The next day, prepare a 1 × solution of stimulation buffer from the 5 × stimulation buffer provided in the IP-One Gq kit.

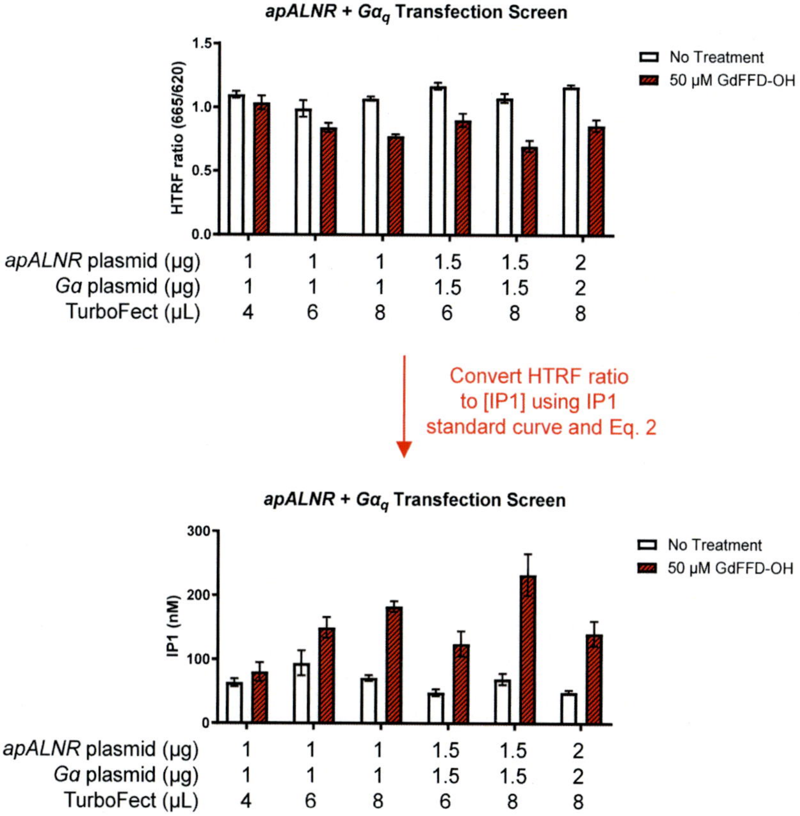

FIG. 7

Example data showing the results of a transfection screen. Amounts of receptor plasmid, promiscuous Gα_q-family protein plasmid, and transfection reagent Turbofect are varied. Activation of the GPCR is then evaluated in the presence or absence of a high concentration of predicted ligand (peptide GdFFD-OH), monitoring IP1 accumulation via HTRF assay. Using an IP1 standard curve (as shown in Fig. 6), HTRF ratio values can be converted to IP1 concentrations to evaluate the optimal transfection conditions. These data screen conditions for expression of the *Aplysia* GPCR *apALNR* (Bauknecht & Jekely, 2015; Checco, Zhang, Yuan, Yu, et al., 2018) in CHO-G5A cells (Bauknecht & Jekely, 2015). Bars represent the mean $\pm$ standard deviation of three replicate wells on the plate.

Note: The amount of 1 × stimulation buffer prepared will vary depending upon the number of transfection conditions to be tested.

17. Prepare a 2 × stock solution of the ligand in 1 × stimulation buffer.

For example: In our assay depicted in Figs. 5–7, 50 µM was chosen as the concentration of ligand tested, therefore, a 100 µM (2 ×) stock was prepared in 1 × stimulation buffer.

18. Remove the 96-well plate seeded with the transfected cells from the incubator and carefully remove the media from each well.

Note: During this step, it is important not to dislodge the cells from the bottom of the well. Using a single-channel pipette, we carefully touch the bottom edge of the well, and slowly remove the media. Use of aspirators can be helpful to speed up the process and reduce labor, but care should be taken to minimize any cell loss during this step.

19. Add 14 μL of 1 × stimulation buffer to each well.

Note: Ensure that the stimulation buffer is added quickly to each well once the media is removed. Allowing the cells to dry reduces their viability and may affect the final IP1 response observed.

20. Add 14 μL of 2 × ligand (for stimulated wells) or 1 × stimulation buffer (for control wells) to each well. Run each condition at least in duplicate.
21. Incubate the 96-well plate at 37 °C in 5% CO_2 for 30–60 min.
22. During this time, prepare a 1 × solution of IP1-d2 and a 1 × solution of anti-IP1-cryptate in lysis buffer provided in the kit.

Note 01. It is advised to prepare fresh stocks of IP1-d2 and anti-IP1-cryptate before each experiment.

Note 02. We have found that using the manufacturer's recommended amount of IP1-d2 and anti-IP1-cryptate (1 × each) provides high sensitivity. However, we have also noted comparable overall results using 0.5 × IP1-d2 and 0.5 × anti-IP1-cryptate (Checco, Zhang, Yuan, Le, et al., 2018; Checco, Zhang, Yuan, Yu, et al., 2018; Do et al., 2018).

23. Prepare serial dilutions of IP1 standard (provided in IP-One Gq kit) in 1 × stimulation buffer. We recommend using a highest concentration of 11,000 nM with five–six 4-fold dilutions.
24. After cells have been stimulated for 30–60 min, remove the plate from incubator and cool to room temperature.
25. Add 28 μL of each prepared IP1 standard solution into empty wells on the plate.
26. Add 6 μL of IP1-d2 in lysis buffer solution to each well (both experimental and those containing IP1 standards). IP1-d2 should be added to each well before anti-IP1-cryptate.
27. Add 6 μL anti-IP1-cryptate in lysis buffer solution to each well (both experimental and those containing IP1 standards).
28. Cover the prepared 96-well plate with aluminum foil and incubate at room temperature for at least 60 min.
29. Read the homogeneous time-resolved fluorescence (HTRF) signal of each well on HTRF-compatible plate reader, using an excitation wavelength of 330 nm and measuring fluorescence emission at both 665 and 620 nm.

Note: The settings of the plate reader will vary with the instrument. Follow guidelines provided with IP-One Gq kit for chosen plate reader.

3.3.4 Data analysis and interpretation

1. Export HTRF data into appropriate format for data processing and conversion (e.g., Microsoft Excel).
2. Calculate ratio of fluorescence emission at 665 nm/620 nm for each well. This is the "HTRF ratio."
3. Using a program suitable for non-linear regression analysis (e.g., GraphPad Prism), plot the IP1 standard curve: log[IP1] vs HTRF ratio (Fig. 6).
4. Fit the IP1 standard curve to a four-parameter logistical regression model (Eq. 1, Fig. 6) (in GraphPad Prism, this is "log(inhibitor) vs response—variable slope (four parameters)") to estimate Bottom, Top, Hill coefficient, and IC_{50} values for the IP1 standard curve.
5. Using Eq. (2) (Fig. 6) and the measured HTRF ratios, calculate the IP1 concentration in each experimental well. These data can be plotted for analysis (Fig. 7).
6. A reproducible increase in IP1 concentration indicates that the ligand activates the chosen receptor. In this case, examination of the different transfection conditions tested can guide the researcher to choosing the optimal conditions for future experiments (e.g., conditions that give the greatest dynamic range).
7. A lack of reproducible IP1 accumulation in transfected cells indicates that the ligand may not be a natural agonist of the chosen receptor. However, a lack of response here may also indicate inefficient transfection, a requirement for co-receptors, or that the chosen promiscuous $G\alpha$ protein does not associate with this receptor. Refer to Section 4 for more discussion.

Note: If the lab is not confident in performing transient transfections, we recommend practicing with a plasmid encoding for GFP. One can practice transfection steps using this plasmid and evaluate relative expression rapidly using fluorescence microscopy or a similar method.

3.4 Determining ligand potency through dose-response curves

Once a ligand-receptor pairing has been validated in a single-point assay and optimal transfection conditions have been established (refer to Section 3.3), ligand potency can be accessed via dose-response experiments. Potency measurements (i.e., EC_{50} values) can then be compared among different ligands. Steps 1–7 in this section are performed in a biosafety hood under sterile conditions. Steps 8–15 can be performed on benchtop and under non-sterile conditions.

3.4.1 Materials and reagents

- CHO-K1 cell line (ATCC, CCL-61), or similar easily maintained and transfectable cell line.
- Plasmid containing receptor of interest in pcDNA3.1(+) vector (generated in Section 3.2).

- pcDNA3.1(+) vector with gene encoding for a promiscuous $G\alpha_q$-family protein (Refer to discussion in Section 3.3.2, as well as additional discussion in Section 4).
- F-12K nutrient medium (ATCC, 30-2004).
- TurboFect Transfection reagent (ThermoFisher, R0531).
- Fetal bovine serum (FBS) (VWR, 97068-085).
- HyClone Penicillin Streptomycin 100 × Solution (10,000 U/mL penicillin, 10,000 μg/mL streptomycin) (Fisher Scientific, SV30010).
- Opti-MEM reduced serum medium (ThermoFisher, 31985070).
- 0.05% Trypsin solution, with 0.53 mM EDTA in HBSS without calcium, magnesium, or sodium bicarbonate (Fisher Scientific, MT25052CI).
- IP-One Gq kit (Cisbio, 62IPAPEC).
- 75 cm^2 tissue culture flasks (T75) (Fisher Scientific, FB012937).
- 35 mm diameter tissue culture-treated dishes (Fisher Scientific, FB012920).
- 96-well, cell culture-treated, flat-bottom, half-area, white microplates (Fisher Scientific, 07-200-309).
- Sterile, single-use 0.2 μm vacuum filter units with media bottles (Fisher Scientific, 09-741-02).
- Disposable sterile serological pipettes (for cell culture).
- Micropipettes and tips capable of accurately dispensing 1–1000 μL.
- Hemocytometer (Fisher Scientific, 02-671-6), or alternative method for cell counting.
- HTRF-compatible microplate reader (e.g., Biotek Synergy Neo2). For more information on HTRF-compatible plate readers, see assay guidance for Cisbio's IP-One Gq kit.

3.4.2 Before you begin

1. Prepare media and culture CHO-K1 cells, as described in Section 3.3.2. Determine optimal transfection conditions, as described in Section 3.3.
2. Note that the concentrations and volumes of reagents used in our protocol are provided as a starting guide for the researcher. The reader can use different dilutions based upon the need of their experiment and the availability of materials and reagents.

3.4.3 Protocol

1. Follow steps 1–15 of Section 3.3.3 to transfect and plate CHO-K1 cells in advance of the experiment. Transfect cells with optimized DNA + Turbofect ratio identified in Section 3.3. Plate transfected cells (20,000 cells/well) in 96-well half-area plates, as described above, and incubate overnight at 37 °C, 5% CO_2.

Note: If the receptor of interest naturally couples to the PLC pathway through endogenous $G\alpha_q$ proteins, the promiscuous $G\alpha$ protein can be omitted during transfections.

This can be tested experimentally by comparing transfection conditions with and without co-expression of promiscuous Gα protein.

2. The next day, prepare a 1 × solution of stimulation buffer from the 5 × stimulation buffer provided in the IP-One Gq kit.

Note: The amount of 1 × stimulation buffer prepared will vary depending upon the number of conditions to be tested.

3. Prepare serial dilutions of ligands at 2 × the intended final concentration (for example, if the desired maximum concentration of ligand is 1 μM, prepare a working stock of 2 μM, and prepare serial dilutions from this working stock) in 1 × stimulation buffer.

Note 01. To obtain accurate EC_{50} values, it is important to use a range of ligand concentrations which elicit the entire response range (allowing for accurate fit of "Top" and "Bottom" of the curve). We generally prepare a six-point dose-response curve using 10-fold serial dilutions. However, a researcher may modify this set-up to increase the number of points in the curve or cover an alternative range. It is not recommended to generate dose-response curves with less than six points.

Note 02. While preparing serial dilutions, it is important to change pipette tips between each dilution to avoid cross contamination.

4. Remove the 96-well plate seeded with the transfected cells from the incubator and carefully remove the media from each well.

Note: During this step, it is important not to dislodge the cells from the bottom of the well. Using a single-channel pipette, we carefully touch the bottom edge of the well, and slowly remove the media. Use of aspirators can be helpful to speed up the process and reduce labor, but care should be taken to minimize any cell loss during this step.

5. Add 14 μL of 1 × stimulation buffer to each well.

Note: Ensure that the stimulation buffer is added quickly to each well once the media is removed. Allowing the cells to dry reduces their viability and may affect the final IP1 response observed.

6. Add 14 μL of each ligand dilution to respective wells in the plate, running each condition at least in duplicate.

Note: A simple example of a plate set-up for testing four different ligands is given in Fig. 8. Plate layouts should be modified to test the desired number of ligands.

7. Incubate the 96-well plate at 37 °C in 5% CO_2 for 30–60 min.
8. During this time, prepare working solutions of IP1-d2 and anti-IP1-cryptate in lysis buffer provided in the kit.

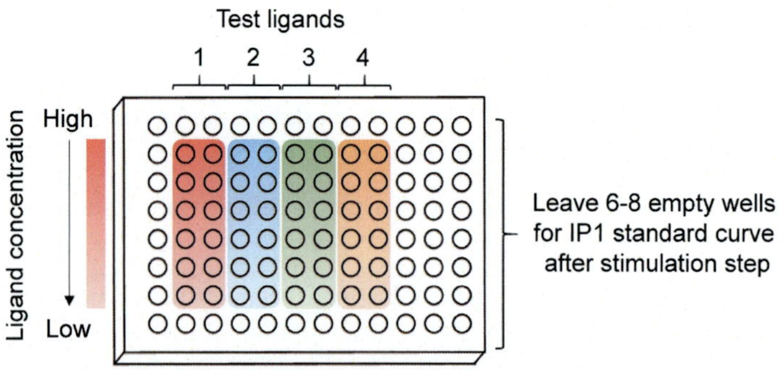

FIG. 8

Example 96-well plate map for evaluating potency of predicted ligands at GPCR of interest. During ligand stimulation, two replicate wells for each ligand concentration are evaluated at 6–8 different ligand concentrations. Empty wells are left on one side of the plate to place the IP1 standards to generate the standard curve.

Note 01. It is advised to prepare fresh stocks of IP1-d2 and anti-IP1-cryptate before each experiment.

Note 02. We have found that using the manufacturer's recommended amount of IP1-d2 and anti-IP1-cryptate (1 × each) provides high sensitivity. However, we have also noted comparable overall results using 0.5 × IP1-d2 and 0.5 × anti-IP1-cryptate (Checco, Zhang, Yuan, Le, et al., 2018; Checco, Zhang, Yuan, Yu, et al., 2018; Do et al., 2018).

9. Prepare serial dilutions of IP1 standard (provided in IP-One Gq kit) in 1 × stimulation buffer. We recommend using a highest concentration of 11,000 nM with five–six 4-fold dilutions.
10. After cells have been stimulated for 30–60 min, remove the plate from incubator and cool to room temperature.
11. Add 28 μL of each prepared IP1 standard solution into empty wells on the plate.

Note: It is critical that an IP1 standard curve is included with every experiment (Burford et al., 2017). 665/620 ratios can differ between experiments. As a result, an IP1 standard curve from an older experiment cannot be used to accurately determine IP1 concentrations for future experiments.

12. Add 6 μL of IP1-d2 in lysis buffer solution to each well (both experimental and those containing IP1 standards). IP1-d2 should be added to each well before anti-IP1-cryptate.

13. Add 6 μL anti-IP1-cryptate in lysis buffer solution to each well (both experimental and those containing IP1 standards).
14. Cover the prepared 96-well plate with aluminum foil and incubate at room temperature for at least 60 min.
15. Read the homogeneous time-resolved fluorescence (HTRF) signal of each well on HTRF-compatible plate reader, using an excitation wavelength of 330 nm and measuring fluorescence emission at both 665 and 620 nm.

Note: The settings of the plate reader will vary with the instrument. Follow guidelines provided with IP-One Gq kit for chosen plate reader.

3.4.4 Data analysis and interpretation

1. Export HTRF data into appropriate format for data processing and conversion (e.g., Microsoft Excel).
2. Calculate ratio of fluorescence emission at 665 nm/620 nm for each well. This is the "HTRF ratio."
3. Using a program suitable for non-linear regression analysis (e.g., GraphPad Prism), plot the IP1 standard curve: log[IP1] vs HTRF ratio (Fig. 9A).
4. Fit the IP1 standard curve to a four-parameter logistical regression model (Eq. 1, Fig. 6) (in GraphPad Prism, this is "log(inhibitor) vs response—variable slope (four parameters)") to estimate Bottom, Top, Hill coefficient, and IC_{50} values for the IP1 standard curve.
5. Using Eq. (2) (Fig. 6) and the measured HTRF ratios, calculate the IP1 concentration in each experimental well.
6. Plot the log[ligand] vs calculated [IP1] for each well in GraphPad Prism. Fit the resulting data to either a three-parameter or four-parameter dose-response model (Fig. 9B).

Note 01. To calculate an accurate EC_{50} value, the dose-response curve must have both a clearly defined top and bottom. Thus, ligands that do not reach maximum response should not be modeled to the dose-response model. In these cases, the EC_{50} values should be reported as > the maximum concentration tested. Ligands that fail to activate the receptor (i.e., show no increase in IP1) should likewise not be fit to the dose-response model.
Note 02. Due to the inherent variability in cell-based assays, it is a good practice to report EC_{50} values obtained from multiple independent experiments (independent transfections, plating, and ligand stimulation performed on different days). Most often, we report the mean and standard deviation for logEC_{50} values from three or more independent experiments. Note that one should not calculate the mean or standard deviation of a non-log transformed EC_{50} value, since EC_{50} values are not normally distributed. We often then convert the mean logEC_{50} back to a non-log value for ease of interpretation.

A

IP1 (nM)	Log[IP1] (nM)	665/620 ratio
11000.0	4.041	0.0653
2750.0	3.439	0.0997
687.5	2.837	0.1938
171.9	2.235	0.4353
43.0	1.633	0.6882
10.7	1.031	0.8472

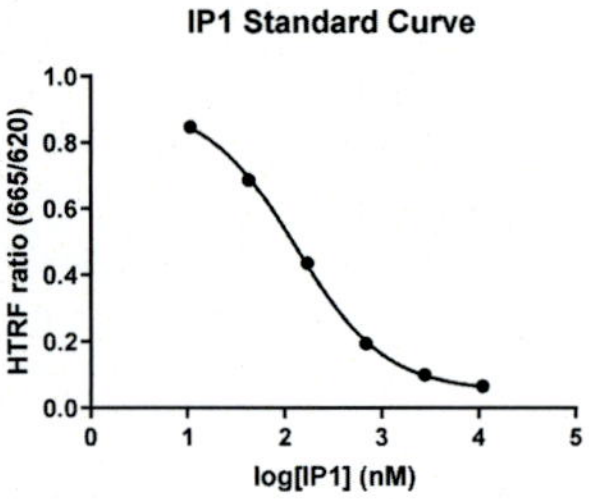

Best-fit values	
Bottom	0.05149
Top	0.9208
$LogIC_{50}$	2.112
Hill	-0.9431
IC_{50} (nM)	129.4

B

		665/520 ratio	
[L-ATRP](μM)	Log[L-ATRP] (μM)	Replicate 1	Replicate 2
10	1	0.2016	0.1902
1	0	0.2172	0.2056
0.1	-1	0.2511	0.2634
0.01	-2	0.5337	0.5287
0.001	-3	0.6482	0.6396
0.0001	-4	0.6229	0.6353

Convert HTRF ratio to [IP1] using IP1 standard curve and Eq. 2

		IP1 (nM)	
[L-ATRP](μM)	Log[L-ATRP] (μM)	Replicate 1	Replicate 2
10	1	682	754
1	0	599	659
0.1	-1	467	430
0.01	-2	102	105
0.001	-3	56	59
0.0001	-4	65	61

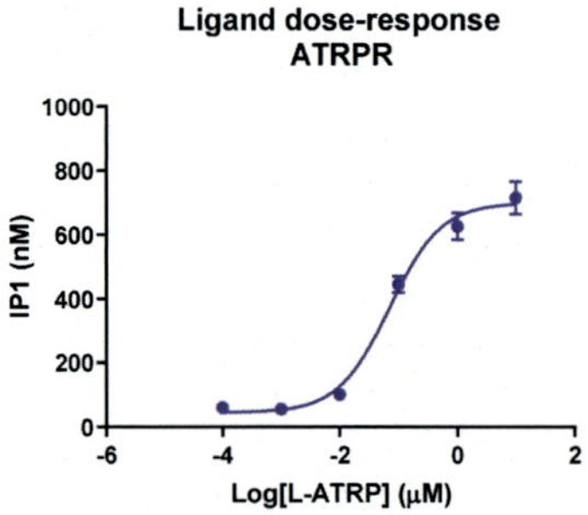

Best-fit values	
Bottom	47.24
Top	701.9
$LogEC_{50}$	-1.150
EC_{50} (μM)	0.07072

FIG. 9

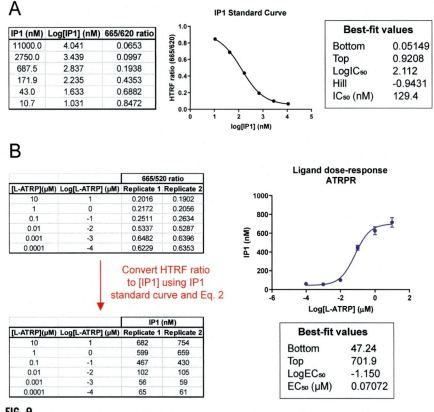

Example data and analysis for activation of *Aplysia* ATRPR by one of its endogenous ligands, peptide L-ATRP (Checco, Zhang, Yuan, Le, et al., 2018). (A) An IP1 standard curve is generated from the experimental plate and fit to a four-parameter logistical equation, as in Fig. 6. (B) Six 10-fold serial dilutions of ligand are prepared and used to stimulate CHO-K1 cells expressing ATRPR. Raw 665/620 ratio (HTRF ratio) is converted to IP1 concentration using the standard curve and Eq. (2). The resulting dose-response curve for L-ATRP is then plotted and fit to a three-parameter logistical equation. Data previously reported in Checco, Zhang, Yuan, Le, et al. (2018), and replotted here.

4 Limitations

As mentioned above, the primary limitation for this method is that gene expression is indirectly confirmed via receptor activation. For receptors with known agonists, functional receptor expression is easily monitored using this method. However, for receptors without verified agonist ligands, a lack of signal in receptor activation assays may be due to the ligand of interest not activating the receptor or due to a lack of appropriate receptor expression on the CHO-K1 cells. To address this limitation, Section 3.3 evaluates a variety of transfection conditions, and recommends controls for evaluating transfection technique. Methods to verify gene expression, such as end-point PCR or qPCR may be helpful to determine if the receptor protein is being expressed. However, in practice we have had trouble with PCR-based assays to confirm expression in recombinant systems, as even minute quantities of plasmid contamination after mRNA extraction can yield false positive results. If aiming to verify gene expression by PCR, care must be taken to ensure full digestion of plasmid by DNase treatment prior to reverse transcription reactions. In this case, it is important to incorporate "No-reverse transcriptase" controls to determine if PCR signal may be arising from plasmid contamination. Alternatively, Western blot can be a useful approach to verify protein expression in transiently transfected cell lines. However, antibodies may not be available for all receptors of interest (especially understudied receptors or those from non-model organisms), limiting the generalizability of the Western blot approach. In these cases, fusion of an epitope tag (e.g., FLAG-tag or HA-tag) to the receptor protein may facilitate Western blot analysis in evaluating receptor protein expression.

Another potential limitation is promiscuity of the Gα protein co-transfected in these studies. Although promiscuous Gα proteins have been shown to associate with a large variety of GPCRs (Conklin et al., 1993; Heydorn et al., 2004; Kostenis, Martini, et al., 2005; Kostenis, Waelbroeck, & Milligan, 2005; Offermanns & Simon, 1995; Wedegaertner et al., 1993) there is no guarantee that a chosen Gα subunit will couple a given GPCR of interest to the PLC pathway. In some cases, it may be beneficial to explore several promiscuous Gα proteins in combination to maximize chances of success. Alternatively, it has been demonstrated that other methods of monitoring receptor function, such as receptor internalization, mass redistribution, and β-arrestin recruitment, can be an effective tool to identifying ligand-receptor interactions, as not all receptors signal through canonical pathways (Foster et al., 2019). Similarly, it is also possible that the GPCR of interest may require co-receptors absent in CHO-K1 cells to signal (Aiyar et al., 1996; Chang, Pearse II, O'Connell, & Rosenfeld, 1993; Flühmann, Muff, Hunziker, Fischer, & Born, 1995). In these cases, it may be beneficial to explore transfection in alternative cell types, or alternative methods of monitoring ligand-receptor interactions (Foster et al., 2019).

5 Summary

This chapter describes our general protocol to develop a new cell-based activity assay for a given GPCR of interest. Steps to isolate the gene of interest from tissue, clone the ORF into an appropriate expression vector, optimize transfection conditions, and generate dose-response experiments are outlined. These methods should allow for evaluation of putative ligand-receptor interactions, and aid laboratories hoping to develop novel GPCR modulators.

Acknowledgments

J.W.C. acknowledges support from the Nebraska Center for Integrated Biomolecular Communication (NIH National Institute of General Medical Sciences P20 GM113126) and from a Nebraska EPSCoR FIRST Award (OIA-1557417). J.W.C. also acknowledges support from the Beckman Institute Postdoctoral Fellowship, provided by the Beckman Foundation as a gift to the Beckman Institute for Advanced Science and Technology at the University of Illinois at Urbana-Champaign.

References

Aiyar, N., Rand, K., Elshourbagy, N. A., Zeng, Z., Adamou, J. E., Bergsma, D. J., et al. (1996). A cDNA encoding the calcitonin gene-related peptide type 1 receptor. *Journal of Biological Chemistry*, *271*, 11325–11329.

Alexander, S. P. H., Battey, J., Benson, H. E., Benya, R. V., Bonner, T. I., Davenport, A. P., et al. (2020). *Class A orphans (version 2020.5) in the IUPHAR/BPS guide to pharmacology database. IUPHAR/BPS Guide to Pharmacology CITE.*

Bauknecht, P., & Jekely, G. (2015). Large-scale combinatorial deorphanization of *Platynereis* neuropeptide GPCRs. *Cell Reports*, *12*, 684–693.

Burford, N. T., Watson, J., & Alt, A. (2017). Standard curves are necessary to determine pharmacological properties for ligands in functional assays using competition binding technologies. *Assay and Drug Development Technologies*, *15*, 320–329.

Chang, C. P., Pearse, R. V., II, O'Connell, S., & Rosenfeld, M. G. (1993). Identification of a seven transmembrane helix receptor for corticotropin-releasing factor and sauvagine in mammalian brain. *Neuron*, *11*, 1187–1195.

Checco, J. W., Zhang, G., Yuan, W. D., Le, Z. W., Jing, J., & Sweedler, J. V. (2018). Aplysia allatotropin-related peptide and its newly identified d-amino acid-containing epimer both activate a receptor and a neuronal target. *Journal of Biological Chemistry*, *293*, 16862–16873.

Checco, J. W., Zhang, G., Yuan, W. D., Yu, K., Yin, S. Y., Roberts-Galbraith, R. H., et al. (2018). Molecular and physiological characterization of a receptor for D-amino acid-containing neuropeptides. *ACS Chemical Biology*, *13*, 1343–1352.

Conklin, B. R., Farfel, Z., Lustig, K. D., Julius, D., & Bourne, H. R. (1993). Substitution of three amino acids switches receptor specificity of Gq alpha to that of Gi alpha. *Nature*, *363*, 274–276.

Do, T. D., Checco, J. W., Tro, M., Shea, J. E., Bowers, M. T., & Sweedler, J. V. (2018). Conformational investigation of the structure-activity relationship of GdFFD and its analogues on an achatin-like neuropeptide receptor of Aplysia californica involved in the feeding circuit. *Physical Chemistry Chemical Physics*, *20*, 22047–22057.

Flühmann, B., Muff, R., Hunziker, W., Fischer, J. A., & Born, W. (1995). A human orphan calcitonin receptor-like structure. *Biochemical and Biophysical Research Communications*, *206*, 341–347.

Foster, S. R., Hauser, A. S., Vedel, L., Strachan, R. T., Huang, X. P., Gavin, A. C., et al. (2019). Discovery of human signaling systems: Pairing peptides to G protein-coupled receptors. *Cell*, *179*, 895–908.e821.

Fricker, L. D., & Devi, L. A. (2018). Orphan neuropeptides and receptors: Novel therapeutic targets. *Pharmacology & Therapeutics*, *185*, 26–33.

Gomes, I., Aryal, D. K., Wardman, J. H., Gupta, A., Gagnidze, K., Rodriguiz, R. M., et al. (2013). GPR171 is a hypothalamic G protein-coupled receptor for BigLEN, a neuropeptide involved in feeding. *Proceedings of the National Academy of Sciences of the United States of America*, *110*, 16211–16216.

Gomes, I., Bobeck, E. N., Margolis, E. B., Gupta, A., Sierra, S., Fakira, A. K., et al. (2016). Identification of GPR83 as the receptor for the neuroendocrine peptide PEN. *Science Signaling*, *9*, ra43.

Hauser, A. S., Attwood, M. M., Rask-Andersen, M., Schiöth, H. B., & Gloriam, D. E. (2017). Trends in GPCR drug discovery: New agents, targets and indications. *Nature Reviews Drug Discovery*, *16*, 829–842.

Heydorn, A., Ward, R. J., Jorgensen, R., Rosenkilde, M. M., Frimurer, T. M., Milligan, G., et al. (2004). Identification of a novel site within G protein alpha subunits important for specificity of receptor-G protein interaction. *Molecular Pharmacology*, *66*, 250–259.

Kostenis, E., Martini, L., Ellis, J., Waldhoer, M., Heydorn, A., Rosenkilde, M. M., et al. (2005). A highly conserved glycine within linker I and the extreme C terminus of G protein alpha subunits interact cooperatively in switching G protein-coupled receptor-to-effector specificity. *Journal of Pharmacology and Experimental Therapeutics*, *313*, 78–87.

Kostenis, E., Waelbroeck, M., & Milligan, G. (2005). Techniques: Promiscuous Galpha proteins in basic research and drug discovery. *Trends in Pharmacological Sciences*, *26*, 595–602.

Laschet, C., Dupuis, N., & Hanson, J. (2018). The G protein-coupled receptors deorphanization landscape. *Biochemical Pharmacology*, *153*, 62–74.

Marchler-Bauer, A., Bo, Y., Han, L., He, J., Lanczycki, C. J., Lu, S., et al. (2017). CDD/SPARCLE: Functional classification of proteins via subfamily domain architectures. *Nucleic Acids Research*, *45*, D200–D203.

Marchler-Bauer, A., Derbyshire, M. K., Gonzales, N. R., Lu, S., Chitsaz, F., Geer, L. Y., et al. (2015). CDD: NCBI's conserved domain database. *Nucleic Acids Research*, *43*, D222–D226.

Marchler-Bauer, A., Lu, S., Anderson, J. B., Chitsaz, F., Derbyshire, M. K., DeWeese-Scott, C., et al. (2011). CDD: A conserved domain database for the functional annotation of proteins. *Nucleic Acids Research*, *39*, D225–D229.

Muratspahić, E., Freissmuth, M., & Gruber, C. W. (2019). Nature-derived peptides: A growing niche for GPCR ligand discovery. *Trends in Pharmacological Sciences*, *40*, 309–326.

Offermanns, S., & Simon, M. I. (1995). G alpha 15 and G alpha 16 couple a wide variety of receptors to phospholipase C. *Journal of Biological Chemistry*, *270*, 15175–15180.

Quiroga Artigas, G., Lapébie, P., Leclère, L., Bauknecht, P., Uveira, J., Chevalier, S., et al. (2020). A G protein–coupled receptor mediates neuropeptide-induced oocyte maturation in the jellyfish Clytia. *PLoS Biology*, *18*, e3000614.

Shiraishi, A., Okuda, T., Miyasaka, N., Osugi, T., Okuno, Y., Inoue, J., et al. (2019). Repertoires of G protein-coupled receptors for Ciona-specific neuropeptides. *Proceedings of the National Academy of Sciences of the United States of America*, *116*, 7847–7856.

Trinquet, E., Fink, M., Bazin, H., Grillet, F., Maurin, F., Bourrier, E., et al. (2006). D-*myo*-inositol 1-phosphate as a surrogate of D-*myo*-inositol 1,4,5-tris phosphate to monitor G protein-coupled receptor activation. *Analytical Biochemistry*, *358*, 126–135.

Wedegaertner, P. B., Chu, D. H., Wilson, P. T., Levis, M. J., & Bourne, H. R. (1993). Palmitoylation is required for signaling functions and membrane attachment of Gq alpha and Gs alpha. *Journal of Biological Chemistry*, *268*, 25001–25008.

Wistrand, M., Kall, L., & Sonnhammer, E. L. (2006). A general model of G protein-coupled receptor sequences and its application to detect remote homologs. *Protein Science*, *15*, 509–521.

CHAPTER

Assays for detecting arrestin interaction with GPCRs

3

Nicole A. Perry-Hauser[a,b], Wesley B. Asher[a,b,c,†], Maria Hauge Pedersen[a,b,d,†], and Jonathan A. Javitch[a,b,c,*]

[a]*Department of Psychiatry, Columbia University, Vagelos College of Physicians and Surgeons, New York, NY, United States*

[b]*Division of Molecular Therapeutics, New York Psychiatric Institute, New York, NY, United States*

[c]*Department of Molecular Pharmacology and Therapeutics, Columbia University, Vagelos College of Physicians and Surgeons, New York, NY, United States*

[d]*NNF Center for Basic Metabolic Research, Section for Metabolic Receptology, Faculty of Health and Medical Sciences, University of Copenhagen, Copenhagen, Denmark*

**Corresponding author: e-mail address: jaj2@cumc.columbia.edu*

Chapter outline

[†]These authors contributed equally.

Methods in Cell Biology, Volume 166, ISSN 0091-679X, https://doi.org/10.1016/bs.mcb.2021.06.007

Abstract

The four vertebrate arrestins play a key role in the desensitization and internalization of G protein-coupled receptors (GPCRs) and also mediate receptor-dependent signaling. Recent work has shown that bias for arrestin vs G protein signaling could offer certain therapeutic advantages (or disadvantages) in different systems, making assays that measure arrestin binding to receptors important for drug discovery efforts. Herein, we briefly review several commonly used techniques for measuring arrestin binding to receptors, as well as provide an in-depth and methodologically focused review of two methods that do not require receptor modification. The first approach measures direct binding between purified arrestin and rhodopsin, and the second measures the recruitment of arrestin to receptors in living cells.

1 Introduction

G protein-coupled receptors (GPCRs) regulate numerous signaling processes and collectively comprise the largest group of therapeutic targets in drug discovery (~27% of the global market) (Hauser et al., 2017). Over 800 different GPCRs mediate intracellular responses through activation of heterotrimeric G proteins or through arrestin-mediated signaling pathways. The arrestins comprise a small family of proteins that were originally discovered for their role in the desensitization and internalization of GPCRs in humans (Gurevich & Gurevich, 2006). Recent experimental efforts have shown that preferentially activating either G protein or arrestin, a mechanism known as biased agonism, can lead to specific signaling effects that are more favorable, i.e., reduce negative side effects (Liu et al., 2017; Rankovic et al., 2016; Slosky et al., 2020; Teoh et al., 2018). For instance, the beta-adrenergic receptor agonist carvedilol has been shown to preferentially recruit arrestin while antagonizing G protein activation and is reported to increase survival rates in patients suffering from heart failure (Kim et al., 2020; Wisler et al., 2007). Therefore, it has become increasingly important to have robust assays that measure arrestin interactions with GPCRs and other intracellular effectors.

The role of arrestins in GPCR signal termination is well established (DeFea, 2011; Luttrell & Lefkowitz, 2002). Termination is initiated by GPCR kinase (GRK), which phosphorylates serines and threonines in the receptor C-terminal tail and, in some cases, the intracellular loops (Gurevich & Gurevich, 2019). The receptor-attached phosphates then bind to the phosphate sensor in arrestin, which primes the molecule for high affinity engagement of the receptor via its activation sensor (Latorraca et al., 2018). Engagement of both sensors is associated with conformational rearrangements of arrestin, including displacement of the C terminus and destabilization of the polar core, that enable high affinity interaction with receptor (Chen et al., 2017; Kim et al., 2013; Shukla et al., 2013; Xiao et al., 2004). As a result, arrestins prefer to interact with the active, phosphorylated receptor, and their binding is weaker if one or the other sensors is not engaged.

In this chapter, we review several experimental approaches for measuring arrestin recruitment to GPCRs. In Section 2 we describe a direct binding assay of arrestin interaction with rhodopsin that uses purified protein, and in Section 3 we briefly review other assays that use purified samples (pull-down and co-immunoprecipitation). In Section 4 we provide a brief review of cellular methods for monitoring GPCR-arrestin interaction and in Section 5 an in-depth analysis of a luminescence approach using unmodified GPCRs developed in our laboratory.

2 Direct binding assay between purified arrestin and rhodopsin

2.1 Direct binding assay with rhodopsin

The prototypical GPCR rhodopsin, a protein involved in visual phototransduction, is one of the best studied GPCR systems (Palczewski, 2006). Rhodopsin is activated via light, which enables activation of its cognate G protein, transducin. Active receptor is subsequently deactivated by GRK phosphorylation, which triggers arrestin recruitment. *In vitro* transcription and translation of radiolabeled arrestin has proved a useful system for efficient characterization of arrestin binding to purified rhodopsin (Fig. 1). This is due to the ease of detecting radiolabeled protein after gel filtration using a scintillation counter. This method can be used to screen numerous arrestin

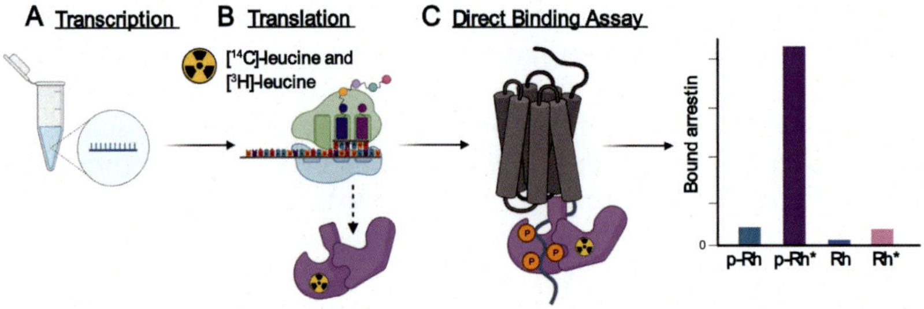

FIG. 1

Overview of direct binding assay using rhodopsin. (A) A plasmid with arrestin cDNA is linearized downstream of the stop codon and transcribed using SP6 polymerase. (B) Arrestin mRNA is translated in the presence of [^{14}C]-leucine and [^{3}H]-leucine to generate radiolabeled arrestin. (C) Arrestin affinity for various purified rhodopsin states is tested using a direct binding assay with radiolabeled protein. The far-right panel shows a graphic representing the results of the assay inspired by Gurevich and Benovic (1995), showing the quantification of arrestin binding to all functional forms of rhodopsin. Functional forms of rhodopsin are labeled as dark phosphorylated (P-Rh), light-activated phosphorylated (P-Rh*), dark unphosphorylated (Rh), and light-activated unphosphorylated (Rh*). Created with https://biorender.com/.

mutants without the requirement for full-scale arrestin purification (Gurevich & Benovic, 1993) and uses purified native rhodopsin (McDowell, 1993) and untagged arrestin, decreasing the possibility of binding artifacts.

In brief, this system uses four different functional forms of rhodopsin purified from bovine retinas: dark phosphorylated (P-Rh), light-activated phosphorylated (P-Rh*), dark unphosphorylated (Rh), and light-activated unphosphorylated (Rh*) (Gurevich & Benovic, 2000). Arrestin preferentially binds to activated and phosphorylated rhodopsin, yielding at least a 10-fold higher binding than to other forms, but can also recognize the activated or phosphorylated state of rhodopsin. Arrestin binding to inactivate unphosphorylated Rh is near zero. Therefore, this assay can be used to study high-affinity arrestin interactions with P-Rh* in addition to lower-affinity interactions with P-Rh and Rh*.

2.2 Materials and equipment

2.2.1 Transcription reaction

- Transcription mix (final concentrations) (Gurevich, 1996): 120mM 4-(2-hydroxyethyl)-1-piperazine-1-ethanesulfonic acid (HEPES)-KOH, pH 7.5; 2mM spermidine-HCl; 16mM $MgCl_2$; 40mM dithiothreitol (DTT); 3mM each of adenosine triphosphate (ATP), guanosine triphosphate (GTP), cytidine triphosphate (CTP), and uridine triphosphates (UTP) (NTP mix); 100μg/mL Ac-BSA, 2.5U/mL of inorganic pyrophosphatase; 200U/mL RNasin; and 1500U/mL SP6 RNA polymerase

- 9 M LiCl
- 70% (v/v) ethanol
- 3 M sodium acetate, pH 5.0
- 100% ethanol
- Nuclease-free water

2.2.2 Translation reaction

- Translation mix (Gurevich & Benovic, 2000) (prepare on ice, in the following order): 140 μL 70% (v/v) rabbit reticulocyte lysate (RRL) (Jackson & Hunt, 1983), 10 μL of 19 unlabeled amino acids (final concentration 50 μM each), 6 μL 1 M creatine phosphate (30 mM final), 2 μL protease inhibitor mix (0.1 μg/mL pepstatin, 0.1 μg/mL leupeptin, 0.1 mg/mL soybean trypsin inhibitor), 5 μL 0.2 M cAMP (5 mM final), 11 μL 2 M potassium acetate (110 mM final), 0.8 μL [^{14}C]-leucine (14,000–35,000 dpm/mL, 30–35 μM final), if using [^{3}H]-leucine, add 2.2 μL (800,000–1,000,000 dpm/mL) (amino acid-free RRL usually supplies 5–15 μM cold leucine), 1 μL RNasin (200 U/mL), and 2 μL creatine kinase (200 μg/mL)
- ATP+GTP mix (40 mM each)
- 10% (w/v) trichloroacetic acid (TCA)
- 50 mM KOH, 1% SDS
- Scintillation fluid

2.2.3 Direct binding assay

- Binding buffer: 50 mM Tris-HCl, pH 7.5, 50 mM potassium acetate, 0.5 M $MgCl_2$, and 0.5 mM DTT
- Equilibration buffer: 10 mM Tris-HCl, pH 7.5 and 100 mM NaCl (ice-cold)
- 2 mL Sephadex G-75 column

2.3 Step-by-step method details

2.3.1 In vitro transcription

Generation of arrestin mRNA for use in *in vitro* translation (typical yields range from 1500 to 3000-nucleotide-long transcripts). This is an abbreviated protocol from the more detailed *Use of Bacteriophage RNA Polymerase in RNA synthesis* (Gurevich, 1996).

2.3.1.1 DNA linearization—Timing: Variable

1. Purify and linearize the cDNA after the stop codon where the arrestin coding sequence is under the control of an SP6 promoter (see Note 1). The timing of linearization will depend on the chosen restriction enzymes; however, most reactions are incubated in a 37 °C water bath for 1 h.

2.3.1.2 Transcription reaction—Timing: 2h incubation at 37 °C

2. Thaw the NTP mix on ice and allow the other components to reach room temperature. Keep RNA polymerase, RNase inhibitor, and PPase at −20 °C until needed.
3. Incubate 30 µg of linearized DNA at 38 °C for 2h in 1 mL transcription mix.

2.3.1.3 mRNA purification—Timing: 35–40 min

4. Add 0.4 volume 9 M LiCl to the transcription mix, vortex, and incubate on ice for 5–10 min.
5. Pellet the mRNA by centrifugation at 8000 *g* for 10 min at 4 °C.
6. Wash the pellet with 1 mL ice-cold 2.5 M LiCl.
7. Wash the pellet with 1 mL room-temperature 70% ethanol and allow to air dry for 5–10 min.
8. Dissolve the pellet in 1 volume nuclease-free distilled water. Remove a 10 µL aliquot to quantify mRNA via absorption at 260 nm.
9. Reprecipitate the mRNA by addition of 0.1 volume 3 M sodium acetate, pH 5.0, and 3.3 volumes of ethanol. Vortex and incubate for 10 min on ice (see Note 2).

2.3.1.4 Notes

Note 1. Other RNA polymerases may be used for this protocol, including T7 and T3. However, the buffer concentrations and optimal reaction conditions will vary.
Note 2. Purified mRNA suspension is stable at −80 °C for several years and can be refrozen up to 30 times.

2.3.2 In vitro translation

The *in vitro* translation system permits rapid expression of 10–20 radiolabeled arrestin mutants simultaneously. This is an abbreviated protocol from the more detailed *Arrestin: Mutagenesis, Expression, Purification, and Functional Characterization* (Gurevich & Benovic, 2000).

2.3.2.1 mRNA prep—Timing: 15 min

1. Pellet 50–150 µg/mL mRNA from suspension by centrifugation for 5–10 min.
2. Wash with 1 mL 70% ethanol (v/v) by vortexing and recentrifugation.
3. Air-dry the pellet for 5–10 min.
4. Dissolve the pellet in 20 µL nuclease-free distilled water.

2.3.2.2 Translation reaction—Timing: 2h incubation at 22.5 °C

5. Add 180 µL translation mix to the mRNA and incubate for 2 h at 22.5 °C (see Note 1).

2.3.2.3 Post-translation purification—Timing: 1.5 h

6. Add 4 µL of a solution containing 40 mM ATP and GTP (1 mM final) to the mix and incubate for 7 min at 37 °C (ribosome “run-off”).

7. Cool the samples on ice prior to centrifugation at 100,000 rpm for 60 min at 4 °C (TLA 100.1 rotor, Beckman, Fullerton, CA) (see Note 2).
8. Carefully remove the supernatant, aliquot, and freeze at −80 °C until ready to use (see Note 3).

2.3.2.4 Calculation of protein yield

9. Dilute a 2 µL aliquot of protein 10-fold with 18 µL water.
10. Apply 5 µL aliquots to Whatman 3MM paper in triplicate (1 cm × 1 cm square) and dry for 3–5 min.
11. Immerse the paper in ice-cold 10% TCA for 10–15 min and then in boiling 5% TCA for 10 min (see Note 4).
12. Rinse the paper briefly in ethanol and diethyl ether to remove water and TCA, respectively.
13. Dry the paper and cut into individual squares.
14. Place each square in a separate scintillation vial containing 500 µL of 50 mM KOH, 1% SDS for 30–50 min at room temperature.
15. Add 5 mL water-miscible scintillation fluid, shake briefly, and quantify protein-incorporated radioactivity (Note 5).

2.3.2.5 Notes

Note 1. When preparing the translation mix, vortex the solution briefly before adding RNase inhibitor and creatine kinase. As a negative control, prepare a sample of translation mix without mRNA.
Note 2. This procedure pellets ribosomes and aggregated protein. The supernatant contains free [^{14}C]- and [^{3}H]-leucine and [^{14}C]- and [^{3}H]-labeled arrestin. These samples can be used either directly for the binding assay or run through gel filtration for further purification. Translated arrestins are stable for several months at −80 °C.
Note 3. Translated arrestin can be used for 3–5 thaw cycles before observing a reduction in activity.
Note 4. Immersion in cold TCA removes radiolabeled leucine while hot TCA hydrolyzes aminoacyl-tRNA.
Note 5. The total concentration of leucine in the translation mix should be in the 30–50 µM range. Typical yields for wild-type bovine visual arrestin are in the 70–150 fmol/µL range.

2.3.3 Direct binding assay

The four different functional forms of rhodopsin are used to measure arrestin selectivity and specificity for receptor. High-affinity binding requires that rhodopsin be both activated and phosphorylated to engage the activation-recognition site and phosphorylation-recognition site of arrestin.

2.3.3.1 Direct binding assay—Timing: 30–45 min

1. Dilute the translated arrestin samples to 100 fmol (2 nM final in 50 µL binding buffer) with 50 mM Tris-HCl, pH 7.5, 50 mM potassium acetate, 0.5 mM $MgCl_2$, and 0.5 mM DTT and add 25 µL of the solution to an Eppendorf tube.
2. Dilute Rh and P-Rh under red light to 12 µg/mL and distribute 25 µL to each tube (see Note 1).
3. Depending on the functional form of rhodopsin, incubate the samples for 5 min in the dark or with illumination at 37 °C (Note 2).
4. Cool the samples on ice.
5. Separate bound from free arrestin at 4 °C on a 2 mL Sephadex G-75 column equilibrated with ice-cold 10 mM Tris-HCl, pH 7.5 and 100 mM NaCl. If using dark rhodopsin, perform this separation in dim red light.
6. After the samples are loaded onto the column, wash with 500 µL of the equilibration buffer.
7. Elute the samples into scintillation vials using 600 µL of the equilibration buffer (Note 3).
8. Add 5 mL of scintillation fluid to each vial and count the radioactivity.

2.3.3.2 Notes

Note 1. Please see the referenced protocol for a comprehensive review of rhodopsin preparation (McDowell, 1993). Rhodopsin can be used in the following membrane systems: native disc membranes (Malmerberg et al., 2015), bicelles (Chen et al., 2015), and nanodiscs (Azevedo et al., 2015; Bayburt et al., 2011; Singhal et al., 2013; Vishnivetskiy et al., 2013). This assay can also be adjusted for use with other GPCRs that are purified and reconstituted in nanodiscs (Cai et al., 2017; Goddard et al., 2015; Knepp et al., 2011; Leitz et al., 2006).
Note 2. Timing is critical due to the rapid decay of light-activated rhodopsin.
Note 3. The membrane-containing fraction elutes between 0.5 and 1.1 mL.

2.4 Quantification and statistical analysis

Data collected from the direct binding assay are typically analyzed using one-way analysis of variance (ANOVA) with correction for multiple comparisons.

2.5 Limitations

All arrestins (arrestin-1, -2, -3, and -4) express in cell-free translation, but the yields for arrestin-1 are higher than for other subtypes.

2.6 Optimization and troubleshooting

2.6.1 High nonspecific binding

Most nonspecific binding encountered in this assay is due to aggregation. This can be counteracted by decreasing the incubation temperature to 20–25 °C or increasing the ionic strength of the column buffer.

2.6.2 Nonfunctional arrestin

Occasionally the translated arrestin will be nonfunctional due to misfolding, denaturation, or proteolysis. To detect whether the protein is degraded, the translated product can be run on an SDS gel (10% acrylamide in the running gel).

3 Brief overview of other assays using purified components

3.1 Affinity pull-down assay

The *in vitro* affinity pull-down assay uses purified proteins or protein fragments to measure direct protein-protein interactions between two or more proteins. For this assay, one or both protein partners are tagged with an affinity label, which permits "bait" protein immobilization on an appropriate matrix. This is dissimilar from immunoprecipitation which uses an antibody as the bait protein. Successful affinity tags that have been used with arrestin have included N-terminal maltose binding protein (MBP) (Perry et al., 2019a; Zhan et al., 2016), glutathione-S-transferase (GST) (Thomsen et al., 2016), and the polyhistidine-tag (Mayer et al., 2019). Additionally, GPCRs are also amenable to several affinity tags including N-terminal Flag tags and C-terminal poly-histidine tags (Kobilka, 1995). In the pull-down assay, the bait protein is first immobilized on resin by its affinity tag. Then the "prey" protein, which can be either purified protein, crude lysate, or another protein source, is incubated with the bait protein. Of note, crude lysate opens the possibility that an additional protein or co-factor may assist in the interaction and, consequently, this format of pull-down cannot distinguish between direct and indirect interactions. Finally, the protein complex is eluted according to its affinity ligand, which can range from adding a high concentration of competing affinity ligand to changing buffer pH. The eluted proteins are then analyzed, most commonly by SDS-PAGE and Western analysis.

Pull-down assays are powerful tools for confirming protein-protein interactions and discovering novel interacting partners. In a structural study that investigated the arrestin-rhodopsin interaction, a variation of the pull-down assay was used to determine binding between rhodopsin and several arrestin variants (Kang et al., 2015). Here, the investigators tagged rhodopsin at its N-terminus with MBP and immobilized the fusion on amylose beads. They then incubated the immobilized rhodopsin with *in vitro* translated arrestin-1 labeled with ^{35}S and assayed for binding using SDS-Page and visualization on a PhosphorImager. As a prototypical GPCR, it is quite common to see rhodopsin used in pull-down assays to test arrestin function (Latorraca et al., 2018; Mayer et al., 2019); however, other GPCRs can also be used. In addition to using pull-down assays for detecting GPCR-arrestin interaction, there are also many examples of pull-down assays for testing arrestin interaction with effectors, which often require stimulation of GPCRs or peptide mimetics to induce complex formation (Brady et al., 2004; Perry et al., 2019a; Smith & Rajagopal, 2016; Zhan et al., 2016).

The greatest advantage of the pull-down method is the ability to detect direct protein interactions. Limitations to the method include the necessity for the protein-protein interaction to have a relatively high affinity and the need to purify interacting partners with a suitable affinity tag.

3.2 Co-immunoprecipitation

Like the *in vitro* pull-down assay, co-immunoprecipitation (co-IP) immobilizes a target protein as "bait"; however, this is accomplished using protein-specific antibodies. As the focus of co-IP is to not only to capture and purify the bait protein, but also to capture native interactions with the bait protein, this method must be verified using a more direct technique, such as the pull-down assay mentioned above. The first step in this assay is to incubate antibodies directed toward a protein of interest with a crude lysate or other protein source to permit antigen-antibody coupling. Typically, antibodies are chosen against an antigen tag (i.e., FLAG-tag, HA-tag, etc.), but it is also possible to use antibodies directly against the target protein. In the case of GPCRs and arrestins, groups have been successful using Flag M2 beads (Flag-arrestin-2) (Shukla et al., 2014; Thomsen et al., 2016; Xiao et al., 2007), anti-HA antibodies (Milano et al., 2006; Perry et al., 2019b), and anti-arrestin antibodies (Lee et al., 2019).

After incubation with the appropriate antibody, protein A or protein G Sepharose beads are added to immobilize the protein-antibody complex. Alternatively, a resin pre-coupled to an antibody can be used for the immobilization step. A recent study highlighted co-IP of the nonvisual arrestins with angiotensin type I receptor at the plasma membrane using protein G-Sepharose beads for monoclonal (mouse) antibodies and protein A-Sepharose beads for polyclonal (rabbit or goat) antibodies to immobilize the primary antibodies for the GPCRs of interest (Wertz et al., 2019). This bead/antibody conjugate was then incubated with lysate for at least 4h prior to the wash step. After the beads were washed, the proteins were eluted by boiling the samples in SDS loading buffer and analyzed via Western blot.

As seen with the example above, co-IP is commonly employed to detect GPCR interaction with arrestin and heterotrimeric G proteins (Lee et al., 2008; Wertz et al., 2019; Zheng et al., 2019). More recently, co-IP was used to investigate the formation of megaplexes (Thomsen et al., 2016), which consist of flag-tagged β2V2 receptor complexed with both arrestin-2 and heterotrimeric Gs when stabilized by Fab30 and stimulated by the high-affinity agonist BI-167107.

4 Overview of common cell-based arrestin recruitment assays

In addition to assays that use purified proteins as discussed above, cell-based assays for measuring arrestin-receptor interactions are widely employed for screening and drug discovery. In this section, we provide an overview of select non-imaging-based fluorescence and luminescence assays that are commonly reported to detect arrestin-receptor interactions in living cells (Fig. 2). An in-depth comparison of these assays including advantages and limitations of each technique is described elsewhere (Donthamsetti et al., 2015; Wang et al., 2004). Of note, all assays described in this section require that C-terminal fusion tags be added to the receptor of interest. These receptor modifications could potentially impact binding of arrestin or other effector proteins and have undesired effects on signaling outcomes.

4.1 Protease-mediated and complementation-based assays

Assays that use protease tags for generating a luminescent or fluorescent readout are commonly used for detecting GPCR-arrestin interactions in cells. The Tango-GPCR assay system is a protease-activated reporter gene assay (Barnea et al., 2008; Kroeze et al., 2015) now also commercially available from ThermoFisher Scientific (Fig. 2A). In this assay, the C-terminus of the GPCR of interest is fused to a transcription factor (TF) by a linker containing a specific protease-cleavage site and arrestin is tagged with the protease. When arrestin is translocated to the membrane and binds activated receptor, the protease tag is brought into close contact with the protease site and cleaves the TF from the receptor, allowing it to enter the nucleus. The TF can then regulate the transcription of a gene coding a beta-lactamase reporter construct, which cleaves a substrate labeled with fluorophores resulting in changes in fluorescence that can be readily monitored. Another assay that relies on protease tags is the LinkLight assay from BioInvenu (Fig. 2A) that uses a modified inactive luciferase attached to arrestin and a protease attached to the C-terminus of the receptor (Eishingdrelo et al., 2011). When arrestin binds receptor, the protease cleaves a site within the modified luciferase, which activates it and allows for light to be produced and measured as a readout.

Assays based on enzyme and luciferase fragment complementation have also been used to measure arrestin-receptor interactions and include DiscoverX's Pathfinder assay (McGuinness et al., 2009) and Promega's NanoBit assay (Dixon et al., 2016), respectively (Fig. 2B). In the Pathfinder assay, split fragments of the enzyme β-galactosidase (β-Gal) are attached to arrestin and the C-terminus of the receptor. Upon arrestin interaction with receptor, the two fragments complement and form a functional enzyme that can then hydrolyze its substrate, generating a chemiluminescent signal. Similarly, the NanoBit assay can be used where large and small split fragments, referred to as LgBiT and SmBiT, of the luciferase NanoLuc are attached to arrestin and the C-terminus of the receptor, respectively.

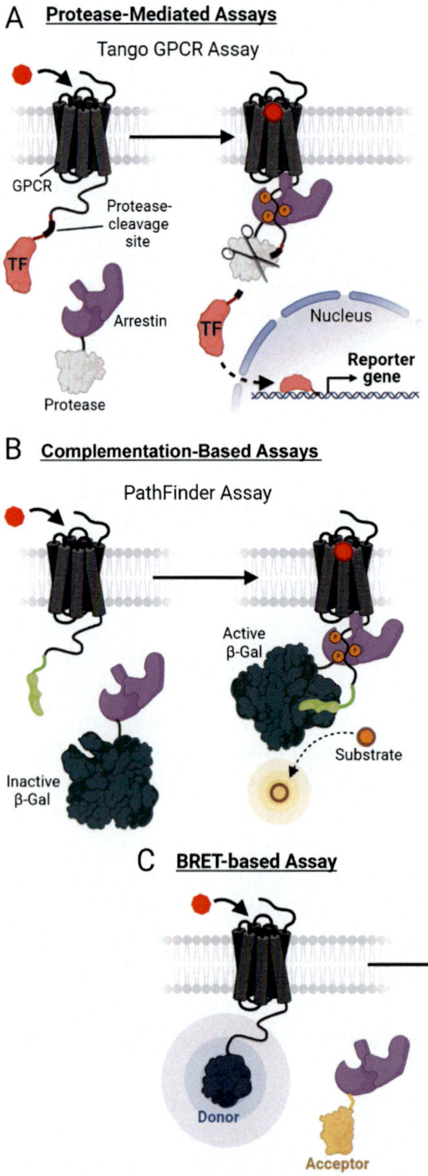

FIG. 2

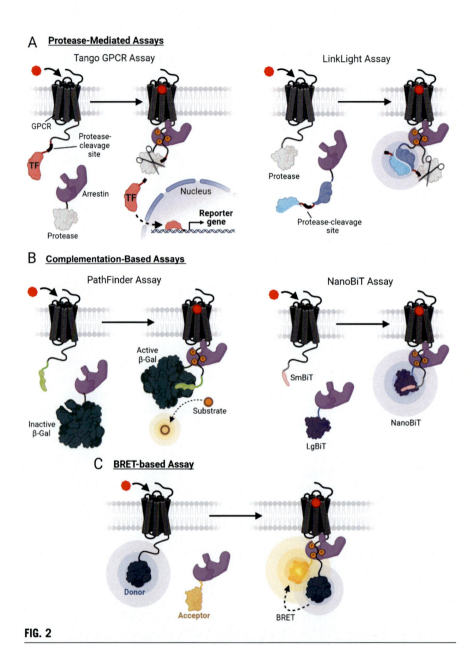

Schematic overview of the commonly used live cell-based assays used to detect arrestin-GPCR interactions. (A) Protease-mediated assays used to monitor arrestin recruitment include the Tango GPCR assay (left) and LinkLight assay (right). (B) Complementation-based assays include the Pathfinder (left) and NanoBit (right) assays. See the main text for detailed descriptions of each assay. (C) BRET-based assay used to monitor direct recruitment of arrestin. Created with https://biorender.com/.

Upon arrestin-receptor binding, complementation of the two components occurs, forming functional NanoLuc, which generates a luminescent readout.

4.2 Bioluminescence resonance energy transfer assays

Bioluminescence Resonance Energy Transfer (BRET) assays have long been used to measure direct receptor-arrestin interactions in living cells (Barak et al., 1997). To measure recruitment, arrestin is tagged with a fluorescent protein (FP) acceptor and the GPCR is fused at the C-terminus with a luciferase donor (or vice versa) that emits light upon reaction with a substrate. Upon agonist stimulation, arrestin is recruited from the cytosol and binds to the receptor, bringing the two fusion tags into proximity where energy is transferred from the luciferase donor to the FP acceptor (Fig. 2C). This process generates sensitized emission from the acceptor at longer wavelengths relative to emission from the donor. BRET is calculated from the ratio of these two signals (FP acceptor emission/luciferase donor emission) and used to detect arrestin-receptor interactions. Variants of *Renilla* luciferase (RLuc) are most often used for the donor in this assay in combination with variants of green fluorescent protein (GFP) as acceptors. Most studies use GFP-tagged or Venus-tagged arrestins and place the tag on the extreme N- or C-terminus (Salahpour et al., 2012). More in depth experimental details including discussion of different FP acceptors can be found elsewhere (Kocan & Pfleger, 2011; Pfleger et al., 2006).

5 Cell-based arrestin recruitment assays to unmodified GPCRS

5.1 Arrestin recruitment to unmodified GPCRS

The assays described in Section 4 require that fusion tags be attached to the C-terminus of GPCRs of interest. This modification has the potential to affect receptor function by altering GRK or arrestin binding and could thereby impact downstream signaling. Therefore, there is a clear benefit to using assays that avoid such receptor modification. We previously developed a modified BRET assay in which RLuc is lipid anchored to the membrane instead of attached to the receptor (Donthamsetti et al., 2015). This assay measures "bystander" BRET between arrestin tagged with the FP acceptor and the RLuc donor at the membrane and is thus able to measure arrestin recruitment to unmodified receptors without C-terminal tags (Fig. 3A). To enhance the dynamic range of this assay, we added "helper peptides" derived from Sp1 and SH3 (Grünberg et al., 2013) to the donor and acceptor components of the assay to enhance their interaction affinity (Donthamsetti et al., 2015).

More recently we established a second membrane arrestin recruitment assay that uses unmodified GPCRs and a novel split NanoLuc as the functional luciferase (Fig. 3B) (Pedersen et al., 2020). One fragment of NanoLuc is attached to a membrane anchor and the other is attached to the N-terminus of arrestin. This assay provides a

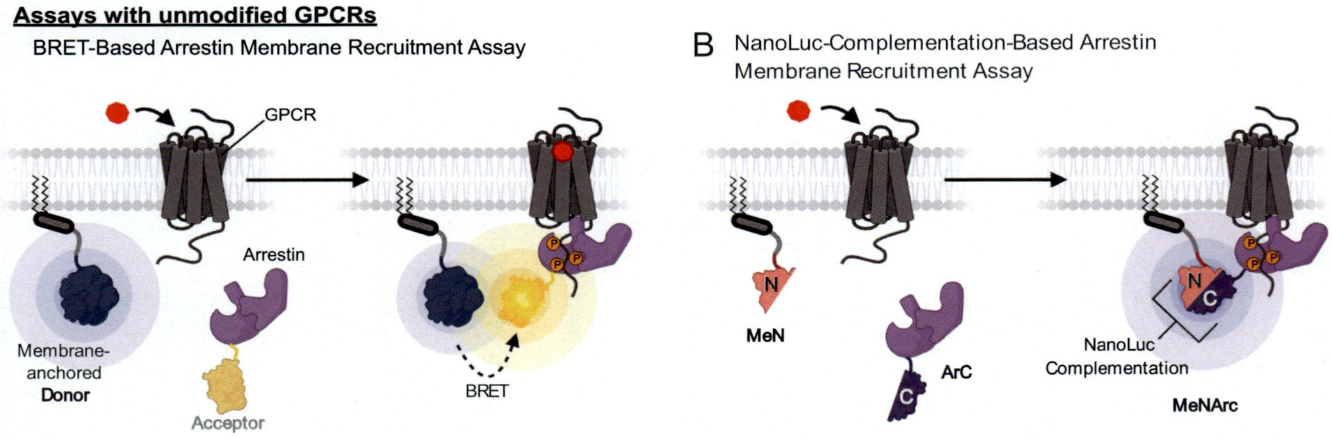

FIG. 3

Overview of arrestin recruitment assay to unmodified GPCRs. (A) A BRET-based assay to measure indirect recruitment of arrestin. RLuc is lipid anchored to the cell membrane and measures "bystander" BRET upon recruitment of fluorescently-tagged arrestin to the unmodified GPCR. (B) The N-terminal fragment (amino acids 1–102) of NanoLuc is attached to a doubly palmitoylated fragment of GAP43 and is anchored in the plasma membrane (membrane tethered N-terminal split, MeN). The C-terminal fragment (amino acids 103–172) is attached to the N-terminus of arrestin-2 or arrestin-3 (arrestin tethered C-terminal split, ArC). In the absence of GPCR stimulation, arrestin is diffuse in the cytosol and the split NanoLuc is unable to complement. Upon GPCR stimulation, arrestin is recruited to the membrane and the split NanoLuc undergoes complementation.

luminescence rather than a BRET readout and does not require the use of the helper peptides that we engineered into the BRET assay. This assay is also now available commercially from Multipsan. In this section, we provide methodological details on the luminescence complementation assay.

5.2 Materials and equipment

5.2.1 Mammalian expression plasmids

- pCDNA3.1-Mem-Link-N1 (MeN)
- pCDNA3.1-N2-Link-Arr3 (ArC)
- pCDNA3.1-N2-Link-Arr2 (ArC)
- pCDNA3.1-GRK2
- pCDNA3.1-N2-Link-FYVE (EeN)
- pCDNA3.1 (empty vector)

5.2.2 Cell culture

- HEK293 cells
- DMEM
- 10% fetal bovine serum
- Penicillin-streptomycin, 100 U/mL
- Dulbecco's Phosphate buffered saline (DPBS)
- Glucose stock (0.5 M)
- 0.05% Trypsin
- Preferred transfection reagent (i.e., polyethyleimine (PEI), lipofectamine, etc.)

5.2.3 Complementation assay

- Black/White 96-well Isoplate
- Coelenterazine H or Fumirazine in absolute ethanol, 5 mM stock
- Repeater pipettor
- Plate reader for luminescence, e.g., Pherastar FS, BMG

5.3 Step-by-step method details

5.3.1 Cell culture

5.3.1.1 Cell line maintenance—Timing: Variable

1. Maintain HEK293 cells in DMEM + GlutaMax™-I with 10% fetal bovine serum and 1% penicillin/streptomycin at 37 °C and 5% CO_2 (see Note 1).
2. To prepare for transfection, seed cells at such a density that they can reach 90–100% confluency within 24 h. For a 10 cm tissue culture plate, seed 3–4 × 10^6 cells in a total volume of 10 mL (see Note 2).

5.3.1.2 Transfection—Timing: 30 min

3. After 24 h, prepare plasmids for transient transfection in a 1.5 mL microcentrifuge tube. The following DNA concentrations were optimized for a 10 cm plate: 1.5 μg receptor, 1.5 μg MeN, 1.5 μg ArC, and if desired, 4.8 μg GRK2 (see Note 3). Adjust the total amount of plasmid DNA to 15 μg using empty vector (see Note 4).
4. In a tissue culture hood, add 500 μL non-supplemented DMEM (or alternatively, OptiMem) to each transfection tube.
5. In a separate 1.5 mL microcentrifuge tube, prepare a 500 μL solution of non-supplemented DMEM with polyethylenimine (PEI) for each transfection. First, add the non-supplemented DMEM to each tube. Then, add PEI (stock 1 μg/μL) directly to the solution at a ratio of 2:1 PEI:DNA (see Note 5).
6. Vortex the DMEM/PEI solution and add it directly to the DNA solution. Vortex the final mixture and incubate at room temperature for 15 min.
7. Drip pipette the DNA/PEI mixture onto the confluent 10 cm plate and gently mix the cells with the solution. Return the plate to the incubator for 24 h.
8. Remove the media and replace with 10 mL complete DMEM. Incubate for an additional 24 h.

5.3.1.3 Notes

Note 1. This assay should be compatible with most transfectable cell lines. In addition to HEK293 cells, the authors have successfully used a CHO-k1 cell line in this format. Cell maintenance will change depending on cell type.

Note 2. To detach cells from the cell culture plate, we use 1 mL trypsin (10-cm plate) and incubate in the cell incubator for 3–5 min. To stop the reaction, we add 9 mL complete DMEM and gently pipet the cells for even distribution. We then count the cells using either a hemocytometer or another analogous method.

Note 3. GRK2 (G protein-coupled receptor kinase 2) phosphorylates active GPCRs to promote arrestin recruitment. Co-transfection of the GRK2 plasmid permits more efficient arrestin recruitment to activated GPCRs, but it may not be necessary depending on the receptor being studied.

Note 4. It may be necessary to optimize transfection values depending on the efficiency of transfection or signal-to-noise ratio. The provided values can also be scaled down to fit smaller transfection setups. Additionally, using other transfection reagents such as lipofectamine or Fugene may increase transfection efficiency.

Note 5. PEI stock can vary in activity depending on the manufacturer. Therefore, it may be necessary to optimize the DNA:PEI ratio to improve transfection efficiency. A brief explanation for assessing PEI activity can be found in our previous publication (Donthamsetti et al., 2015).

5.3.2 Arrestin complementation assay

The complementation assay does not require aseptic technique and can be performed outside of the tissue culture hood.

5.3.3 Dose response/kinetics assay—Timing: 1 h

1. Prepare coelenterazine-H or fumirazine at a 50 μM concentration in DPBS (see Note 1).
2. Aspirate the media from the transfected cells.
3. Wash the cells with 5 mL DPBS. Aspirate the DPBS.
4. Add 7.5 mL DPBS with 5 mM glucose to the plate and gently pipette the cells into suspension.
5. Using a multichannel pipette, aliquot 50 μL of cells into each well of a 96-well black/white isoplate (see Note 1).
6. Supplement the cells with 10 μL coelenterazine-H or fumirazine and incubate for 5 min at room temperature with a foil cover.
7. Add 40 μL of ligand at 2.5 times the desired final concentration to the cells and measure luminescence after 40 min for dose response curves or sequentially after agonist stimulation for kinetics (see Note 3). Alternatively, to measure the ability of ligands to antagonize receptor activation, add 20 μL agonist at 5 times the EC80 concentration 20 min prior to the addition of 20 μL antagonist. Then measure luminescence after 20 min incubation (see Note 4).

5.3.4 Notes

Note 1. It is possible to use either fumirazine or coelenterazine H as the substrate for the luciferase.
Note 2. This should yield a cell count of 25–40,000 cells/well.
Note 3. We recommend using a V-bottom compound plate and preparing the added compound prior to the experiment to make the assay flow smoothly.
Note 4. To study class B GPCR's that are known to bind arrestin more tightly, it is necessary to add antagonist prior to agonist.

5.4 Quantification and statistical analysis

Calculate the fold change in luminescence counts after agonist addition by comparing to an untreated control. Import the data into the preferred data analysis software and generate an XY plot with a minimum of three replicate values per treatment. Fit the data to a non-linear regression curve using log(agonist) vs response for agonist curves and log(inhibitor) vs response for antagonist curves. To calculate the fold change of agonist stimulation, transform the Y-values using the $Y = Y/K$, where the K-value is the bottom fit of the non-linear regression.

5.5 Advantages

Unlike other luminescence assays, this experimental setup does not require C-terminal receptor modification and therefore avoids potential artifacts related to such alterations. This advantage also allows for measurement of arrestin recruitment to a range of GPCR types without the need for additional cloning or construct optimization. In principle, this assay can be used to study endogenous receptors across cell types, although this will depend on the level of receptor expression. A variant of the bystander constructs can also be used to measure arrestin recruitment to early endosomes (Pedersen et al., 2020).

NanoLuc has the advantages of being smaller and brighter than other established luciferases (Hall et al., 2012). Of note, the complementation of split NanoLuc is reversible and can consequently be used to investigate a range of compounds including agonists, partial agonist, antagonists, and inverse agonists.

5.6 Limitations

The output for this assay is luminescence and is susceptible to differences in transient transfection. This variability requires that collected data be normalized and shown as fold-change. It is possible to avoid this normalization by using stable cell lines in which the expression level of the complementing components is consistent from experiment to experiment.

6 Summary

GPCRs are the main interacting partners for the arrestins, a family of proteins that act to desensitize and internalize GPCRs. The GPCR-arrestin interaction is frequently used in drug development as an indirect readout for ligand binding and studies have shown that GPCR-dependent arrestin-mediated intracellular signaling cascades lead to biological responses that are distinct from GPCR-dependent G protein signaling. Therefore, assays that measure arrestin recruitment are necessary for assessing GPCR activity. Here, we focused on two assays that can be used for measuring arrestin recruitment to unmodified GPCRs using either purified proteins or in cell approaches. These assays require little specialized equipment and can therefore be adapted to most laboratories.

7 Key resources table

Note that not all resources will be used in every protocol.

Reagent or resource	Source	Identifier
Biological samples		
Rabbit reticulocyte lysate	Commercial, or prepared as described in Jackson and Hunt (Jackson & Hunt, 1983)	
Purified rhodopsin	Native disc membranes, bicelles (Chen et al., 2015), and nanodiscs (Azevedo et al., 2015; Bayburt et al., 2011; Singhal et al., 2013; Vishnivetskiy et al., 2013)	

—cont'd

Reagent or resource	Source	Identifier
Chemicals, peptides, and recombinant proteins		
1 M HEPES-KOH (pH 7.5)	Sigma	
1 M $MgCl_2$	Sigma	
1 M spermidine-HCl	Sigma	
1 M DTT	Sigma	
1 M KOH	Sigma	
9 M LiCl (stored at −10 °C)	Sigma	
2.5 M LiCl (stored at −10 °C)	Sigma	
70% ethanol, 100% ethanol	Sigma	
50 mM NTP mix (stored at −80 °C up to 2 years)	Sigma	
Acetylated Bovine Serum Albumin (Ac-BSA)	Promega or Boehringer	
Inorganic pyrophosphatase (PPase)	Sigma	I4503
RNase inhibitor from human placenta (RNasin)	Boehringer Manehem or Promega	
SP6 RNA Polymerase	Promega	P108B
3 M NaOAc (sodium acetate), pH 5.0	Sigma	
Nuclease-free water	Sigma	
20 × mix of 19 amino acids (-Leu)	Prepared as described in Jackson and Hunt (Jackson & Hunt, 1983)	
1 M creatine phosphate	Sigma	
100 × mix of protease inhibitors (10 μg/mL pepstatin and leupeptin, 10 mg/mL soybean trypsin inhibitor)	Sigma	
0.2 M cAMP	Sigma	
2 M Potassium acetate, pH 7.4	Sigma	
10 mCi/mL [^{14}C]Leucine	New England Nuclear (NEN)	
5 mCi/mL [^{3}H]Leucine	NEN	
20 mg/mL in 50% glycerol creatine phosphokinase (CPK)	Sigma	CPK Type I
Trichloroacetic acid (TCA)	Sigma	T6399
Diethyl ether	Sigma	309966
Sodium dodecyl sulfate (SDS)	Sigma	436143
Scintillation fluid		
1 M Tris-HCl, pH 7.5	Sigma	T2319
5 M NaCl	Sigma	
DMEM	Gibco	11965-092
OptiMem	Gibco	31985088
GlutaMAX™-I	Gibco	35050061
Fetal bovine serum	Corning	#35-010

Continued

—cont'd

Reagent or resource	Source	Identifier
Penicillin/streptomycin	Corning	#30-002
Dulbecco's Phosphate buffered saline (DPBS)	Cellgro	21-031-CV
0.5 M Glucose	Sigma	
Trypsin	Cellgro	25-052-CI
1 μg/μL polyethyleimine	Polysciences	23966-2
Coelenterazine-H in ethanol	Dalton	#50909-86-9
Fumirazine in ethanol	Promega	#N1120
Experimental models: cell lines		
HEK293 cells	ATTC	CRL-1573
Recombinant DNA		
pCDNA3.1+-Mem-Link-N1 (MeN)	Available upon request	
pCDNA3.1+N2-Link-Arr3 (ArC)	Available upon request	
pCDNA3.1+-N2-Link-Arr2 (ArC)	Available upon request	
pCDNA3.1+-GRK2	Available upon request	
pCDNA3.1+-N2-Link-FYVE (EeN)	Available upon request	
pCDNA3.1+	Life Technologies	

Acknowledgments

The authors would like to thank Dr. Vsevolod V. Gurevich and Dr. Sergey A. Vishnivetskiy for their input on the rhodopsin binding assay.

References

Azevedo, A. W., et al. (2015). C-terminal threonines and serines play distinct roles in the desensitization of rhodopsin, a G protein-coupled receptor. *eLife*, *4*, e05981.

Barak, L. S., et al. (1997). A beta-arrestin/green fluorescent protein biosensor for detecting G protein-coupled receptor activation. *The Journal of Biological Chemistry*, *272*, 27497–27500.

Barnea, G., et al. (2008). The genetic design of signaling cascades to record receptor activation. *Proceedings of the National Academy of Sciences of the United States of America*, *105*, 64–69.

Bayburt, T. H., et al. (2011). Monomeric rhodopsin is sufficient for normal rhodopsin kinase (GRK1) phosphorylation and arrestin-1 binding. *The Journal of Biological Chemistry*, *286*, 1420–1428.

Brady, A. E., et al. (2004). Study of G-protein-coupled receptor-protein interactions using gel overlay assays and glutathione-S-transferase-fusion protein pull-downs. *Methods in Molecular Biology*, *259*, 371–378.

Cai, Y., et al. (2017). Purification of family B G protein-coupled receptors using nanodiscs: Application to human glucagon-like peptide-1 receptor. *PLoS One*, *12*, e0179568.

Chen, Q., et al. (2015). The rhodopsin-arrestin-1 interaction in bicelles. *Methods in Molecular Biology*, *1271*, 77–95.

Chen, Q., et al. (2017). Structural basis of arrestin-3 activation and signaling. *Nature Communications*, *8*, 1427.

DeFea, K. A. (2011). Beta-arrestins as regulators of signal termination and transduction: How do they determine what to scaffold? *Cellular Signalling*, *23*, 621–629.

Dixon, A. S., et al. (2016). Nanoluc complementation reporter optimized for accurate measurement of protein interactions in cells. *ACS Chemical Biology*, *11*, 400–408.

Donthamsetti, P., et al. (2015). Using bioluminescence resonance energy transfer (BRET) to characterize agonist-induced arrestin recruitment to modified and unmodified G protein-coupled receptors. *Current Protocols in Pharmacology*, *70*, 2.14.1–2.14.14.

Eishingdrelo, H., et al. (2011). A cell-based protein-protein interaction method using a permuted luciferase reporter. *Current Chemical Genomics*, *5*, 122–128.

Goddard, A. D., et al. (2015). Reconstitution of membrane proteins: A GPCR as an example. *Methods in Enzymology*, *556*, 405–424.

Grünberg, R., et al. (2013). Engineering of weak helper interactions for high-efficiency FRET probes. *Nature Methods*, *10*, 1021–1027.

Gurevich, V. V. (1996). [21] Use of bacteriophage RNA polymerase in RNA synthesis. In *275. Viral polymerases and related proteins* (pp. 382–397). Elsevier.

Gurevich, V. V., & Benovic, J. L. (1993). Visual arrestin interaction with rhodopsin. Sequential multisite binding ensures strict selectivity toward light-activated phosphorylated rhodopsin. *The Journal of Biological Chemistry*, *268*, 11628–11638.

Gurevich, V. V., & Benovic, J. L. (1995). Visual arrestin binding to rhodopsin. Diverse functional roles of positively charged residues within the phosphorylation-recognition region of arrestin. *The Journal of Biological Chemistry*, *270*, 6010–6016.

Gurevich, V. V., & Benovic, J. L. (2000). [29] Arrestin: Mutagenesis, expression, purification, and functional characterization. In *315. Vertebrate phototransduction and the visual cycle, part A* (pp. 422–437). Elsevier.

Gurevich, V. V., & Gurevich, E. V. (2006). The structural basis of arrestin-mediated regulation of G-protein-coupled receptors. *Pharmacology & Therapeutics*, *110*, 465–502.

Gurevich, V. V., & Gurevich, E. V. (2019). GPCR signaling regulation: The role of grks and arrestins. *Frontiers in Pharmacology*, *10*, 125.

Hall, M. P., et al. (2012). Engineered luciferase reporter from a deep sea shrimp utilizing a novel imidazopyrazinone substrate. *ACS Chemical Biology*, *7*, 1848–1857.

Hauser, A. S., et al. (2017). Trends in GPCR drug discovery: New agents, targets and indications. *Nature Reviews. Drug Discovery*, *16*, 829–842.

Jackson, R. J., & Hunt, T. (1983). [4] Preparation and use of nuclease-treated rabbit reticulocyte lysates for the translation of eukaryotic messenger RNA. In *96. Biomembranes part J: Membrane biogenesis: assembly and targeting (general methods, eukaryotes)* (pp. 50–74). Elsevier.

Kang, Y., et al. (2015). Crystal structure of rhodopsin bound to arrestin by femtosecond X-ray laser. *Nature*, *523*, 561–567.

Kim, Y. J., et al. (2013). Crystal structure of pre-activated arrestin p44. *Nature*, *497*, 142–146.

Kim, J., et al. (2020). The β-arrestin-biased β-adrenergic receptor blocker carvedilol enhances skeletal muscle contractility. *Proceedings of the National Academy of Sciences of the United States of America*, *117*, 12435–12443.

Knepp, A. M., et al. (2011). Direct measurement of thermal stability of expressed CCR5 and stabilization by small molecule ligands. *Biochemistry*, *50*, 502–511.

Kobilka, B. K. (1995). Amino and carboxyl terminal modifications to facilitate the production and purification of a G protein-coupled receptor. *Analytical Biochemistry*, *231*, 269–271.

Kocan, M., & Pfleger, K. D. G. (2011). Study of GPCR-protein interactions by BRET. *Methods in Molecular Biology*, *746*, 357–371.

Kroeze, W. K., et al. (2015). PRESTO-Tango as an open-source resource for interrogation of the druggable human GPCRome. *Nature Structural & Molecular Biology*, *22*, 362–369.

Latorraca, N. R., et al. (2018). Molecular mechanism of GPCR-mediated arrestin activation. *Nature*, *557*, 452–456.

Lee, C., et al. (2008). Site-specific cleavage of G protein-coupled receptor-engaged beta-arrestin. Influence of the AT1 receptor conformation on scissile site selection. *The Journal of Biological Chemistry*, *283*, 21612–21620.

Lee, S., et al. (2019). Nedd4 E3 ligase and beta-arrestins regulate ubiquitination, trafficking, and stability of the mGlu7 receptor. *eLife*, *8*, e44502.

Leitz, A. J., et al. (2006). Functional reconstitution of Beta2-adrenergic receptors utilizing self-assembling Nanodisc technology. *BioTechniques*, *40*, 601–602. 604, 606, passim.

Liu, C.-H., et al. (2017). Arrestin-biased AT1R agonism induces acute catecholamine secretion through TRPC3 coupling. *Nature Communications*, *8*, 14335.

Luttrell, L. M., & Lefkowitz, R. J. (2002). The role of beta-arrestins in the termination and transduction of G-protein-coupled receptor signals. *Journal of Cell Science*, *115*, 455–465.

Malmerberg, E., et al. (2015). Conformational activation of visual rhodopsin in native disc membranes. *Science Signaling*, *8*, ra26.

Mayer, D., et al. (2019). Distinct G protein-coupled receptor phosphorylation motifs modulate arrestin affinity and activation and global conformation. *Nature Communications*, *10*, 1261.

McDowell, J. H. (1993). Preparing rod outer segment membranes, regenerating rhodopsin, and determining rhodopsin concentration. In *Vol. 15. Photoreceptor cells* (pp. 123–130). Elsevier.

McGuinness, D., et al. (2009). Characterizing cannabinoid CB2 receptor ligands using DiscoveRx PathHunter beta-arrestin assay. *Journal of Biomolecular Screening*, *14*, 49–58.

Milano, S. K., et al. (2006). Nonvisual arrestin oligomerization and cellular localization are regulated by inositol hexakisphosphate binding. *The Journal of Biological Chemistry*, *281*, 9812–9823.

Palczewski, K. (2006). G protein-coupled receptor rhodopsin. *Annual Review of Biochemistry*, *75*, 743–767.

Pedersen, M. H., et al. (2020). A novel luminescence-based β-arrestin membrane recruitment assay for unmodified GPCRs. *BioRxiv*. https://doi.org/10.1101/2020.04.09.034520.

Perry, N. A., et al. (2019a). Using in vitro pull-down and in-cell overexpression assays to study protein interactions with arrestin. *Methods in Molecular Biology*, *1957*, 107–120.

Perry, N. A., et al. (2019b). Arrestin-3 interaction with maternal embryonic leucine-zipper kinase. *Cellular Signalling*, *63*, 109366.

Pfleger, K. D. G., et al. (2006). Bioluminescence resonance energy transfer (BRET) for the real-time detection of protein-protein interactions. *Nature Protocols*, *1*, 337–345.

Rankovic, Z., et al. (2016). Biased agonism: An emerging paradigm in GPCR drug discovery. *Bioorganic & Medicinal Chemistry Letters*, *26*, 241–250.

Salahpour, A., et al. (2012). BRET biosensors to study GPCR biology, pharmacology, and signal transduction. *Frontiers in Endocrinology*, *3*, 105.

Shukla, A. K., et al. (2013). Structure of active β-arrestin-1 bound to a G-protein-coupled receptor phosphopeptide. *Nature*, *497*, 137–141.
Shukla, A. K., et al. (2014). Visualization of arrestin recruitment by a G-protein-coupled receptor. *Nature*, *512*, 218–222.
Singhal, A., et al. (2013). Insights into congenital stationary night blindness based on the structure of G90D rhodopsin. *EMBO Reports*, *14*, 520–526.
Slosky, L. M., et al. (2020). β-arrestin-biased allosteric modulator of NTSR1 selectively attenuates addictive behaviors. *Cell*, *181*, 1364–1379.e14.
Smith, J. S., & Rajagopal, S. (2016). The β-arrestins: Multifunctional regulators of G protein-coupled receptors. *The Journal of Biological Chemistry*, *291*, 8969–8977.
Teoh, J.-P., et al. (2018). β-arrestin-biased agonism of β-adrenergic receptor regulates dicer-mediated microRNA maturation to promote cardioprotective signaling. *Journal of Molecular and Cellular Cardiology*, *118*, 225–236.
Thomsen, A. R. B., et al. (2016). GPCR-G protein-β-arrestin super-complex mediates sustained G protein signaling. *Cell*, *166*, 907–919.
Vishnivetskiy, S. A., et al. (2013). Constitutively active rhodopsin mutants causing night blindness are effectively phosphorylated by GRKs but differ in arrestin-1 binding. *Cellular Signalling*, *25*, 2155–2162.
Wang, T., et al. (2004). Measurement of β-arrestin recruitment for GPCR targets. In G. S. Sittampalam, et al. (Eds.), *Assay guidance manual* Eli Lilly & Company and the National Center for Advancing Translational Sciences.
Wertz, S. L., et al. (2019). Co-IP assays for measuring GPCR-arrestin interactions. *Methods in Cell Biology*, *149*, 205–213.
Wisler, J. W., et al. (2007). A unique mechanism of beta-blocker action: Carvedilol stimulates beta-arrestin signaling. *Proceedings of the National Academy of Sciences of the United States of America*, *104*, 16657–16662.
Xiao, K., et al. (2004). Activation-dependent conformational changes in β-arrestin 2. *The Journal of Biological Chemistry*, *279*, 55744–55753.
Xiao, K., et al. (2007). Functional specialization of beta-arrestin interactions revealed by proteomic analysis. *Proceedings of the National Academy of Sciences of the United States of America*, *104*, 12011–12016.
Zhan, X., et al. (2016). Peptide mini-scaffold facilitates JNK3 activation in cells. *Scientific Reports*, *6*, 21025.
Zheng, C., et al. (2019). Critical role of the finger loop in arrestin binding to the receptors. *PLoS One*, *14*, e0213792.

CHAPTER

BRET-based assay to specifically monitor β_2AR/GRK2 interaction and β-arrestin2 conformational change upon βAR stimulation

4

Warisara Parichatikanond[a], Ei Thet Htar Kyaw[b], Corina T. Madreiter-Sokolowski[c], and Supachoke Mangmool[d,*]

[a]*Department of Pharmacology, Faculty of Pharmacy, Mahidol University, Bangkok, Thailand*

[b]*Pharmacology and Biomolecular Science Graduate Program, Faculty of Pharmacy, Mahidol University, Bangkok, Thailand*

[c]*Gottfried Schatz Research Center, Molecular Biology and Biochemistry, Medical University of Graz, Graz, Austria*

[d]*Department of Pharmacology, Faculty of Science, Mahidol University, Bangkok, Thailand*

**Corresponding author: e-mail address: supachoke.man@mahidol.ac.th*

Chapter outline

Methods in Cell Biology, Volume 166, ISSN 0091-679X, https://doi.org/10.1016/bs.mcb.2021.06.005

Abstract

The β-adrenergic receptors (βARs) are members of G protein-coupled receptor (GPCR) family and have been one of the most important GPCRs for studying receptor endocytosis and signaling pathway. Agonist binding of βARs leads to an activation of G proteins and their canonical effectors. In a parallel way, βAR stimulation triggers the termination of its signals by receptor desensitization. This termination process is initiated by G protein-coupled receptor kinase (GRK)-induced βAR phosphorylation that promotes the recruitment of β-arrestins to phosphorylated βAR. The uncoupled βARs which formed a complex with GRK and β-arrestin subsequently internalize into the cytosol. In addition, GRKs and β-arrestins also act as scaffolding proteins and signal transducers in their own functions to modulate various downstream effectors. Upon translocation to the βAR, β-arrestin is believed to undergo an important conformational change in the structure that is necessary for its signal transduction. The bioluminescence resonance energy transfer (BRET) technique involves the fusion of donor (luciferase) and acceptor (fluorescent) molecules to the interested proteins. Co-expression of these fusion proteins enables direct detection of their interactions in living cells. Here we describe the use of our established BRET technique to track the interaction of βAR with both GRK and β-arrestin. The assay described here allows the measurement of the BRET signal for detecting the interaction of β_2AR with GRK2 and the conformational change of β-arrestin2 following βAR stimulation.

1 Introduction

G protein-coupled receptors (GPCRs) are a conserved family of seven transmembrane receptors and are one of the largest targets for drug discovery. βARs belong to the GPCR family and are subdivided into 3 subtypes; β_1-, β_2-, and β_3-AR (Salazar, Chen, & Rockman, 2007). Stimulation of βARs elicits receptor coupling with heterotrimeric G proteins and promotes the dissociation of $G\alpha_s$ subunit from $G\beta\gamma$ subunit contributing to a stimulation of adenylyl cyclase (AC) and an elevation of cAMP levels. After that, cAMP interacts and activates its downstream effectors, resulting in the induction of cAMP-dependent signaling pathway

(Post, Hammond, & Insel, 1999). Upon agonist binding, GRKs phosphorylate and activate βARs thereby β-arrestins are subsequently recruited to suppress further interactions of the receptor with G proteins. This process is known as "receptor desensitization" (Ferguson, 2001; Lutrell & Lefkowitz, 2002). In addition, β-arrestins have essential functions in the trafficking of βAR into intracellular compartments located in the cytosol (Ferguson et al., 1996). This process is termed "βAR internalization." Thus, βAR has been one of the most popular experimental systems to study the mechanism of GPCR endocytosis and transmembrane signaling pathways.

1.1 Interaction of $β_2$AR with GRK2

Based on its sequences and structural homology, the mammalian GRK family consists of seven members which are classified into three subfamilies: rhodopsin kinase subfamily (GRK1 and GRK7), βARK subfamily (GRK2 and GRK3), and GRK4 subfamily (GRK4, GRK5, and GRK6) (Penn, Pronin, & Benovic, 2000). The accumulating evidence has directly implicated GRK2 phosphorylation in the desensitization of $β_2$AR. Expression of GRK2 together with $β_2$AR resulted in enhanced desensitization of $β_2$AR, whereas co-expression of a dominant-negative mutant of GRK2 (containing a mutation that suppresses catalytic activity) inhibited this desensitization (Mangmool et al., 2006). In addition, overexpression of GRK2 enhanced the $β_2$AR-mediated translocation of β-arrestin to the plasma membrane and induced $β_2$AR internalization (Mangmool et al., 2006).

GRK2 generally locates in the cytosol without receptor stimulation. Besides translocation to the plasma membrane, GRK2 is detected in endosomal vesicles and shares endocytic mechanisms with $β_2$AR in response to the receptor activation indicating that macromolecular complexes consisted of $β_2$AR, GRK2, and β-arrestins are present in the initial steps of internalization (Ruiz-Gòmez & Mayor, 1997). Current several lines of evidence refer that GRKs function as signal transducers by themselves leading to interactions of substrates to not only GPCR but also other interacting partners and thereby regulate various crucial signaling processes (Mangmool, Parichatikanond, & Kurose, 2018; Penela, Ribas, & Mayor, 2003).

1.2 Stimulation of βAR promotes conformational changes of β-arrestin2

The arrestin family comprises four members. Arrestin1 and arrestin4 (known as visual arrestins) are abundantly identified in the rods and cones of eyes, respectively. The β-arrestin-1 and -2 (called arrestin2 and arrestin3, respectively) are widely distributed throughout mammalian cells and tissues (Ferguson, 2001). The βAR endocytosis and trafficking are sophisticatedly controlled by the key regulators, β-arrestins. After occupying with its agonist, βAR undergoes phosphorylation and interacts with β-arrestins resulting in the suppression of further G protein activation (Ferguson et al., 1996). Beyond their classical regulations, β-arrestins function as crucial modulators linking receptors to many downstream effectors such as MAPK cascades (ERK and JNK), Src, and the ubiquitin ligase Mdm2 (Lefkowitz, Rajagopal, & Whalen, 2006; Lefkowitz & Shenoy, 2005). β-Arrestins are also known to interact and activate Ca^{2+}/calmodulin kinase II (CaMKII) activity following β_1AR stimulation (Mangmool, Shukla, & Rockman, 2010). Thus, arrestins act as scaffold proteins that encompass manifold signaling pathways in responding to cellular functions, highlighting their pivotal roles in cellular physiology and pathophysiology of heart diseases (Mangmool et al., 2018).

Interestingly, a conformational rearrangement of β-arrestin structure accompanies its transition to the high-affinity state for receptor-binding (Xiao, Shenoy, Nobles, & Lefkowitz, 2004). Intramolecular interactions between the N-terminal and C-terminal domains are found in the inactive state of β-arrestin (Charest, Terrillon, & Bouvier, 2005). These interactions are lost in the active state of β-arrestin following receptor stimulation, indicating that the domains inside β-arrestin structure relatively move to each other upon activation (Gurevich & Benovic, 1993; Gurevich & Gurevich, 2003; Hirsch, Schubert, Gurevich, & Sigler, 1999). Thus, the conformational change of β-arrestin reflects its transition from an inactive state to an active state and subsequently follows its initial translocation to activated receptors, and involves the relative movement of the C-terminus towards N-terminus of β-arrestin (Charest et al., 2005; Shukla et al., 2008).

1.3 Bioluminescence resonance energy transfer (BRET) assay

BRET is a phenomenon reflecting an energy transfer between two molecules, a light-emitting molecule (typically a luciferase) and a light-sensitive molecule (typically a fluorescent protein). The BRET-based reaction principally depends on the energy derived from a luciferase reaction which is used to excite a fluorescent protein when the fluorescent protein (acceptor) is in close proximity to the luciferase (donor) (Pfleger & Eidne, 2006). The BRET technology involves the fusion of luciferase domain (donor) and fluorescent domain (acceptor) to proteins of interest. Co-expression of these fusion constructs in living cells enables us to monitor the protein-protein interactions in a quantitative real-time manner (Marullo & Bouvier, 2007). Construction of the two proteins of interest genetically conjugated with luciferase protein and fluorescent protein depends on the chosen BRET approach at either N-terminus or C-terminus, including the applied substrates (Table 1) (Borroto-Escuela, Flajolet, Agnati, Greengard, & Fuxe, 2013). Energy is transferred through a dipole-dipole coupling from a donor to an acceptor when they are in an adjacent distance (approximately within 10 nm), leading to fluorescence emission at a characteristic wavelength (Fig. 1A). The energy emitted by the acceptor relative to that emitted by the donor is known as the BRET ratio (or BRET signal) (Pfleger & Eidne, 2006).

Nowadays, the BRET technique is a potential tool extensively applied for the quantitatively real-time monitoring of protein-protein interactions in live cells particularly a coupling of GPCRs and their regulating proteins, such as G proteins

Table 1 Different BRET methods.

Method	Donor	Acceptor	Substrate	Emission wavelength (nm)	
				Donor	Acceptor
BRET[1]	RLuc	eYFP	Coelenterazine h	480	530
BRET[2]	RLuc	GFP[2]	Coelenterazine 400a (DeepBlueC)	395	510
eBRET[2]	RLuc8	GFP[2]	Coelenterazine 400a (DeepBlueC)	395	510
BRET[3]	Firefly	DsRed	Luciferin	565	583
BRET[3]	RLuc/RLuc8	mOrange	EnduRen/ Coelenterazine h	480	564
QD-BRET	RLuc/RLuc8	QD	Coelenterazine	480	605
NanoBRET	NanoLuc	HaloTag Ligand	Furimazine	460	618

eBRET, *enhanced BRET;* QD-BRET, *quantum dot-BRET;* RLuc, Renilla *luciferase.*

(Maziarz & Garcia-Marcos, 2017) and β-arrestins (Bertrand et al., 2002). Not only protein-protein interactions, BRET can also be used for the detection of conformational rearrangement of β-arrestin molecules (Charest et al., 2005; Mangmool et al., 2010; Shukla et al., 2008). For instance, β-arrestin2 was constructed as the intramolecular BRET-based biosensor, in which β-arrestin2 is sandwiched between the *Renilla* luciferase (RLuc) and the yellow fluorescent protein (YFP) (Fig. 1B). This BRET construction was further named a doubled-brilliance β-arrestin2 (Charest et al., 2005).

The BRET assay utilizes heterologous co-expression of fusion proteins that link two proteins of interest such as βAR and GRK2. βAR linked with a bioluminescent donor, *Renilla* luciferase (named as βAR-RLuc), and GRK2 linked to an acceptor fluorescent protein such as enhanced GFP (GFP^2) (named as GRK2-GFP^2). If βAR-RLuc is in close distance to GRK2-GFP^2, the energy resulting from an oxidation of the coelenterazine substrate by the donor, βAR-RLuc will transfer to the acceptor, GRK2-GFP^2, which in turn fluoresces and subsequently generates the BRET ratio (Fig. 1A).

BRET can be determined in living cells using microplate readers to study protein-protein interactions, especially in the fields of GPCR endocytosis and signal transduction, including receptor dimerization and transactivation (Kocan, See, Seeber, Eidne, & Pfleger, 2008). The ability to study protein-protein interactions in living cells circumvents many of problems related to conventional techniques such as yeast two-hybrid and co-immunoprecipitation. In addition, the high sensitivity of BRET enables the study of protein-protein interactions at physiological conditions and thereby represents a significant advantage of BRET over several techniques that require high protein expression levels (Pfleger & Eidne, 2003).

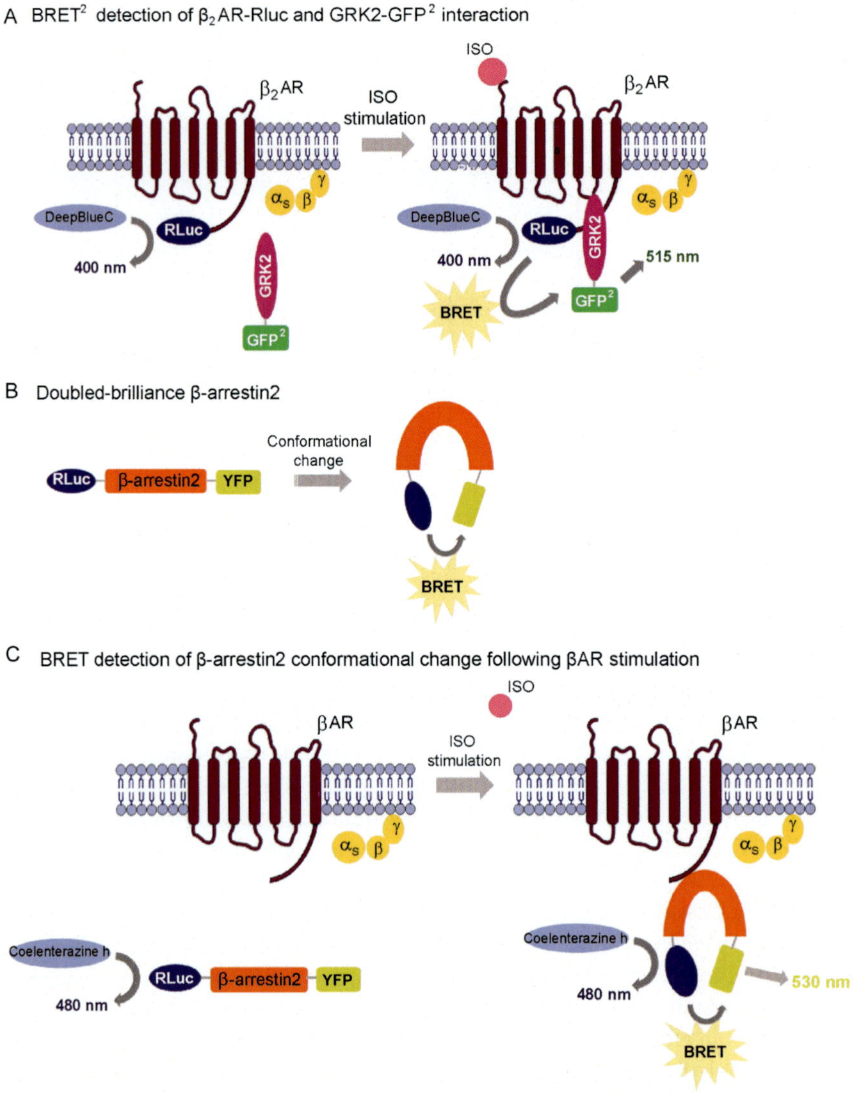

FIG. 1

Schematic representation of BRET-based assays. (A) The BRET^2 detection of β_2AR-RLuc and GRK2-GFP^2 interaction. The advanced BRET (BRET^2) technique is used for monitoring the interaction between RLuc-tagged β_2AR (β_2AR-RLuc) and GFP^2-tagged GRK2. Stimulation of β_2AR with isoproterenol (ISO) induced the interaction of β_2AR with GRK2 that brings the acceptor (GFP^2) closed proximity to the donor (RLuc). The energy is transferred from a donor to an acceptor, leading to the detection of BRET signal. (B) The structure of doubled-brilliance RLuc-β-arrestin2-YFP. The intramolecular BRET-based biosensor detects the BRET signal originated from the interaction between *Renilla* luciferase (RLuc) and yellow fluorescent protein (YFP) tagged β-arrestin2 at its N-terminus and C-terminus, respectively. (C) The BRET detection of β-arrestin2 conformational change. ISO stimulation promotes the conformational rearrangement of β-arrestin2 structure that brings the acceptor (YFP) closed proximity to the donor (RLuc).

2 Materials

2.1 Cells and reagents

1. Human embryonic kidney 293 (HEK-293) cells (ATCC, CRL-1573)
2. Dulbecco's Modified Eagle's Medium (DMEM) (Gibco, 11965-092)
3. Fetal Bovine Serum (FBS) (Gibco, 16000-044)
4. Penicillin-Streptomycin (P/S) solution (Gibco, 11965-092)
5. Dulbecco's phosphate-buffered saline (DPBS) containing 0.1 g/L $CaCl_2$, 0.1 g/L $MgCl_2$ and 1 g/L D-glucose (Gibco, 14287080)
6. 10cm tissue-cultured treated culture dishes (Corning Costar, 430167)
7. Isoproterenol hydrochloride (Sigma aldrich, I6504)
8. FuGENE 6 transfection reagent (Omega, E2692)
9. Coelenterazine 400a (LifeSpan BioSciences, LS-H7214) (also known as DeepBlueC, a trademark of Perkin Elmer)
10. Coelenterazine h (Promega, S2011)
11. White opaque 96-well plates (Perkin Elmer OptiPlate-96, 6005290)
12. Multilabel Reader Mithras LB 940 (Berthold) or similar plate reader suitable for the dual emission luminescence detection

2.2 Plasmids

Plasmids expressing the following constructs are required:

1. Plasmids encoding β_2AR-RLuc and GRK2-GFP^2 were obtained from Prof. H. Kurose (Kyushu University, Japan). (Mangmool et al., 2006)
2. Plasmids encoding RLuc-β-arrestin2-YFP and RLuc-β-arrestin2 were obtained from Prof. M Bouvier (Université de Montréal, Montréal, Canada) (Charest et al., 2005)
3. Plasmids encoding β_1AR and β_2AR were obtained from Prof. R.J. Lefkowitz (Duke University Medical Center, USA) (Mangmool et al., 2010)
4. pcDNA3.1 vector (Invitrogen, V790-20)

3 Methods

3.1 $BRET^2$ detection of β_2AR-RLuc and GRK2-GFP^2 interaction in HEK-293 cells

The following procedure describes the important steps to measure an β_2AR/GRK2 interaction in response to the β_2AR stimulation. The procedure consists of four major parts, including (1) transfection of cells, (2) harvesting cells, (3) $BRET^2$ measurement, and (4) data analysis.

3.1.1 Transfection of cells

(1) Seed HEK-293 cells at a density of 1×10^6 cells/10cm dishes and incubate the cells overnight (16–24h) in a CO_2 incubator at 37°C with 5% CO_2.

(2) After seeding the cells, aspirate the medium in each well and replace with 10mL of fresh DMEM containing 10% FBS plus 1% P/S solution. Maintain the cells in a CO_2 incubator until subsequent experiments.

(3) Start to prepare the transfection reagent and plasmid. For BRET detection of β_2AR/GRK2 interaction, cells are transfected with two conditions, (1) HEK-293 cells are transfected with β_2AR-RLuc alone (donor only) and (2) HEK-293 cells are transfected with 2.5μg of β_2AR-RLuc and 2.5μg of GRK2-GFP^2 (1:1 ratio for amount of DNA plasmid), by using FuGENE 6 transfection reagent. (see Note 1 and 2) For an initial optimization, use FuGENE 6 and DNA plasmid at a ratio of 3:1 (Table 2).

(4) Prepare and label two polystyrene tubes corresponding to each transfection condition. In a separate tube, pipet 580μL of serum-free DMEM (without antibiotics or fungicides) into the tube. Pipet 15μL of the FuGENE6 transfection reagent directly into the medium (without allowing contact with the sides of tube) and mix immediately by pipetting. Incubate for 5min at room temperature.

(5) Add 5μg of total plasmid DNA (β_2AR-RLuc alone or β_2AR-RLuc together with GRK2-GFP^2) into the labeled tube and mix immediately by pipetting. Incubate FuGENE6/DNA mixture for 15min at room temperature (see Note 4).

(6) Add the mixture to the cells (in each 10cm dish) in a drop-wise manner. Swirl the wells to ensure distribution over the entire plate surface. Return the cells to a CO_2 incubator for 48h.

Table 2 Amount of medium, FuGENE 6 transfection reagent, and plasmid DNA for 10cm culture dishes.

	Total volume of medium (per dish) (mL)	Amount of FuGENE6 (per dish) (μL)	Amount of plasmid DNA (per dish)		Total transfection volume (per dish) (μL)
			β_2AR-RLuc (μg)	GRK2-GFP^2	
β_2AR-RLuc+GRK2-GFP^2	10	15	2.5	2.5μg	600
β_2AR-RLuc alone (see Note 3)	10	15	2.5	– (see Note 3)	600

3.1.2 Harvesting cells

(1) Two days after transfection (see Note 5), wash HEK-293 cells with PBS and then detach the cells with PBS/EDTA solution (see Note 6). Collect the cell suspension into centrifuge tubes and wash twice with PBS before counting the cells using a hemocytometer.

(2) After the last wash, centrifuge the tube containing cells at 1,000 rpm for 5 min and carefully aspirate the cell supernatant.

(3) Resuspend the cell pellets in the assay buffer (DPBS containing 0.1 g/L $CaCl_2$, 0.1 g/L $MgCl_2$ and 1 g/L D-glucose) to achieve a density of 10×10^6 cells per mL.

(4) Add 50 μL of cell suspension (approximately 5×10^5 cells per well) into each well of white 96-well plate.

3.1.3 BRET^2 measurement

The advanced BRET (BRET^2) technique tremendously improves the separation of emission spectra between the donor (luminescent; RLuc) and acceptor (fluorescent; GFP^2) molecules compared with classical BRET. The BRET^2 signal is measured by the amount of green light emitted by GFP^2 dividing by the amount of blue light emitted by RLuc. The ratios of green to blue increase when two proteins are in close proximity (Fig. 1A).

(1) Incubate the cells in a white 96-well plate (5×10^5 cells/well) under the presence or absence of 1 μM isoproterenol (ISO; non-selective βAR agonist) in the dark for 2 min at room temperature.

(2) Add Coelenterazine 400a (DeepBlueC) at a final concentration of 5 μM. The BRET^2 signal is detected immediately by using a Multilabel Reader Mithras LB 940 with 400 nm (RLuc) and 515 nm (GFP^2) emission filters over a period of certain time.

3.1.4 Data analysis

(1) Determine the BRET^2 ratio by calculating the ratio of the light emitted signal of the acceptor GRK2-GFP^2 (emission at 515 nm) relative to the signal of the donor β_2AR-RLuc (emission at 400 nm) (see Note 7).

(2) Represent the data as the specific BRET^2 ratio calculated by subtracting from the BRET^2 ratio value above the background BRET^2 ratio (see Note 8).

(3) When analyzing experimental replicates, calculate the mean and standard error of mean (SEM) of BRET^2 ratio between multiple experiments as shown in Fig. 2A (real-time measurement) and Fig. 2B (measurement at 10 min).

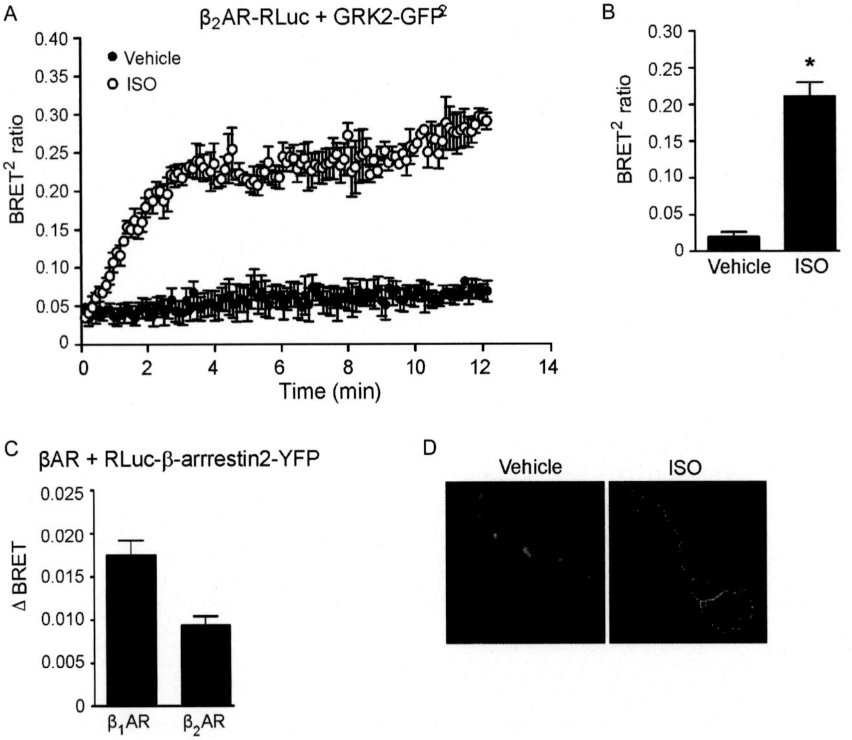

FIG. 2

Measurement of BRET ratios. (A) The real-time BRET measurements of ISO-induced β_2AR-GRK2 interaction. HEK-293 cells were transfected with β_2AR-RLuc and GRK2-GFP^2. Cells were stimulated with ISO or vehicle over a period of time. The BRET^2 ratios were shown as the mean and standard error of mean (SEM) from four different experiments. (B) The BRET^2 ratios were measured at 10 min after ISO stimulation. The BRET^2 ratios were shown as the mean $\pm$ SEM ($n=4$). *, $P<0.05$ vs vehicle. (C) The BRET detection of β-arrestin2 conformational change. HEK-293 cells were transfected with either β_1AR or β_2AR together with RLuc-β-arrestin2-YFP. The BRET ratios indicated the conformational rearrangement of β-arrestin2 were determined. The ΔBRET was defined as the ISO stimulated minus non-stimulated BRET ratio, and data was shown as the mean $\pm$ SEM ($n=4$). (D) HEK-293 cells were transfected with β_2AR and RLuc-β-arrestin2-YFP. Cells were stimulated with ISO for 15 min, and translocation of β-arrestin2 biosensor was monitored by a fluorescent microscopy.

3.2 BRET detection of β-arrestin2 conformational change following βAR stimulation

The approaches applied to detect a real-time change of β-arrestin2 conformations upon the activation of βAR are described in the following procedure comprised of four parts, including (1) transfection of cells, (2) harvesting cells, (3) BRET measurement, and (4) data analysis.

3.2.1 Transfection of cells

(1) Seed HEK-293 cells at a density of 1×10^6 cells/10cm dishes and incubate the cells overnight (16–24h) in a CO_2 incubator at 37°C with 5% CO_2.

(2) After seeding the cells, aspirate the medium in each well and replace with 10mL of fresh DMEM containing 10% FBS plus 1% P/S solution. Maintain the cells in a CO_2 incubator until subsequent experiments.

(3) Start to prepare the transfection reagent and plasmid. For BRET detection of β-arrestin2 conformational change, cells are transfected with four conditions, (1) HEK-293 cells are transfected with 2.5μg of β_1AR and 2.5μg of RLuc-β-arrestin2 alone (donor only); (2) HEK-293 cells are transfected with 2.5 μg of β_2AR and 2.5μg of RLuc-β-arrestin2 alone (donor only); (3) HEK-293 cells are transfected with 2.5μg of β_1AR and 2.5μg of RLuc-β-arrestin2-YFP; and (4) HEK-293 cells are transfected with 2.5μg of β_2AR and 2.5μg of RLuc-β-arrestin2-YFP, by using FuGENE 6 transfection reagent. (see Notes 1 and 2) For an initial optimization, use FuGENE 6 and DNA plasmid at a ratio of 3:1 (Table 3).

(4) Prepare and label four polystyrene tubes corresponding to each transfection condition. In a separate tube, pipet 580μL of serum-free DMEM (without

Table 3 Amount of medium, FuGENE 6 transfection reagent, and plasmid DNA for 10cm culture dishes.

	Total volume of medium (per dish) (mL)	Amount of FuGENE6 (per dish) (μL)	Amount of plasmid DNA (per dish)		Total transfection volume (per dish) (μL)
			RLuc-β-arr2-YFP	βAR	
β_1AR + RLuc-β-arr2-YFP	10	15	2.5μg	2.5μg of β_1AR	600
β_2AR + RLuc-β-arr2-YFP	10	15	2.5μg	2.5μg of β_2AR	600
β_1AR + RLuc-β-arr2 (see Note 9)	10	15	2.5μg of RLuc-β-arr2	2.5μg of β_1AR	600
β_2AR + RLuc-β-arr2 (see Note 9)	10	15	2.5μg of RLuc-β-arr2	2.5μg of β_2AR	600

β-arr2, β-arrestin2.

antibiotics or fungicides) into the tube. Pipet 15 μL of the FuGENE6 transfection reagent directly into the medium (without allowing contact with the sides of tube) and mix immediately by pipetting. Incubate for 5 min at room temperature.

(5) Add the total plasmid DNA (β_1AR + RLuc-β-arrestin2, β_2AR + RLuc-β-arrestin2, β_1AR + RLuc-β-arrestin2-YFP or β_2AR + RLuc-β-arrestin2-YFP) into the labeled tube and mix immediately by pipetting. Incubate the FuGENE6/DNA mixture for 15 min at room temperature (see Note 4).

(6) Add the mixture to the cells (in each 10 cm dish) in a drop-wise manner. Swirl the wells to ensure distribution over the entire plate surface. Return the cells to a CO_2 incubator for 48 h.

3.2.2 Harvesting cells

(1) Two days after transfection (see Note 10), wash HEK-293 cells with PBS and then detach the cells with PBS/EDTA solution (see Note 6). Collect the cell suspension in centrifuge tubes and wash twice with PBS before counting the cells using a hemocytometer.

(2) After the last wash, centrifuge the tube containing cells at 1,000 rpm for 5 min and carefully aspirate the supernatant.

(3) Resuspend the cell pellets in the assay buffer (DPBS containing 0.1 g/L $CaCl_2$, 0.1 g/L $MgCl_2$ and 1 g/L D-glucose) to achieve a density of 10×10^6 cells per mL.

(4) Add 50 μL of cell suspension (approximately 5×10^5 cells per well) into each well of white 96-well plate.

3.2.3 BRET measurement

The BRET signal is measured by the amount of yellow light emitted by YFP (at C-terminus) as compared to the blue light emitted by RLuc (at N-terminus). The change of BRET signal (ratio of yellow to blue) reflects different conformations of RLuc-β-arrestin2-YFP, indicating the rearrangements of β-arrestin2 structure (Fig. 1C).

(1) Incubate the cells in a white 96-well plate (5×10^5 cells/well) with 5 μM of Coelenterazine h for 10 min and subsequently stimulate the cells without or with 1 μM ISO for the indicated time in the dark at room temperature (see Note 11).

(2) Detect the BRET signal immediately by using a Multilabel Reader Mithras LB 940 with 460–500 nm (Luc) and 510–515 nm (YFP) emission filters over a period of certain time.

3.2.4 Data analysis

(1) Determine the BRET ratio by calculating the ratio of the light emitted signal of the acceptor YFP (emission at 528 nm) relative to the signal of the donor RLuc (emission at 460 nm) (see Note 7).

(2) Calculate the BRET ratio by subtracting from the BRET ratio value above the background BRET ratio. The values of BRET ratio are corrected by subtracting the background BRET signal detected when cells expressing the donor alone RLuc-β-arrestin2.

(3) Represent the data as the ΔBRET that defined ISO stimulated minus non-stimulated BRET ratio.

(4) When analyzing experimental replicates, calculate the mean and standard error of mean (SEM) of ΔBRET between multiple experiments as shown in Fig. 2C (measurement at 10 min).

4 Notes

1. The optimal amount of each fusion construct to be used in co-transfection depends on many factors especially the levels of endogenous proteins and the levels at which each construct is expressed. Thus, a simple 1:1 ratio (2.5 μg β_2AR-RLuc per 2.5 μg GRK2-GFP[2]) may not give the best results. Optimization is necessary.
2. We found that Lipofectmine 2000 (ThermoFisher, 11668019) can be used as the transfection reagent instead of FuGENE6 transfection reagent with the similar outcome.
3. The BRET ratio is corrected for the background signal according to the overlap of the donor (RLuc) emission at the acceptor excitation wavelength. Thus, cells expressed the donor alone (β_2AR-RLuc alone) is used as the background BRET signal. The empty vector (pCDNA3.1 vector) is used to maintain equal total amount of DNA per dish during transfection.
4. Total transfection volume (per 10 cm dish) is approximately 600 μL (medium + FuGENE6 + plasmid DNA).
5. For GRK2-GFP[2], an analysis for fluorescence of GRK2-GFP[2] is performed by exciting at 425 nm and examining the emission signal at 515 nm. GRK2-GFP[2] can be examined for the cytosol localization via a fluorescent microscopy. For β_2AR-RLuc, an assay for luminescence of β_2AR-RLuc is performed in the presence of Coelenterazine 400a (DeepBlueC).
6. The PBS/EDTA solution is defined as 0.5 mM EDTA in PBS, pH 7.4
7. The calculation of BRET ratio is performed automatically by using the MikroWin software from Berthold.
8. The values of $BRET^2$ ratio are corrected by subtracting the background $BRET^2$ signal detected when cells expressing the donor β_2AR-RLuc alone. The $BRET^2$ ratio is corrected for the background signal due to the overlap of the donor emission at the acceptor excitation wavelength, which is always determined in parallel for cells expressing the donor alone.
9. The value of BRET is corrected by subtracting the background BRET signals detected in cells expressed donor alone (RLuc-β-arrestin2). Thus, cells expressed

RLuc-β-arrestin2 with either β_1AR or β_2AR (RLuc-β-arrestin2 + β_1AR; RLuc-β-arrestin2 + β_2AR) are used as the background BRET signal.

10. The RLuc-β-arrestin2-YFP can be examined for cytosol localization via a fluorescent microscopy (Fig. 2D) and can be test for luminescence of RLuc-β-arrestin2-YFP in the presence of Coelenterazine h.
11. The HEK-293 cells are stimulated with ISO at approximately 5–10 min for detection of the conformational change of β-arrestin2

Acknowledgments

This work has been supported by grants from the Ministry of Education, Culture, Sports, Science, and Technology of Japan to Prof. Hitoshi Kurose (in Mangmool et al., 2006) and by the National Institutes of Health grants HL56687 and HL-75443 to Prof. Howard A. Rockman (in Mangmool et al., 2010).

Disclosure statement

The authors have nothing to disclose.

References

Bertrand, L., Parent, S., Caron, M., Legault, M., Joly, E., Angers, S., et al. (2002). The BRET²/arrestin assay in stable recombinant cells: a platform to screen for compounds that interact with G protein-coupled receptors (GPCRs). *Journal of Receptor and Signal Transduction Research*, *22*, 533–541.

Borroto-Escuela, D. O., Flajolet, M., Agnati, L. F., Greengard, P., & Fuxe, K. (2013). Bioluminescence resonance energy transfer methods to study G protein-coupled receptor-receptor tyrosine kinase heteroreceptor complexes. *Methods in Cell Biology*, *117*, 141–164.

Charest, P. G., Terrillon, S., & Bouvier, M. (2005). Monitoring agonist-promoted conformational changes of beta-arrestin in living cells by intramolecular BRET. *EMBO Reports*, *6*(4), 334–340.

Ferguson, S. S. (2001). Evolving concepts in G protein-coupled receptor endocytosis: the role in receptor desensitization and signaling. *Pharmacological Reviews*, *53*, 1–24.

Ferguson, S. S., Downey, W. E., Colapietro, A. M., Barak, L. S., Menard, L., & Caron, M. G. (1996). Role of beta-arrestin in mediating agonist-promoted G protein-coupled receptor internalization. *Science*, *271*, 363–366.

Gurevich, V. V., & Benovic, J. L. (1993). Visual arrestin interaction with rhodopsin. Sequential multisite binding ensures strict selectivity toward light-activated phosphorylated rhodopsin. *The Journal of Biological Chemistry*, *268*, 11628–11638.

Gurevich, V. V., & Gurevich, E. V. (2003). The new face of active receptor bound arrestin attracts new partners. *Structure*, *11*, 1037–1042.

Hirsch, J. A., Schubert, C., Gurevich, V. V., & Sigler, P. B. (1999). The 2.8 A crystal structure of visual arrestin: A model for arrestin's regulation. *Cell*, *97*, 257–269.

Kocan, M., See, H. B., Seeber, R. M., Eidne, K. A., & Pfleger, K. D. (2008). Demonstration of improvements to the bioluminescence resonance energy transfer (BRET) technology for the monitoring of G protein-coupled receptors in live cells. *Journal of Biomolecular Screening*, *13*(9), 888–898.

Lefkowitz, R. J., Rajagopal, K., & Whalen, E. J. (2006). New roles for beta-arrestins in cell signaling: Not just for seven-transmembrane receptors. *Molecular Cell*, *24*, 643–652.

Lefkowitz, R. J., & Shenoy, S. K. (2005). Transduction of receptor signals by β-arrestins. *Science*, *308*, 512–517.

Lutrell, L. M., & Lefkowitz, R. J. (2002). The role of β-arrestins in the termination and transduction of G-protein-coupled receptor signals. *Journal of Cell Science*, *115*, 455–465.

Mangmool, S., Haga, T., Kobayashi, H., Kim, K. M., Nakata, H., Nishida, M., et al. (2006). Clathrin required for phosphorylation and internalization of β_2-adrenergic receptor by G protein-coupled receptor kinase 2 (GRK2). *The Journal of Biological Chemistry*, *281*, 31940–31949.

Mangmool, S., Parichatikanond, W., & Kurose, H. (2018). Therapeutic targets for treatment of heart failure: Focus on GRKs and β-arrestins affecting βAR signaling. *Frontiers in Pharmacology*, *2018*(9), 1336.

Mangmool, S., Shukla, A. K., & Rockman, H. A. (2010). β-Arrestin-dependent activation of Ca^{2+}/calmodulin kinase II after β_1-adrenergic receptor stimulation. *The Journal of Cell Biology*, *189*, 573–587.

Marullo, S., & Bouvier, M. (2007). Resonance energy transfer approaches in molecular pharmacology and beyond. *Trends in Pharmacological Sciences*, *28*(8), 362–365.

Maziarz, M., & Garcia-Marcos, M. (2017). Rapid kinetic BRET measurements to monitor G protein activation by GPCR and non-GPCR proteins. *Methods in Cell Biology*, *142*, 145–157.

Penela, P., Ribas, C., & Mayor, F., Jr. (2003). Mechanisms of regulation of the expression and function of G protein-coupled receptor kinases. *Cellular Signalling*, *15*, 973–981.

Penn, R. B., Pronin, A. N., & Benovic, J. L. (2000). Regulation of G protein-coupled receptor kinases. *Trends in Cardiovascular Medicine*, *10*, 81–89.

Pfleger, K. D., & Eidne, K. A. (2003). New technologies: bioluminescence resonance energy transfer (BRET) for the detection of real time interactions involving G-protein coupled receptors. *Pituitary*, *6*(3), 141–151.

Pfleger, K. D., & Eidne, K. A. (2006). Illuminating insights into protein-protein interactions using bioluminescence resonance energy transfer (BRET). *Nature Methods*, *3*(3), 165–174.

Post, S. R., Hammond, H. K., & Insel, P. A. (1999). β-Adrenergic receptors and receptor signaling in heart failure. *Annual Review of Pharmacology and Toxicology*, *39*, 343–360.

Ruiz-Gòmez, A., & Mayor, F., Jr. (1997). β-Adrenergic receptor kinase (GRK2) colocalizes with β-adrenergic receptors during agonist-induced receptor internalization. *The Journal of Biological Chemistry*, *272*, 9601–9604.

Salazar, N. C., Chen, J., & Rockman, H. A. (2007). Cardiac GPCRs: GPCR signaling in healthy and failing hearts. *Biochimica et Biophysica Acta*, *1768*, 1006–1018.

Shukla, A. K., Violin, J. D., Whalen, E. J., Gesty-Palmer, D., Shenoy, S. K., & Lefkowitz, R. J. (2008). Distinct conformational changes in β-arrestin report biased agonism at seven-transmembrane receptors. *Proceedings of the National Academy of Sciences of the United States of America*, *105*, 9988–9993.

Xiao, K., Shenoy, S. K., Nobles, K., & Lefkowitz, R. J. (2004). Activation dependent conformational changes in β-arrestin 2. *The Journal of Biological Chemistry*, *279*, 55744–55753.

CHAPTER

5

Cannabinoid receptor CB_1 and CB_2 interacting proteins: Techniques, progress and perspectives

Caitlin R.M. Oyagawa[a,b,c] and Natasha L. Grimsey[a,b,c,*]

[a]*Department of Pharmacology and Clinical Pharmacology, School of Medical Sciences, Faculty of Medical and Health Sciences, University of Auckland, Auckland, New Zealand*

[b]*Centre for Brain Research, Faculty of Medical and Health Sciences, University of Auckland, Auckland, New Zealand*

[c]*Maurice Wilkins Centre for Molecular Biodiscovery, Auckland, New Zealand*

**Corresponding author: e-mail address: n.grimsey@auckland.ac.nz*

Chapter outline

Methods in Cell Biology, Volume 166, ISSN 0091-679X, https://doi.org/10.1016/bs.mcb.2021.06.011

Abstract

Cannabinoid receptors 1 and 2 (CB_1 and CB_2) are implicated in a range of physiological processes and have gained attention as promising therapeutic targets for a number of diseases. Protein-protein interactions play an integral role in modulating G protein-coupled receptor (GPCR) expression, subcellular distribution and signaling, and the identification and characterization of these will not only improve our understanding of GPCR function and biology, but may provide a novel avenue for therapeutic intervention. A variety of techniques are currently being used to investigate GPCR protein-protein interactions, including Förster/fluorescence and bioluminescence resonance energy transfer (FRET and BRET), proximity ligation assay (PLA), and bimolecular fluorescence complementation (BiFC). However, the reliable application of these methodologies is dependent on the use of appropriate controls and the consideration of the physiological context. Though not as extensively characterized as some other GPCRs, the investigation of CB_1 and CB_2 interacting proteins is a growing area of interest, and a range of interacting partners have been identified to date. This review summarizes the current state of the literature regarding the cannabinoid receptor interactome, provides commentary on the methodologies and techniques utilized, and discusses future perspectives.

1 Introduction

Cannabinoid receptors 1 and 2 (CB_1 and CB_2), class A G Protein-Coupled Receptors (GPCRs) named due to their interactions with components of *Cannabis sativa*, are both in hot pursuit as therapeutic targets. CB_1 is predominantly expressed in the central nervous system, with particularly high levels in the basal ganglia, hippocampus, cortex and cerebellum, where it is involved in the regulation of motor function, mood, memory, and pain (Fernández-Ruiz & González, 2005; Manzanares, Julian, & Carrascosa, 2006; Morena & Campolongo, 2014; Witkin, Tzavara, & Nomikos, 2005). CB_1 is most frequently expressed at the presynaptic terminals of both GABAergic and glutamatergic neurons, where activation inhibits neurotransmitter release (Castillo, Younts, Chávez, & Hashimotodani, 2012; Domenici et al., 2006; Irving et al., 2000; Szabo & Schlicker, 2005). Dysfunction and/or dysregulation of CB_1 has been implicated in a number of neurodegenerative disorders, including Huntington's disease, Alzheimer's disease, and Parkinson's disease (Stampanoni Bassi, Sancesario, Morace, Centonze, & Iezzi, 2017; Tanveer, McGuinness, Daniel, Gowran, & Campbell, 2012). In contrast, CB_2 distribution is largely peripheral, with high expression in immune tissues and cells such as the spleen, tonsils, and leukocyte populations (Bouaboula et al., 1993; Galiègue et al., 1995), though it is also expressed in the resident immune cells of the brain, the microglia (Carlisle, Marciano-Cabral, Staab, Ludwick, & Cabral, 2002; Ehrhart et al., 2005). Unsurprisingly, a number of CB_2-mediated functions are immune-related, for example, inhibiting leukocyte migration and chemokine mediated-chemotaxis, and reducing macrophage secretion of pro-inflammatory mediators (Ghosh, Preet, Groopman, & Ganju, 2006; Nilsson, Fowler, & Jacobsson, 2006; Persidsky et al., 2015). As such, CB_2 has been suggested to be a promising therapeutic target in inflammatory diseases such as rheumatoid arthritis, atherosclerosis, inflammatory bowel disease and neuroinflammation (Benito et al., 2008; Turcotte, Blanchet, Laviolette, & Flamand, 2016; Wright, Duncan, & Sharkey, 2008). Both cannabinoid receptors have also been implicated in various cancers, though whether their activity is pro- or anti-oncogenic appears to be context-dependent (Śledziński, Zeyland, Słomski, & Nowak, 2018).

Endogenous cannabinoids (endocannabinoids), the most prominent of which are anandamide and 2-arachidonoyl glycerol (2-AG), are the endogenous ligands that bind to the orthosteric binding site of cannabinoid receptors. Unlike most neurotransmitters, these are synthesized "on demand" from precursors present in the plasma membrane and typically act as retrograde messengers (Lu & Mackie, 2016; Zou & Kumar, 2018). Despite being structurally similar, anandamide and 2-AG are synthesized and degraded by distinct enzymes. Since the discovery of the endocannabinoid system (comprised of the cannabinoid receptors, endocannabinoids, and the enzymes that mediate their synthesis and degradation), numerous synthetic cannabinoid orthosteric agonists and inverse agonists have been developed. In recent years, a number of allosteric modulators of CB_1 activity have been identified, and are being explored as an alternative therapeutic approach (reviewed in Mielnik, Lam, & Ross, 2021). The identification of allosteric modulators of CB_2 is comparatively in its infancy, as only few have been discovered to date (e.g., Pandey, Roy, & Doerksen, 2020).

Protein-protein interactions underpin the ability of GPCRs to induce signaling responses, as well as being critical in controlling subcellular distribution and expression levels. It is increasingly being recognized that the interplay between all these factors give rise to the specific physiological impacts that are unique to each receptor type, in the particular context in which they are activated. In the GPCR field in general, interactions with cognate signaling G proteins ($G\alpha/\beta/\gamma$), GPCR kinases (GRKs) and arrestins are extremely well established, have been demonstrated via a wide range of techniques, and are relatively straightforward to validate in that interactions are promoted and/or regulated by ligands binding to the receptor. Interactions between receptors to form dimers or higher order oligomers, and between receptors and adaptor/effector proteins to regulate subcellular localization and signaling, are also widely accepted. The best-validated of these interactions are again those that are influenced by ligand, and/or that produce stark alterations to subcellular localization—such as the $GABA_B$ receptor which functions as an obligate dimer, and the requirement of receptor activity modifying proteins (RAMPs) for cell surface expression of the calcitonin receptor-like receptor (CRL) (Evenseth, Gabrielsen, & Sylte, 2020; Hay, Garelja, Poyner, & Walker, 2018). The availability and ready application of proximity-reporting and protein complementation tools have facilitated considerable expansion of research in this area, and so too, the apparent prevalence of such interactions.

However, the reliable application of these techniques call for extreme care and scrutiny in terms of utilization of controls and consideration of the physiological context. Commonly utilized methods include immunoprecipitation, Förster/fluorescence and bioluminescence resonance energy transfer (FRET and BRET), proximity ligation assay (PLA), and bimolecular fluorescence complementation (BiFC). Each of these techniques, though powerful and useful when utilized carefully, can readily produce false-positive (or false-negative) results. Immunoprecipitation is often not designed to differentiate direct protein-protein interactions from complexes which may involve intermediates, and is critically dependent on the fidelity of antibodies utilized to precipitate and detect the proteins of interest, as well as on careful sample preparation to avoid non-specific protein aggregation (Szidonya, Cserzo, & Hunyady, 2008; Zvonok et al., 2007). Proximity-based resonance energy transfer (RET), PLA and BiFC are well-documented to be able to work in a "bystander" fashion, where positive signals are obtained from proteins that occur in close proximity but do not directly interact (Alsemarz, Lasko, & Fagotto, 2018; Donthamsetti, Quejada, Javitch, Gurevich, & Lambert, 2015; Kodama & Hu, 2012; Lan, Liu, Li, Wu, & Lambert, 2012; Szidonya et al., 2008; Tiulpakov et al., 2016). Good evidence for saturable interactions that are unlikely to be due to bystander effects can be obtained by interrogating a range of expression levels/ratios of the partner; for example, by knocking down one of the interacting partners or transfecting in different amounts of DNA to alter expression levels. However, it is extremely important that the cellular (potentially even sub-cellular) expression levels of the proteins of interest are measured directly, as opposed to simply modifying the DNA ratios transfected which will not necessarily correlate linearly with expression and/or may alter the proportion of cells co-expressing protein rather than the expression level/ratio per cell (Lan et al., 2015). Furthermore, although many studies utilize alternative receptors/partners as negative controls, smaller signals can be produced due to differences in tag or antibody conformations and/or orientations when attached to different proteins rather than truly reflecting a different proximity of the proteins of interest (e.g., Cannaert, Storme, Franz, Auwärter, & Stove, 2016). Functional implications, such as alterations in subcellular distribution and/or signaling responses, are useful in providing additional evidence for interactions as well as contributing to understanding the biological relevance. These experiments must also be carefully controlled to avoid misinterpretation, and in some cases functional interaction could reflect cross-talk, competition for signaling proteins or molecules, and/or feedback regulation rather than direct physical interaction (Chabre, Deterre, & Antonny, 2009; Lambert & Javitch, 2014). Furthermore, to date in the majority of cases the primary evidence for interaction has been obtained from over-expression models. It is important to consider whether the interacting protein concentration and stoichiometry assayed are reflective of the *in vivo* system of interest; even when assays are well designed and controlled, high concentrations and/or abnormal stoichiometry of the interacting partners may produce interactions that are *possible* but would not necessarily occur *in vivo* (Szidonya et al., 2008; Terrillon & Bouvier, 2004).

Although the CB_1 and CB_2 interactome has not been studied in as much depth as some other GPCRs, this is a growing area of interest and robust knowledge of receptor interactions may well be important in the development and optimization of therapeutic drugs. Here, we review the current state of the literature regarding the cannabinoid receptor interactome, providing commentary on the methodologies utilized and perspectives for future research.

2 Canonical G protein signaling interactions

The GPCR conformational state(s) promoted by agonist binding facilitate the canonical interaction with the $G\alpha/\beta/\gamma$ heterotrimer and exchange of guanosine triphosphate (GTP) for guanosine diphosphate (GDP), resulting in release of the heterotrimer and induction of downstream signaling. It is well established that both CB_1 and CB_2 primarily couple to the $G\alpha_{i/o}$ family, wherein upon activation adenylate cyclase is inhibited and cyclic adenosine monophosphate (cAMP) concentrations decrease.

Constitutive interaction of CB_1 with all three $G\alpha_i$ subtypes and $G\alpha_o$ was confirmed in early work by immunoprecipitation from rat brain, while interactions with $G\alpha_s$, $G\alpha_q$ and $G\alpha_z$ were not detected (Mukhopadhyay & Howlett, 2001; Mukhopadhyay, McIntosh, Houston, & Howlett, 2000). Interestingly, subtype selectivity in the receptor-G protein interaction point was indicated, wherein CB_1 intracellular loop 3 acted as a substrate for $G\alpha_{i1}$ and $G\alpha_{i2}$ binding, whereas a juxtamembrane portion of the CB_1 C-terminal tail (401–417) was implicated in $G\alpha_{i3}$ and $G\alpha_o$ binding (Mukhopadhyay & Howlett, 2001; Mukhopadhyay et al., 2000). Coupling to these $G\alpha_{i/o}$ subtypes has since been corroborated in a variety of contexts (e.g., Diez-Alarcia et al., 2016) and the interaction interfaces have been further refined via static and dynamic structural modeling, mutagenesis, mass spectrometry, and cryoelectron microscopy (cryo-EM) (Hua et al., 2020; Krishna Kumar et al., 2019; Mnpotra et al., 2014; Shim, Ahn, & Kendall, 2013).

CB_1 has additionally been described as a "promiscuous" receptor, with coupling to both $G\alpha_s$ and $G\alpha_q$ also reported, as interpreted via assessment of signaling responses and effects of inhibitors, knock-ins and knock-outs (e.g., Bonhaus, Chang, Kwan, & Martin, 1998; Finlay et al., 2017; Lauckner, Hille, & Mackie, 2005). However, this coupling is not always present and may well be context-dependent (e.g., Avet et al., 2020; Diez-Alarcia et al., 2016; Glass & Northup, 1999). Regulation of coupling has been linked both to GPCR conformational states via ligand-induced bias, and stoichiometry of receptor to signaling effectors, implying that CB_1 activation could have considerably different effects on signaling cascades depending upon the ligand encountered and cellular context.

The interaction of CB_2 with specific $G\alpha_i$ subtypes has not undergone in-depth study, though CB_2 has been shown to couple to purified $G\alpha_i$ (comprised of a mixture of $G\alpha_{i1}$, $G\alpha_{i2}$, and $G\alpha_{i3}$ subunits) via GTPγS binding assays (Glass & Northup, 1999). The same study also found that, unlike with $G\alpha_i$, activation of GTPγS with increasing concentrations of $G\alpha_o$ was not saturable under equivalent assay conditions, indicating a low affinity for CB_2-$G\alpha_o$ coupling. Furthermore, CB_2 did not catalyze the activation of $G\alpha_q$ nor $G\alpha_t$. In two recent large screens of human GPCR libraries for interactions with $G\alpha$ subunit C termini chimerized to reporter systems, agonist-stimulated CB_2 reached the "positive coupling" threshold for $G\alpha_{i1/2}$, $G\alpha_{i3}$, $G\alpha_o$ and $G\alpha_z$ (Avet et al., 2020; Inoue et al., 2019). CB_2 coupling with $G\alpha_q$ and $G\alpha_s$ has also been reported, though only rarely to date, and these interactions seem considerably more context-dependent than the more widely-observed pleiotropic coupling observed for CB_1 (Brailoiu et al., 2014; Glass & Northup, 1999; Saroz, Kho, Glass, Graham, & Grimsey, 2019).

The second intracellular loop (ICL) of CB_2 has been suggested to play a crucial role in CB_2 G protein coupling, as substitution of ICL2 with the corresponding region from CB_1 resulted in a twofold increase in basal activity, and a pertussis toxin-insensitive increase in forskolin-stimulated cAMP in response to agonist (Zheng et al., 2013). Within this region, P139 was identified as a key residue, replacement of which with hydrophobic amino acids resulted in agonist-induced stimulation of cAMP synthesis. Recent studies have utilized single-particle cryo-EM to obtain CB_2-$G\alpha_{i1}$-bound structures in a nucleotide-free state (i.e., an intermediate state in which GDP has been released from the G protein prior to GTP binding). Despite having different agonists bound (AM12033, Hua et al., 2020; WIN55,212-2, Xing et al., 2020), both structures indicate that the major interaction interface occurs between the Ga_{i1} C-terminal $\alpha5$ helix and CB_2 transmembrane domains (TM) 2, 3, 5, and 6, and ICL2. These data are also consistent with an earlier study which experimentally identified three specific cross-link sites between CB_2 and $G\alpha_{i1}$ using liquid chromatography with tandem mass spectrometry (LC-MS/MS) and electrospray ionization-MS/MS, all of which are plausible in both cryo-EM structures (Mnpotra et al., 2014).

Functional selectivity, or biased agonism, is a topic of widespread interest in the GPCR field currently. This concept, wherein a single receptor type may initiate different signaling response complements depending on the activating ligand, has been demonstrated for both CB_1 and CB_2 in heterologous expression systems (e.g., Khajehali et al., 2015; Oyagawa et al., 2018; Soethoudt et al., 2017). At the molecular level this phenomenon is thought to arise from ligand stabilization of different complements of receptor conformations which then facilitates interactions with different G proteins and/or other signaling effectors. As yet, little is known about the specific molecular mechanisms of CB_1 or CB_2 functional selectivity. An early study utilizing re-constitution of receptors with G proteins indicated that different CB_1 ligands could alter the propensity to couple to $G\alpha_{i1/2/3}$ vs $G\alpha_o$ (Glass & Northup, 1999). More recently, scintillation proximity detection of immunoprecipitated [^{35}S]GTPγS-labeled G proteins has been utilized to measure activation of various G protein subtypes by three cannabinoid ligands in brain tissue from wild-type and CB_1/CB_2 knockout mice. Interestingly, G protein coupling was seemingly mutually exclusive between CB_1 and CB_2, and the ligands produced differing patterns of coupling; for example, delta-9-tetrahydrocannabinol (Δ^9-THC) induced coupling to $G\alpha_{i1}$ and $G\alpha_q$ via CB_1, and $G\alpha_o$ and $G\alpha_z$ via CB_2, while WIN55,212-2 activated all G proteins as for Δ^9-THC and additionally $G\alpha_{i3}$ and $G\alpha_{12/13}$ via CB_1. Surprisingly, none of the ligands tested detectably activated $G\alpha_{i2}$ or $G\alpha_s$, though note positive controls were not provided which may have confirmed sensitivity of the assay for these pathways (Diez-Alarcia et al., 2016). The emergence of new RET-based techniques to assay coupling to specific G protein subtypes at high throughput provide tools to more readily investigate the G protein subtypes involved in cannabinoid receptor signaling and bias, as well as contributions of coupling kinetics and potentially receptor subcellular distribution (e.g., Inoue et al., 2019; Olsen et al., 2020). Attainment of the aforementioned crystal and cryo-EM structures of CB_1 and CB_2, with additional structures in alternative conformational states likely to follow, promise to support further understanding of biased agonism and accelerate the ability to rationally design ligands with specific bias profiles.

3 Non-G protein signaling mediators and modulators

3.1 G protein coupled receptor kinases (GRKs) and arrestins

Two of the most widely-studied protein types which modulate signaling via direct interaction with GPCRs are G Protein Coupled Receptor Kinases (GRKs) and arrestins. Following receptor activation, GRKs phosphorylate the cytoplasmic domains of GPCRs in particular patterns which are determined by the accessibility of phosphorylatable receptor residues, complement of GRK subtype expression, and GRK affinity for particular receptor conformations and other proximal interacting proteins. This "barcode" of phosphorylation enhances the affinity for arrestins which are recruited to GPCRs and act initially to "arrest" G protein-mediated signaling—essentially by sterically hindering further interaction with G proteins—as well as promoting internalization (Kim & Chung, 2020). Arrestins may also mediate the induction of further signaling either when interacting with receptor at the plasma membrane, subsequent to internalization, or as a result of persistent conformational change after dissociating from receptor (Appleton et al., 2019; Thomsen, Jensen, Hicks, & Bunnett, 2018). Four arrestin isoforms are known, two of which are found exclusively in the visual system, while the other two are expressed ubiquitously. The latter two, termed arrestins 2 and 3 (also known as β-arrestins 1 and 2), seem to be able to interact with the majority of GPCRs in either a transient or prolonged manner, these phenotypes being delineated as "class A" and "class B," respectively.

While a considerable amount is known about GRK and arrestin interactions with GPCRs in general, relatively little has been directly studied for CB_1 and CB_2. Over-expression of GRK3 and arrestin 3 is sufficient to enable CB_1 desensitization following activation with agonist WIN55,212-2, with phosphorylation at either S426 or S430 being implicated (Jin et al., 1999). Mutation of these sites in mice resulted in resistance to tolerance for Δ^9-THC and WIN55,212-2 but not CP55,940, raising the possibility of ligand-dependence of GRK3-mediated phosphorylation and desensitization (Nealon, Henderson-Redmond, Hale, & Morgan, 2019). GRK3 was also implicated in downstream extracellular signal-regulated kinase 1/2 (ERK1/2) phosphorylation, with knockdown reducing activation of this pathway whereas reducing expression of GRKs 2 and 4–6 had no effect (Delgado-Peraza et al., 2016). In contrast, GRK2 has been found to participate in CB_1-stimulated ERK1/2 phosphorylation via focal adhesion kinase (FAK) and growth factor receptors in a neuroblastoma cell line, though it was hypothesized that this was via interaction of GRK2 with Gβγ rather than with CB_1 (Dalton, Carney, Marshburn, Norford, & Howlett, 2020). Expression of a dominant negative form of GRK2 in rat hippocampal neurons reduced desensitization of CB_1-mediated effects (Kouznetsova, Kelley, Shen, & Thayer, 2002). GRK2 over-expression also slightly potentiated CB_1-stimulated arrestin 2 and 3 translocation in a heterologous cell line, though the effect was subtle in comparison with GRK2-enhanced recruitment to a dopamine receptor in the same system (Ibsen et al., 2019).

In regard to CB_2/GRK interactions, GRK5 has been shown to phosphorylate CB_2 and play a role in the downstream agonist-induced interaction between arrestin 3 and ERK1/2 in a neuronal cell line that endogenously expresses CB_2, as confirmed by shRNA knockdown of CB_2 and/or GRK5 (Franklin & Carrasco, 2013). Over-expression of GRK2 in a heterologous cell line did not significantly affect CB_2-stimulated translocation of arrestin 2 nor 3, despite translocation efficacy being comparatively much lower than CB_1 and other control receptors whose arrestin translocation was enhanced following GRK2 introduction (Ibsen et al., 2019). Moreover, in another heterologous expression system, agonist-induced internalization of CB_2 was unaffected by the presence of a GRK2/3 inhibitor, indicating an apparent lack of involvement of these GRKs in CB_2 internalization and desensitization (Udoh, Santiago, Devenish, McGregor, & Connor, 2019). Interestingly, a CB_2-selective agonist, JWH-015, restored spinal cord expression of GRK2 in a rat model of bone cancer pain (Lu et al., 2017). This correlated with the alleviation of bone cancer-induced allodynia and ambulatory pain, however, it was not clear whether the changes in GRK2 expression had any homologous effect on CB_2 responses.

Both CB_1 and CB_2 are typically described as "class A" arrestin interactors and accordingly tend to have short-lived arrestin interaction with greater propensity for recruiting arrestin 3 rather than 2. For CB_1, this trend has been measured via immunoprecipitation (Delgado-Peraza et al., 2016), and for both CB_1 and CB_2 indirectly via a BRET-based assay detecting arrestin proximity to the plasma membrane (Ibsen et al., 2019). A range of studies, usually focusing on arrestin 3 and most frequently utilizing a BRET-based assay detecting proximity of arrestin to a luciferase-tagged receptor, have reported ligand-dependence of arrestin recruitment to CB_1 and CB_2, presumably a manifestation of ligand bias and with potential functional consequences (e.g., Dhopeshwarkar & Mackie, 2016; Ibsen et al., 2019; Laprairie, Bagher, Kelly, & Denovan-Wright, 2016; Soethoudt et al., 2017).

Despite the general consensus that CB_1 and CB_2 can transiently recruit arrestins, there are some discrepancies in specific results and relative ligand effectiveness between studies. Some of these may be due to the cellular context and availability of interacting proteins, or specific protocols such as timepoint assayed, however, we caution that the majority of methods that are routinely applied should be interpreted carefully. Firstly, any assays involving tagged receptor or arrestin should consider the possibility of the tag influencing biology and interactions. Further, whenever the receptor of interest is required to be tagged with an assay component it is difficult to control for components of the apparent positive signal that are not mediated by the receptor of interest (i.e., arrestin recruitment to sites proximal to the receptor but not representing direct interaction) (Ibsen et al., 2019). Conversely, methods measuring arrestin proximity to the plasma membrane allow the use of unmodified receptor, and controlling for non-receptor-mediated effects is usually straightforward, but some such methods do not facilitate measurement of post-endocytic arrestin interactions which may mediate distinct signaling responses and/or have implications for trafficking (Appleton et al., 2019; Nogueras-Ortiz et al., 2017). We also note that the vast majority of studies have utilized non-human arrestins which, when studying human receptors and cell models, has additional potential to introduce artifact/misinterpretation (Ibsen et al., 2019). Furthermore, RET proximity-based assays are indirect methods and conformational changes that influence the RET signal can be incorrectly interpreted as changes in recruitment/interaction (though any such conformational change may in itself be interesting/relevant). Determination of arrestin translocation and/or formation of vesicles via immunocytochemistry avoids some of these drawbacks, but is a less sensitive and more labor-intensive assay. Immunoprecipitation with western blotting represents the most direct method for assessing interaction, however, is prone to artifact, and only semi-quantitative. We have noted with interest new arrestin conformational biosensors which would be interesting to apply to the cannabinoid receptors (Lee et al., 2016). Each of the methods described has potential utility in better understanding GPCR-arrestin interactions; indeed, a direct comparison between the various approaches may well be illuminating. However, also important to note that all studies of CB_1/CB_2 arrestin interaction to date have relied upon receptor and arrestin over-expression, which will clearly have abnormal stoichiometry—likely to each other but also to any other signalsome components, in comparison with endogenous expression systems. Animal models such as knockouts have utility in advancing understanding of the physiological consequences of arrestin interactions. This approach has been applied to further understanding of the consequence of CB1 and arrestin 3 interactions, with knockout of arrestin 3 from mouse inducing subtle consequences for Δ^9-THC-induced analgesia and hypothermia (e.g., Nguyen et al., 2012). However, widespread and chronic knock-outs can give rise to network and/or adaptive changes, and some findings in the opioid field that were based on arrestin knockouts have recently been disputed (Gillis et al., 2020).

3.2 Cannabinoid receptor interacting proteins (CRIPs)

The Cannabinoid Receptor Interacting Proteins (CRIPs) are plasma membrane-associated but non-transmembrane-spanning proteins discovered via a yeast two-hybrid interaction assay screen of a human brain cDNA library using the last 55 residues of the CB_1 C-terminal tail as bait (Niehaus et al., 2007). The two proteins are splice variants of the same gene; "1a" having 164 amino acids with orthologues across vertebrates and detected in brain, heart, lung and intestinal tissue (and elsewhere), whereas "1b" is only found in primates with residues 1–110 identical to CRIP1a but the last 53 residues being replaced by a unique 18 amino acid sequence. CRIP1a incorporates a PDZ domain (a commonly-occurring structural domain frequently involved in protein-protein interaction and trafficking) and predicted palmitoylation site which are not present in CRIP1b. In the initial report, interactions were validated by co-immunoprecipitation of bacterially-expressed CRIPs 1a and 1b with the CB_1 (but not CB_2) C-terminal tail, and for CRIP1a with CB_1 from rat brain homogenate (Niehaus et al., 2007). CRIP1a did not influence WIN55,212-2-induced inhibition of voltage-gated Ca^{2+} channels but reduced constitutive CB_1-mediated inhibition without influencing inverse agonist affinity nor overall CB_1 expression (surface expression was not specifically measured), whereas CRIP1b had no detectable effect on the parameters measured (Niehaus et al., 2007). In most subsequent reports CRIP1a has tended to suppress CB_1 activity, reducing agonist-induced G protein coupling, phospho-ERK stimulation, constitutive internalization and agonist-induced internalization (Blume, Eldeeb, Bass, Selley, & Howlett, 2015; Blume et al., 2016, 2017; Mascia et al., 2017; Smith et al., 2015), though effects were not always consistent between studies which may to some degree involve ligand-dependence (Blume et al., 2016), or otherwise likely implies context-dependence. One study in hippocampal neurons instead found that CB_1 agonist-induced G protein activation was enhanced (Guggenhuber et al., 2016), though whether this could have been a consequence of lower constitutive activity is not clear. Interestingly, CRIP1a may facilitate a switch in G protein coupling from predominantly $G\alpha_{i3}$ and $G\alpha_o$ with low CRIP1a to $G\alpha_{i1}$ and $G\alpha_{i2}$ in high CRIP1a-expressing cells (Blume et al., 2015). This again highlights the importance of context and stoichiometry of interaction partners when studying receptor pharmacology; indeed, differential CRIP1a expression has been hypothesized as a reason for contrasting CB_1 trafficking results obtained from different cell lines (Daigle, Kwok, & Mackie, 2008). The CB_1-CRIP1a interaction has been proposed to be important in a variety of physiological and pathological contexts; these have been comprehensively reviewed recently (Oliver et al., 2020).

The interaction site for CRIP1a was originally reported to be the final nine amino acids of the CB_1 C-terminal tail (Niehaus et al., 2007) and this was more recently refined to five amino acids as the minimal motif (Mascia et al., 2017). Phosphorylation at threonine-467 within this motif inhibited interaction with CRIP1a and seemingly facilitated exchange in affinity for arrestin 3; accordingly, competition for binding between CRIP1a and arrestins is a viable mechanism for CRIP1a reducing internalization (Blume et al., 2017). The distal CB_1 C-terminal tail has also been implicated in G protein coupling, indicating a possible mechanism for the aforementioned G protein coupling switch induced by CRIP1a (Blume et al., 2015). Interestingly, a similar five amino acid motif is present in another pre-synaptically expressed receptor, metabotropic glutamate receptor (mGlu8R), and CRIP1a reduced its constitutive internalization in a similar manner as for CB_1 (Mascia et al., 2017). Immunoprecipitation with peptide competition and modeling both further supported the importance of the distal CB_1 C-terminal tail, but also implicated the central C-terminal tail as a CRIP1a contact site (Ahmed, Kellogg, Selley, Safo, & Zhang, 2014; Blume et al., 2017; Singh, Ganjiwale, Howlett, & Cowsik, 2019).

CRIP1b has undergone extremely little study, perhaps owing to its lack of expression in widely-used rodent models and that the discovering report suggested lack of functional effect (Niehaus et al., 2007). It has been hypothesized that the lack of PDZ domain and palmitoylation site in CRIP1b might facilitate function as a dominant negative modulator of CRIP1a activity (Smith, Sim-Selley, & Selley, 2010). *In silico* methods have been utilized to predict the structure of CRIP1b and interaction points with CB_1, but these have yet to be tested experimentally (Singh, Ganjiwale, et al., 2019; Singh, Ganjiwale, Howlett, & Cowsik, 2017).

3.3 Src homology 3-domain growth factor receptor-bound 2-like (endophilin) interacting protein 1 (SGIP1)

Src homology 3-domain growth factor receptor-bound 2-like (endophilin) interacting protein 1 (SGIP1) is a proposed CB_1 interacting protein which modulates clathrin-mediated endocytosis of a range of plasma membrane proteins (e.g., Uezu et al., 2007) and has been implicated in energy homeostasis (e.g., Trevaskis et al., 2005). SGIP1 is expressed primarily in the CNS, including pre-synaptically in rat primary cortical neurons where it co-localizes with CB_1 (Hájková et al., 2016). Similarly to CRIP1a, the interaction of SGIP1 with CB_1 was discovered via a yeast two-hybrid screen of the CB_1 C-terminal tail with subsequent validation by immunoprecipitation and BRET (Hájková et al., 2016); to our knowledge this is the only evidence to date for a direct protein-protein interaction. SGIP1 over-expression in a heterologous cell line had no significant effect on G protein coupling, but ERK phosphorylation was markedly reduced. This was correlated with a reduction in both constitutive and agonist-induced internalization, whereas arrestin 3 recruitment was enhanced slightly (Hájková et al., 2016). A SGIP1 knockout mouse exhibited alterations of cannabinoid tetrad behaviors and augmented anti-nociceptive responses to CB_1 agonists and morphine (Dvorakova et al., 2021).

3.4 Calneuron-1 and neuronal calcium sensor 1 (NCS1)

Two further proposed CB_1 interacting proteins, though reported in only one publication to date, are calcium binding proteins calneuron-1 and neuronal calcium sensor 1 (NCS1, also known as frequenin). Evidence for the interactions was obtained via BRET saturation assays in transfected cells, modification of BRET signals via mutations to CB_1 and interacting proteins (implicating the CB_1 C-terminal tail for both interactions, and CB_1 IL3 for NCS1 only), and competition between NCS1 and calneuron-1 for binding to CB_1. Lack of interaction with calcium binding protein caldendrin was also indicated, though this negative result could have been a result of non-permissive conformational arrangement of the BRET tags. BRET in transfected cells and *in situ* PLA in mouse primary striatal neurons indicated a switch in the calcium binding protein predominantly binding to CB_1 at different intracellular calcium levels, the NCS1 interaction being predominant at basal calcium levels. These states correlated with CB_1 exhibiting $G\alpha_i$-mediated signaling with basal calcium, but $G\alpha_s$-like signaling when cytoplasmic calcium was enhanced with ionomycin (Angelats et al., 2018).

3.5 Heat shock protein 90 (Hsp90)

Hsp90 is well established as a molecular chaperone which facilitates the maturation and appropriate folding of a number of substrates (Schopf, Biebl, & Buchner, 2017). Direct interaction between CB_2 and Hsp90 has been demonstrated via co-immunoprecipitation in transfected cells, and in HL60 cells that endogenously express CB_2 (He et al., 2007). Geldanamycin, a specific Hsp90 inhibitor (Ochel, Eichhorn, & Gademann, 2001), reduced the amount of Hsp90 immunoprecipitated with CB_2 without affecting the amount of Hsp90 or CB_2 in membrane fractions, validating the interaction. Furthermore, inhibition and siRNA knockdown of Hsp90 attenuated the extent of 2-AG-induced cell migration, indicating the involvement of Hsp90 in this process. This interaction was later observed in trabecular meshwork cells (via co-immunoprecipitation), disruption of which inhibited CB_2-mediated phosphorylation of ERK1/2 and actin cytoskeleton remodeling (He, Kumar, & Song, 2012).

3.6 Post-translational modifications

Post-translational modifications such as glycosylation, palmitoylation and SUMOylation have also been detected for CB_1 and/or CB_2, have the potential to influence receptor subcellular distribution and function, and may themselves undergo dynamic regulation (e.g., Gowran, Murphy, & Campbell, 2009; Oddi et al., 2012; Song & Howlett, 1995; Zhang et al., 2007). Introduction (and potential regulation) of these modifications presumably involve recognition of consensus sequences on the receptors and transient interaction with enzymes. To date, little is known about these processes for the cannabinoid receptors.

4 Receptor oligomerization

Whereas GPCRs were once considered to function as singular units, a plethora of literature now implicates many GPCRs as functioning in complex with receptors of same or different types. These are often described as "dimers," with some evidence pointing to this dual topology but other evidence implying the potential interaction of three or more units. GPCR oligomerization has the potential to influence ligand binding and receptor coupling to signaling pathways (both constitutively and in response to ligand), as well as subcellular localization both basally and following ligand interaction (Pin, Kniazeff, Prézeau, Liu, & Rondard, 2019).

Both homo-oligomerization and hetero-oligomerization with a range of other GPCRs have been reported for both CB_1 and CB_2. As noted in the introduction, studying protein-protein interaction is inherently challenging, and the study of GPCR oligomerization has been particularly fraught as has been the subject of a number of commentaries (e.g., Lambert & Javitch, 2014; Milligan, 2013). Given that, to date, a number of the suggested cannabinoid receptor interactions have been reported by only one study and/or laboratory, we have made particular note of the techniques and controls applied in each.

4.1 Homo-dimer/oligomerization

Both CB_1 and CB_2 have been suggested to form homo-oligomers, though it is not typically claimed that oligomerization is *required* for function. Western blot analysis frequently indicates the presence of dimers and/or higher order oligomers on the basis of size, though it is not often clarified whether these species are truly homo-oligomers or represent interaction of the receptors with other proteins (e.g., Mukhopadhyay et al., 2000; Wager-Miller, Westenbroek, & Mackie, 2002). In more direct biochemical evidence for homodimer formation, mass spectromic analysis of an SDS-resistant protein species of a size consistent with homodimer was confirmed to represent only CB_1 (Xu et al., 2005), though in this study the receptor had been expressed to high levels in insect cells. As also addressed in the same study, care must be taken when interpreting SDS-PAGE protein gels and western blots given that receptor protein is known to be prone to aggregation in lysates, particularly when heated (Esteban et al., 2020). At least three *in vitro* studies have contributed evidence supporting CB_1 homodimerization utilizing FRET and BRET (Jäntti, Mandrika, & Kukkonen, 2014; Ward, Pediani, & Milligan, 2011; Zou, Somvanshi, & Kumar, 2017). Most recently, BiFC with peptide competition implicated TM4 as a putative

CB_1 homodimer interface (Köfalvi et al., 2020). Whether disruption of CB_1 homodimer by peptide competition had any functional consequence was not investigated.

Limited evidence for mammalian *in vivo* CB_1 homodimer formation has been provided through the use of a "dimer-specific" antibody directed against the CB_1 C-terminus which detected CB_1 in the rat brain in a very similar pattern to that observed with a typical "monomeric" N-terminally directed CB_1 antibody (Wager-Miller et al., 2002). Assuming reliability of this antibody, this implies that CB_1 dimers or multimers are present *in vivo*, in similar locations to the monomer. However, it is important to note that this antibody was selected for its ability to detect dimer on the basis of western blotting, which may well present antigens in considerably different conformations and availability in comparison with immunohistochemistry. Although this study also noted that CB_1 homodimers in rat brain were unlikely to be stabilized via disulfide bonds as presence of the monomeric species was not enhanced under reducing conditions, a subsequent study found putative dimers of human CB_1 from prefrontal cortex membranes to be sensitive to reducing conditions (De Jesús, Sallés, Meana, & Callado, 2006).

On the basis of this evidence, and the general assumption that most GPCRs probably homodimerize to some degree, at least two groups have sought to design bivalent CB1 ligands with the intention of specifically targeting the homodimer. In both cases, two molecules of a well characterized high affinity orthosteric ligand were connected by a linking moiety and affinity assessed for different linker lengths to produce a bivalent ligand. The first-discovered of these had higher affinity for CB_1 than the monovalent control ligand, which on simple interpretation might be considered to contribute evidence toward CB_1 homodimerization (Fernández-Fernández et al., 2013; Zhang et al., 2010). However, subsequent study of similar compounds found that monovalent ligands acted with higher potency than divalent ligands (Huang et al., 2014), and interrogation of these compounds by modeling into multiple potential homodimer conformations indicated that the likelihood of these compounds being physically large enough to bind to both CB_1 homodimer binding sites simultaneously was extremely small (if not impossible) (Glass et al., 2016; Pérez-Benito et al., 2018). The enhanced affinity observed was hypothesized to instead be a result of an allosteric effect of the bivalent molecule via the monomer or across the dimer, or of increased "statistical binding" wherein binding of one pharmacophore increases the local concentration of ligand.

Less evidence of CB_2 homodimerization has been reported, with only two studies carried out by the same laboratory group proposing that CB_2 forms homodimers when overexpressed in Sf21 insect cells (Filppula et al., 2004; Zvonok et al., 2007). MALDI-TOF MS confirmed that an SDS-resistant protein species observed via western blot was comprised of CB_2 only. However, the authors also noted that as with CB_1, CB_2 tended to form aggregates when lysates were incubated at higher temperatures and/or had undergone a dual-step purification regimen, and the apparent oligomerization could have at least in part been artefactual.

In our opinion, to date there is a lack of definitive direct evidence to support the existence of CB_1 and CB_2 homodimers, let alone any insight into the potential functional relevance of this putative dimerization. Although their presence seems a reasonable likelihood based on evidence from other class A GPCRs, it would be useful to apply some of the approaches applied to other GPCRs in order to verify this hypothesis and begin to interrogate potential functional relevance, including the prevalence of dimerization vs oligomerization both in individual cells and between cell/organ types (Parker et al., 2008), and whether there is therapeutic value in targeting either state. For example, a widely-accepted method for establishing homodimerization is to assess the effects of non-trafficking receptor mutants on wild-type receptor (e.g., Lee et al., 2000). A more simplistic and more readily accessible approach would be to co-express CB_1 or CB_2 transmembrane-spanning domains and interrogate any alteration in receptor function and/or complexed state via western blot or protein interaction/proximity assay (e.g., Hebert et al., 1996). Sophisticated live cell imaging techniques such as dual color fluorescence recovery after photobleaching (dcFRAP) with immobilization of one dimer partner to assess influence on mobility of a freely mobile partner and fluorescence cross-correlation spectroscopy (FCCS), also have strong potential to lend further insight as to the existence and importance of oligomeric receptor states (Youker & Voet, 2020).

4.2 Hetero-dimer/oligomerization

Hetero-oligomerization of CB_1 and CB_2 has received considerably more attention than homo-oligomerization, perhaps partially due to the somewhat more straightforward design of assays and the greater likelihood that heterodimer/oligomers will present therapeutic opportunities via restricted co-expression and/or unique functional properties.

4.2.1 CB_1-CB_2

The possibility of CB_1 and CB_2 heterodimerizing has been investigated in only a few studies to date. As well as being demonstrated in transfected cells (BRET), CB_1-CB_2 heteromers were identified in rat brain, specifically in pinealocytes and neurons of the pineal gland, by *in situ* proximity ligation assay and were found to produce negative signaling crosstalk in the phospho-Akt pathway to influence neurite outgrowth (Callén et al., 2012). The putative presence of CB_2 in the healthy brain (particularly in neurons) remains a point of contention in the field (e.g., Atwood & MacKie, 2010), and it is interesting to speculate whether the presence of CB_1-CB_2 heteromers could have somehow obfuscated detection of CB_2 in some studies and/or produce a functional effect of neuronal CB_2 even if expressed at extremely low levels. A caution regarding this specific study is that there is considerable divergence between the human and rat CB_2 primary amino acid sequences which have been reported to influence CB_2 molecular pharmacology (e.g., Mukherjee et al., 2004), and a secondary splice variant exists for rat CB_2 which is not present for human CB_2 (Brown, Wager-Miller, & Mackie, 2002), both of which have potential to modify the CB_2 expression and dimerization profiles. It will be interesting to continue to investigate whether CB_1-CB_2 heterodimers exist in other contexts where co-expression has been noted, for example, in some immune cell types (Kaplan, 2013), platelets (Catani et al., 2010), some cancers (Laezza et al., 2020), and spermatozoa (Agirregoitia et al., 2010). CB_1-CB_2 heteromerization has since been demonstrated by BRET and PLA in a microglial cell line, with the interaction enhanced by the presence of pro-inflammatory stimuli and in a rat 6-hydroxy-dopamine model of Parkinson's disease (Navarro et al., 2018).

4.2.2 CB_1-D_2

The most rigorously established and best-characterized dimer pairing for a cannabinoid receptor is between CB_1 and dopamine receptor 2 (D_2). Direct interactions between these receptors were first indicated by a striking signaling switch from inhibition of adenylate cyclase when the receptors were stimulated separately, to stimulation of adenylate cyclase when both receptors were activated; presumably a result of a change in preference from $G\alpha_i$ to $G\alpha_s$-coupling (Glass & Felder, 1997). Subsequent studies have provided evidence to suggest this signaling switch is a result of direct physical interaction between CB_1 and D_2, including immunoprecipitation (e.g., Kearn, Blake-Palmer, Daniel, Mackie, & Glass, 2005; Khan & Lee, 2014), FRET (e.g., Marcellino et al., 2008), BRET (e.g., Bagher, Laprairie, Kelly, & Denovan-Wright, 2016), BiFC (e.g., Przybyla & Watts, 2010) and *in situ* PLA in mouse globus pallidus (Bagher et al., 2020). Furthermore, CB_1-D_2 dimerization can influence ligand affinity, presumably via allosteric modification across the dimer interface; specifically, a CB_1 agonist reduced the affinity of a D_2 orthosteric ligand in rat striatal membranes and tissue sections (Marcellino et al., 2008). Pre-treatment (but not co-incubation) with a D_2 antagonist can also influence CB_1-G protein coupling and induction of downstream signaling cascades (Bagher et al., 2016). Immunoprecipitation with receptor fragments and peptide competition has indicated that dimerization involves direct interaction between the CB_1 carboxy-terminal (cytoplasmic) tail and the D_2 third intracellular loop (Khan & Lee, 2014). To our knowledge, the molecular mechanism for the signaling switch has not been studied for CB_1 and D_2 heterodimers, however, studies of other dimer pairs indicate that conformational changes via the dimer interface can allosterically promote re-arrangement of the G protein binding cleft which could feasibly alter affinity for G protein subtypes either at the individual dimer constituents or via formation of an asymmetric dimer where only one receptor in the dimer can bind G protein (reviewed in Maurice, Kamal, & Jockers, 2011).

Given the striking impact on signaling, and that co-expression of CB_1 and D_2 in the pre-synaptic membrane of basal ganglia neurons is well established (e.g., Pickel, Chan, Kearn, & Mackie, 2006), it seems likely that this dimer pairing plays an important physiological role. Furthermore, both CB_1 and D_2 have altered expression in various pathologies (e.g., Glass, Dragunow, & Faull, 2000) indicating the potential for dimer-mediated signaling to be disrupted and contribute to pathogenesis.

4.2.3 CB_1-MOR/DOR/KOR

Cannabinoids are well-established to have interactions with opioid-mediated pain and reward pathways via CB_1. While some such interactions likely occur via signaling crosstalk and/or neuronal pathway integration, the co-expression of CB_1 with the mu and delta opioid receptors (MOR and DOR) in various neuronal cell types indicated the possibility of direct physical interaction between these receptors (reviewed in Sierra, Gomes, & Devi, 2017). The first study to investigate this utilized BRET and observed that CB_1 heteromerized in an agonist-independent manner with MOR, DOR and also the kappa opioid receptor (KOR), but not with the CCR5 chemokine receptor (Rios, Gomes, & Devi, 2006). However, note that saturation experiments were not performed and the chemokine receptor negative control could reflect a difference in BRET tag location/conformation rather than truly controlling for lack of bystander BRET in the assay. Only the CB_1-MOR dimer was interrogated for functional interactions; concurrent activation of the dimer constituents reduced G protein cycling as measured by GTPγS incorporation in cell line and rat striatal tissue-derived membranes with endogenous CB_1 and MOR expression (Rios et al., 2006). Again, these are intriguing observations that may have been induced by dimerization, but could also be explained by other means of crosstalk (Christie, 2006). Further evidence toward CB_1-MOR dimerization has since been reported in the form of co-immunoprecipitation and FRET, though only limited control conditions were implemented (Hojo et al., 2008). Bivalent ligands intended to bind to both constituents of CB_1-MOR dimers produced potent *in vivo* effects (Le Naour et al., 2013), but subsequent modeling and structural analysis of the spatial requirements for concurrent binding indicated that this is unlikely to be physically capable, and the high potency may be a result of other mechanisms (Glass et al., 2016; Pérez-Benito et al., 2018; see also discussion above in Section 4.1).

Further evidence supporting CB_1-DOR dimerization has been obtained via co-immunoprecipitation from the Neuro2a (mouse) cell line with endogenously-expressed CB_1 and transfected DOR (though validation of the CB_1 antibody was not demonstrated; Rozenfeld et al., 2012), and by development and application of a CB_1-DOR heteromer antibody which indicated increased presence of dimers in a rat neuropathic pain model (Bushlin, Gupta, Stockton, Miller, & Devi, 2012). Interestingly, in unmodified Neuro2a cells (or those transfected with an unrelated transmembrane protein) CB_1 could not be detected at the plasma membrane, but transfection of DOR permitted CB_1 cell surface expression. This re-distribution correlated with an increase in the ability of CB_1 to immunoprecipitate AP-2 (an adaptor protein involved in endocytosis), and a decrease in immunoprecipitation of AP-3 (involved in lysosomal sorting) (Rozenfeld et al., 2012). Signaling cross-influence was also detected; cortical membranes from DOR knockout mice produced seemingly greater CB_1 basal G protein turnover (as measured by [^{35}S]GTPγS) as well as slightly enhanced potency of response to a CB_1 agonist in comparison with wild-type mice. In a cell line model, introduction of DOR modified the subcellular distribution of CB_1 agonist-induced pERK, facilitated induction of apoptosis-promoting signaling mediator pBAD, and enhanced interaction of CB_1 with arrestin 3. Both the results in mice and in the cell line could imply enhanced desensitization and/or arrestin-mediated signaling when CB_1 and DOR are co-expressed (Rozenfeld et al., 2012). To our knowledge CB_1-KOR dimerization has not been revisited since the study mentioned above.

4.2.4 CB_1-OX1/OX2

CB_1 and orexin (also known as hypocretin) receptors (OX1 and OX2) both have recognized roles in appetite regulation, the latter being neuropeptide receptors selectively expressed in the lateral hippocampus where CB_1 is also expressed. As with other dimer pairs, there is a range of evidence for functional interaction of these receptor systems, some of which is suggestive of direct physical interaction. The first reported evidence for OX1 dimerization is quite compelling in that heterologous effects on receptor subcellular distribution are apparent. Without applying ligand, co-expression of CB_1 modifies OX1 subcellular localization to increase cytoplasmic/vesicular expression (but has no effect on MOR subcellular distribution) (Ellis, Pediani, Canals, Milasta, & Milligan, 2006). Enhancement of surface expression of one receptor with a cognate inverse agonist produced surface upregulation of the other putatively interacting receptor, implying that the receptors were maintained in a constitutive dimer interaction with inhibition of constitutive internalization of either protomer able to prevent constitutive internalization of the other (Ellis et al., 2006). Although these initial studies were undertaken at one expression level and with large fluorescent tags on the receptor C-termini, these limitations were largely resolved in a follow-up study by the same lab which found that agonist stimulation of OX1 produced co-internalization of CB_1 (Ward et al., 2011). Likely physical interaction was also indicated by co-immunoprecipitation, FRET (using relatively small SNAP and CLIP tag labels) and BRET (Jäntti et al., 2014; Ward et al., 2011).

BRET assays have also provided preliminary evidence for CB_1-OX2 interaction, but to our knowledge this potential dimer pair has not yet been studied further (Jäntti et al., 2014).

4.2.5 CB_1-A_{2A}

CB_1 and adenosine receptor type 2A (A_{2A}) are co-expressed pre-synaptically in cortico-striatal projection neurons. Both play a critical role in modulating glutamatergic and GABAergic neurotransmission, and a range of studies have indicated degrees of crosstalk and interaction between these receptor types (reviewed in Tebano, Martire, & Popoli, 2012). CB_1 and A_{2A} were co-immunoprecipitated from rat striatum (although antibody specificity was seemingly not validated/confirmed) (Carriba et al., 2007; Ferreira et al., 2015) and had interaction indicated by BRET in a transfected cell line (Carriba et al., 2007). Ten years later, this dimer pairing was re-visited in a detailed and carefully controlled study utilizing conditional receptor knockouts observing CB_1-A_{2A} heteromers by *in situ* PLA in mouse striatum (Moreno et al., 2018). Dimers were detected in GABAergic but not glutamatergic neurons, and had a peri-nuclear rather than plasma membrane localization. BiFC was paired with peptide competition to demonstrate that CB_1 TM5 and TM6, but not 7, were important for heterodimerization and to provide direct evidence toward dimerization being responsible for cross-antagonism and G protein switching of signaling responses (Moreno et al., 2018). Most recently, the quaternary structure of the CB_1-$A2_A$ interaction was again evaluated with BiFC and peptide competition. This confirmed the importance of CB_1 TM5 and TM6, while also implicating A_{2A} TM5 and TM6 as well as the A_{2A} C-terminal tail. This study went on to model the structural arrangements of multiple A_{2A}-containing dimers as well as demonstrate the roles of various interface domains on dimer-modulated cAMP signaling responses (Köfalvi et al., 2020).

4.2.6 CB_1-β2A

Despite being co-expressed in various tissues including the cardiovascular system and brain (e.g., cortex, hippocampus, hypothalamus) (Bylund, 2007; Miller & Devi, 2011), to date only one study has contributed direct evidence toward the existence of CB_1 and β2-adrenergic receptor (β2A) dimers (Hudson, Hébert, & Kelly, 2010). An interaction was firstly indicated by BRET (saturation and competition with unlabeled receptor), and dimer was unmodified by agonists but enhanced by a CB_1 inverse agonist (within 15 min, so it is unlikely that considerable changes in surface CB_1 expression due to inhibition of constitutive internalization would have occurred). Presence of β2A (but not control transmembrane proteins HERG and mGluR6) seemed to prevent CB_1 constitutive internalization, leading to greater cell surface expression, and a β2A agonist produced internalization of CB_1 only when β2A was also present. Various signaling interactions were also noted in both HEK cells artificially expressing the receptors (aided by an inducible expression system) and in a cell line with endogenous CB_1 and β2A expression, though genuine dimer-mediated effects could not be discriminated from signaling crosstalk.

4.2.7 CB_1-AT1

Unlike the majority of other CB_1 dimer pairs here discussed, dimerization between CB_1 and angiotensin II (AngII) receptor 1 (AT1) is thought most likely to occur in the periphery, specifically in the liver in hepatic stellate cells where AT1 is expressed normally and CB_1 is upregulated in liver fibrosis (Teixeira-Clerc et al., 2006). Putative dimerization was first demonstrated by BRET (saturation study and competition with untagged receptor) and cross-antagonism of signaling responses which was correlated with profibrogenic effects of AngII (Rozenfeld et al., 2011). Reminiscent of the interaction with OX1, CB1 was found predominantly intracellularly when expressed alone, but surface expression was enhanced when AT1 was co-expressed. A heterodimer-selective antibody was also developed and could disrupt dimer-associated signaling and functional fingerprints. To our knowledge this dimer pair has not undergone further study subsequent to this first report.

Interestingly, AT1 activation can stimulate endocannabinoid 2-AG synthesis (this may be a general property of $G\alpha_q$-coupled receptors), adding another dimension for potential feedback and regulation for these two receptors (Gyombolai et al., 2012).

4.2.8 CB_1-GPR55

GPR55 is an orphan GPCR which is expressed in the CNS, immune system and elsewhere, and whose most widely accepted putative endogenous agonist is lysophosphatidylinositol (LPI). GPR55 has also been proposed as a candidate cannabinoid receptor due to some overlap in ligand activity, including Δ^9-THC and an endocannabinoid (anandamide), though classification remains controversial due to some discrepancies between studies (Liu, Song, Jones, & Persaud, 2015; Reggio & Shore, 2015). CB_1 and GPR55 dimerization and cross-influences on signaling might assist in explaining some of the disparities. The potential for dimerization was first demonstrated by co-immunoprecipitation, though no cross-influence on either constitutive or agonist-induced internalization was detected which may imply the dimer interaction is transient (Kargl et al., 2012). Nonetheless, co-expression of CB_1 and GPR55 had marked effects on signaling responses; most strikingly, the presence of CB_1 completely abrogated GPR55-mediated induction of nuclear factor of activated T-cells (NFAT) and serum response element (SRE) in a manner that was dependent on the CB_1 expression level (but was not affected by co-expression of a control receptor). Subsequently, dimerization of this pair has been supported by BRET (saturable interaction) and their close proximity in both cells and *in situ* in the striatum of a non-human primate (cynomolgus monkey) by PLA (Martínez-Pinilla et al., 2014). While exciting to see these results, we note that the PLA technique is critically reliant on robust antibodies and the specific lots were seemingly not validated at the time of the study in the cells and tissue under observation; in particular, non-specific binding may have accounted for the apparent presence of dimers throughout the nucleus and cytoplasm and absence from the plasma membrane in the cell line assay. The same technique and antibodies were applied to in a cynomolgus monkey methyl-4-phenyl-1,2,3,6-tetrahydropyridine (MPTP) model of Parkinson's disease (PD; Martínez-Pinilla et al., 2020). CB_1-GPR55 PLA signal increased in PD for all three studied regions, the caudate nucleus, putamen, and nucleus accumbens (the same was observed for CB_2-GPR55, discussed below). Unfortunately, again, staining patterns for the individual antibodies were not provided and it is therefore difficult to gauge the fidelity of detection, as well as infer whether the increase in proximity may have been due to increased expression or altered localization of one or all receptor types, or a result of modified regulation of the dimer pair.

4.2.9 CB_1-$5HT_{2a}$

Motivated by numerous reports of functional interactions between the cannabinoid and serotonin systems, putative dimerization between CB_1 and serotonin receptor 2a ($5HT_{2a}$) was first demonstrated in heterologous cell lines by a saturable BRET signal,

BiFC (with control experiments indicating lack of signal with non-interacting dopamine receptor 1), and PLA (though considerable signal was apparent in the cytoplasm and nucleus, with little at the plasma membrane) (Viñals et al., 2015). Bi-directional cross-antagonism was observed, and co-expression of CB_1 TM5 or TM6 (but not TM7) was able to block this effect (without influencing receptor expression) as well as disrupt detection of dimers by PLA, lending further support for the existence of a functionally-relevant dimer and suggesting these CB_1 domains contribute to the dimer interface. This same study subsequently applied PLA to detect dimer in the mouse cortex, hippocampus and striatum, but not in the nucleus accumbens, despite CB_1 and $5HT_{2a}$ both being expressed in all these regions. This finding corresponds with observations of CB_1 signaling having different characteristics and beneficial vs adverse outcomes in different brain regions, and highlights the potential for pharmacologically targeting dimers (Viñals et al., 2015). Specifically in this case, CB_1-$5HT_{2a}$ dimerization seemed to be responsible for adverse effects, including memory impairments. Disruption of this dimer interaction may therefore be therapeutically useful. A potential approach was investigated *in vitro* by designing peptides which mimic CB_1 transmembrane domains, incorporating a cell-penetrating peptide and introducing aryl-carbon-stapling with the aim of stabilizing secondary structure and improving cellular uptake and stability. Using a recently-developed NanoBiT complementation assay to sensitively and dynamically monitor dimer, one of the resultant peptides disrupted pre-existing CB_1-$5HT_{2a}$ dimer with micromolar potency and reaching maximum efficacy after around 5 min (Botta, Bibic, Killoran, McCormick, & Howell, 2019). It will certainly be interesting to learn whether this peptide (and/or a more potent derivative thereof) will be efficacious *in vivo*. If so, this would represent direct evidence to support the physiological relevance of CB_1 heterodimers and a promising therapeutic strategy worthy of continued pursuit, though the likely involvement of CB_1 TM domains in various dimer interactions imply that specificity to the dimer pair of interest may be difficult to obtain via this approach.

4.2.10 CB_1-SST_5

To our knowledge the most recently-reported CB_1 dimer partner is somatostatin receptor 5 (SST_5). CB_1 and SST_5 were co-localized in and co-immunoprecipitated from the rat brain cortex, striatum and hippocampus. In transfected cells, constitutive dimer (as measured by photobleaching FRET) was disrupted by treatment with SST_5 agonist (but not CB_1 agonist); SST_5 agonist also promoted SST_5 homodimer, while CB_1 agonist dissociated CB_1 homodimer. Further, the presence of SST_5 appeared to disrupt the ability of CB_1 to modulate cAMP levels (Zou et al., 2017). A follow-up study indicated that the CB_1-SST_5 heterodimer was present in "normal" and Huntington's disease striatal neuronal cell line models, wherein stimulation of either receptor was protective against quinolinic acid-induced excitotoxicity but applying agonists for both receptors simultaneously had a diminished protective effect (Zou & Kumar, 2019).

4.2.11 CB_2-CXCR4

CB_2 and the chemokine receptor CXCR4 are both upregulated in various tumor tissues, and the increased expression of these has been associated with poorer disease outcomes in a number of cancers (Fraguas-Sánchez, Martín-Sabroso, & Torres-Suárez, 2018; Zhao et al., 2015). Evidence of signaling crosstalk between these two receptors was first observed in breast cancer cells wherein the CB_2-selective agonist, JWH-015, inhibited agonist-induced CXCR4-mediated cell migration and ERK1/2 phosphorylation (Nasser et al., 2011). A physical interaction between these receptors in response to co-stimulation with CB_2 and CXCR4 agonists was later confirmed in prostate cancer cells by co-immunoprecipitation (Coke et al., 2016). In the same study, PLA was also used to visualize these heteromers in both prostate and breast cancer cells, where they were again induced by agonist co-stimulation. Moreover, dimerization was induced by both exogenous and endogenous agonists, as validated by siRNA knockdown of endogenous SDF1α (a CXCR4 agonist). Consistent with the findings in Nasser et al. (2011), dimerization of CB_2 and CXCR4 also reduced agonist-induced ERK1/2 phosphorylation and cell migration. More recently, the reduction of cell migration, invasion and adhesion by the induced heterodimer was shown to be via inhibition of $G\alpha_{13}$/Rho-A signaling (Scarlett et al., 2018). Collectively, these studies indicate that the formation of CB_2-CXCR4 heteromers reduce CXCR4-signaling, and may be a promising target for metastasis intervention.

4.2.12 CB_2-GPR55

CB_2-GPR55 heteromerization has also been demonstrated in transfected cells via BRET (Balenga et al., 2014; Moreno et al., 2014) and co-immunoprecipitation (Balenga et al., 2014), and these receptors have been shown to co-internalize following stimulation with the endogenous GPR55 agonist, lysophosphatidylinositol (LPI; Balenga et al., 2014). Furthermore, these heteromers have been visualized using PLA in both transfected cancer cells and a glioblastoma tumor cell line (T98G) that endogenously expresses both receptors (Moreno et al., 2014), with antibody specificity validated by shRNA knockdown of the individual protomers. More recently, *in situ* PLA has revealed CB_2-GPR55 heteromers in post-mortem human tissue from the dorsolateral prefrontal cortex (García-Gutiérrez et al., 2018) and in monkey caudate and putamen tissue (Martínez-Pinilla et al., 2020), however, as previously mentioned (see CB_1-GPR55 above) no antibody validation controls were carried out. Moreover, the anti-CB_2 antibody used in these studies has been reported to lack specificity for CB_2, particularly in brain tissue (Baek, Darlington, Smith, & Ashton, 2013; Li & Kim, 2015; Marchalant, Brownjohn, Bonnet, Kleffmann, & Ashton, 2014).

Evidence of signaling cross-talk between CB_2 and GPR55 and/or cross-antagonism was apparent in all studies that carried out functional assays, and has been observed in multiple cell types including transfected HEK293 cells (Balenga et al., 2014), mast cells (Cruz et al., 2018), and tumor cells (Moreno et al., 2014), indicating that these heteromers may be relevant in a number of contexts.

4.2.13 CB_2-GPR18

GPR18 is an orphan GPCR that is predominantly expressed in the testis, spleen, peripheral bloody leukocytes, and lymph nodes—suggesting a role in immune system regulation (Gantz et al., 1997). It has also been considered a putative cannabinoid receptor as some studies have reported GPR18-mediated signaling in response to cannabinoids (e.g., Δ^9-THC and anandamide; Console-Bram, Brailoiu, Brailoiu, Sharir, & Abood, 2014; McHugh, Page, Dunn, & Bradshaw, 2012), though this remains controversial as other studies have shown no cannabinoid-mediated effects across multiple GPR18-expressing cell types (Finlay, Joseph, Grimsey, & Glass, 2016; Lu, Puhl, & Ikeda, 2013). CB_2 has been reported to form dimers with GPR18, as indicated in transfected cells by BRET and via *in situ* PLA in both healthy primary murine microglial cultures and those derived from a transgenic mouse model of Alzheimer's disease (Reyes-Resina et al., 2018). Note, however, that the anti-CB_2 antibody utilized has previously been shown to bind non-specifically in mouse hippocampus (Li & Kim, 2015). Nevertheless, bi-directional cross-antagonism of ERK phosphorylation was indicated in both resting and lipopolysaccharide (LPS) plus interferon gamma (IFN-γ) stimulated microglia, as well as in the transgenic microglia, suggesting that these heteromers are likely functional and may be of relevance in neurodegenerative diseases (Reyes-Resina et al., 2018).

4.2.14 CB_2-A_{2A}

CB_2 and A_{2A} are co-expressed in microglia, and are both upregulated in activated cells (e.g., Carlisle et al., 2002; Ehrhart et al., 2005; Orr, Orr, Li, Gross, & Traynelis, 2009). In a recent study, CB_2 and A_{2A} were shown to interact in a heterologous expression system as indicated by a saturable BRET signal (Franco et al., 2019). These heteromers were functional as the receptors exhibited signaling crosstalk and partial cross-antagonism, wherein A_{2A} antagonism potentiated CB_2-mediated inhibition of cAMP and partially reduced ERK1/2 phosphorylation. Functional heteromers were also observed in resting and primary microglial cells, and those derived from a transgenic mouse model of Alzheimer's disease ($APP_{Sw,Ind}$). Additionally, the authors visualized these heteromers via *in situ* PLA in primary microglial cultures from both control and transgenic mice, and observed an increase in the number of cells positive for CB_2-A_{2A} heteromers in the $APP_{Sw,Ind}$ microglia. However, appropriate negative controls were not included, and like a number of commercially-available anti-CB_2 antibodies, the one used in this study has previously been shown to be non-specific in mouse brain (Li & Kim, 2015).

4.2.15 CB_2-HER2

In addition to dimerizing with other GPCRs, CB_2 has been reported to dimerize with the receptor tyrosine kinase human epidermal growth factor receptor 2 (HER2; also known as ERBB2). CB_2 expression appears to be upregulated in HER2+ breast tumors, and this correlates with poor patient prognosis (Blasco-Benito et al., 2019; Pérez-Gómez et al., 2015). Dimerization of CB_2 and HER2 has been shown in transfected cells via co-immunoprecipitation, and in endogenously expressing breast cancer cells using PLA, wherein shRNA knockdown of CB_2 was used to validate the anti-CB_2 antibody used (Pérez-Gómez et al., 2015). In the same study, the pro-oncogenic activity associated with CB_2-HER2 was linked to signaling of the tyrosine kinase, c-Src. In a follow-up study conducted by the same research group, the interaction was confirmed in transfected cells via BRET, and *in situ* PLA indicated that higher expression of CB_2-HER2 dimers in HER2+ breast cancer biopsies was associated with greater metastases and lower disease-free patient survival (Blasco-Benito et al., 2019). In breast cancer cells, CB_2-HER2 dimers were disrupted following Δ^9-THC stimulation, which in turn reduced HER2 activation, as indicated by a decrease in phosphorylated HER2. Co-immunoprecipitation assays in HEK293 cells using peptide competition suggested a potential interaction between TMs 1, 3, 4 and 5 of CB_2 and HER2. Additionally, a reduction in BiFC between full-length CB_2 and HER2 was apparent when TM4 and TM5 constructs were co-transfected, and a specific peptide targeting TM5 disrupted dimer formation via BiFC and PLA, indicating that this region is involved in the interaction between CB_2 and HER2 (Blasco-Benito et al., 2019).

4.2.16 Hetero-oligomerization

Hetero-oligomerization of more than two receptor protomers, including of more than two receptor types, is also plausible (Ferré et al., 2014). The design of many studies restricts interrogation to assume and/or detect a dimer pair, but this often does not preclude that higher order multimerization (and/or other protein-protein interactions) could be present and functionally important.

To our knowledge, only one hetero-oligomer involving the cannabinoid receptors has been reported to date; CB_1-D_2-A_{2A}. This was first demonstrated by sequential BRET-FRET (SRET) (Carriba et al., 2008) and subsequently the same group has expanded the evidence for this interaction by interrogating the trimer quaternary structure (Navarro et al., 2010). Importantly, some evidence for this trimer interaction has been obtained in the primate brain by *in situ* PLA, though additional controls would have been required to have full confidence in antibody specificity (Bonaventura et al., 2014).

For these receptors (and most other suggested dimer pairs), little is currently known about the proportional existence of homo-oligomer vs hetero-dimer and hetero-oligomer, nor regulation of dimerization in health and disease (though a potential role for CB_1-D_2-A_{2A} has been suggested in schizophrenia, Parkinson's disease and drug addiction; Fuxe et al., 2010). These will be important questions to address as more robust evidence for cannabinoid receptor interactions with other GPCRs is gathered, in particular regarding the presence and consequences of oligomers *in vivo* at native expression levels.

5 Interactions influencing subcellular distribution

CB_1 and CB_2 are both typically found to be expressed on the cell surface, however, have also been described as having "intracellular" expression (e.g., Brailoiu et al., 2014; Castaneda, Harui, Kiertscher, Roth, & Roth, 2013; Grimsey, Graham,

Dragunow, & Glass, 2010; Kleyer et al., 2012; Rozenfeld & Devi, 2008). Unlike some other GPCRs where it has been found that dimerization and/or accessory proteins are critical for enabling surface expression, to date there are no specific interacting proteins identified that regulate this basal distribution. Similarly, although some intracellularly-expressed GPCRs are recognized as being functional and playing important and unique roles in signaling/physiology (Plouffe, Thomsen, & Irannejad, 2020), very little is known about the functional role of intracellular CB_1 and CB_2 other than this population may induce distinct signaling responses from cell surface receptors. Control of basal distribution within plasma membranes is likely to be particularly important for CB_1, for example, in establishing pre-synaptic localization (Leterrier et al., 2006; McDonald, Henstridge, Connolly, & Irving, 2007).

More is understood about CB_1 and CB_2 trafficking post-activation, with both receptors undergoing clathrin-dependent internalization. Following endocytosis, CB_1 has been described to undergo either recycling or degradation, with determination of this fate likely involving protein-protein interactions (e.g., Grimsey et al., 2010; Leterrier, Bonnard, Carrel, Rossier, & Lenkei, 2004; Martini et al., 2007; Wu et al., 2008). Regional differences in the development of tolerance and subsequent recovery of responsiveness following chronic cannabinoid exposure might be indicative of differential CB_1 regulation and protein-protein interactions between cell types (González, Cebeira, & Fernández-Ruiz, 2005). In the majority of contexts studied to date, CB_2 recycles following internalization and is refractory to agonist-induced degradation (Grimsey, Goodfellow, Dragunow, & Glass, 2011; Kleyer et al., 2012).

In Sections 3 and 4 we already noted some cannabinoid receptor interacting proteins that influence CB_1 and/or CB_2 subcellular localization and trafficking, namely the CRIPs, SGIP1, and various receptors. This section discusses additional interacting proteins which interact with CB_1 and/or CB_2 primarily in the context of modulating subcellular distribution and/or expression.

GPCR-associated sorting protein 1 (GASP1) is known to interact with a variety of GPCRs, usually via the receptor C-terminal tail, and promotes sorting of the receptor to lysosomes for subsequent degradation (Heydorn et al., 2004; Simonin, Karcher, Boeuf, Matifas, & Kieffer, 2004; Whistler et al., 2002). GASP1 interaction with CB_1 has been demonstrated by pulldown to immobilized CB_1 C-terminal tail, and co-immunoprecipitation from transfected cells and rodent brain homogenate (Martini et al., 2007; Tappe-Theodor et al., 2007). The expression level of GASP1 can influence the post-endocytic fate of various GPCRs, for example, expression of a dominant negative GASP1 fragment (cGASP) could re-route DOR from a degradative to a recycling pathway (Whistler et al., 2002). cGASP expression or GASP knockdown also prevented CB_1 degradation in cell and mouse models and reduced behavioral tolerance toward chronic agonist administration, raising the possibility that this protein-protein interaction may be worthy of targeting therapeutically (Martini, Thompson, Kharazia, & Whistler, 2010; Tappe-Theodor et al., 2007). Based on GST pulldown with the human CB_2 C-terminal tail and surface plasmon resonance of CB_2 purified from a yeast expression system, CB_2 might also interact with GASP1 via a central region of GASP1 containing a repeated 15 amino acid motif (Bornert et al., 2013).

Association of CB_1 with adaptor protein 3 (AP-3) heterotetrameric complex has also been reported (co-immunoprecipitation, confocal microscopy and cell fractionation) and, given the primary recognized role of AP-3 in directing cargo from early to late endosomes and lysosomal-related organelles, again supports a degradative phenotype of CB_1 in some contexts (Park & Guo, 2014; Rozenfeld & Devi, 2008). Similarly to the supposed influence of GASP1, depletion of the AP-3 δ subunit by siRNA prevented increased CB_1 plasma membrane expression. Whether one or more subunits of AP-3 interact directly with CB_1 has not been investigated, though by analogy with other interacting receptors this was speculated to occur via dileucine- or tyrosine-containing motifs in the CB_1 cytoplasmic domains.

6 Putative interactors with as yet undefined function

As protein-protein interaction assay technology becomes more accessible and amenable to high throughput screening, efforts to define cellular interactomes are increasing and have the potential to rapidly accelerate our understanding of cell biology. However, such screens will always have inherent false-negative and false-positive rates, which can be challenging to meaningfully assess. Having undertaken a screen, validation will typically begin with the strongest hits, but interrogation of all potential interactors constitutes a major effort and investment.

The yeast two-hybrid approach has twice been successfully applied to CB_1, using a receptor C-terminal fragment as bait and identifying interacting proteins CRIP1a and SGIP1 as detailed above (Hájková et al., 2016; Niehaus et al., 2007). Unfortunately, neither paper provided any indication as to the other putative interacting proteins identified in these screens.

More recently, the CB_2 interactome was investigated in HEK293 cells stably expressing Strep-hemagglutinin (HA)-tagged human CB_2. Affinity purification in combination with mass spectroscopy, identified 83 proteins as highly probable interactors (identified by a minimum of two unique peptides) (Sharaf et al., 2019). Of these, six were present in CB_2-expressing cells only; delta(24)-sterol reductase (DHCR24), dehydrogenase/reductase SR family member 7 (DHRS7), glutathione hydrolase 7 (GGT7), E3 ubiquitin-protein ligase HECTD3 (HECTD3), phospholipid scramblase 1 (PLSCR1), and an uncharacterized protein (KIAA2013). The known functions of two of these proteins suggest roles in plasma membrane organization; PLSCR1 is an enzyme involved in trans-bilayer redistribution of phospholipids (Bassé, Stout, Sims, & Wiedmer, 1996; Zhou et al., 1997) and DHCR24 is an enzyme involved in cholesterol biosynthesis (Luu, Hart-Smith, Sharpe, & Brown, 2015). Unsurprisingly, molecular chaperones and proteins involved in protein folding and/or translocation were identified as putative interactors (e.g., calnexin, SEC62, protein disulfide isomerase TMX3). The interaction of CB_2 and another probable interactor, p62 (also known as SQSTM1 or Sequestosome 1), was confirmed via co-immunoprecipitation and co-localization. Deletion of individual CB_2 intracellular loops or the C-terminus did not appear to alter the interaction, however, mutant expression levels were comparatively lower than wild-type, and it would seem surprising if folding and subcellular localization of mutant receptors were not influenced to some degree. Though no functional consequences were assessed, p62 has been reported to act as a cargo protein in autophagy (Fujita, Maeda, Xiao, & Srinivasula, 2011; Kwon et al., 2018; Lin et al., 2013), and is involved in the regulation of NF-κB (Kim & Ozato, 2009; Sanz, Sanchez, Lallena, Diaz-Meco, & Moscat, 1999), and therefore may be involved in the clearance of CB_2 or perhaps influence downstream signaling.

An impressive yeast two-hybrid human ORFeome screen included hits for CB_2 interactions (though not for CB_1) (Luck et al., 2020). Twenty binary interactors, none of which had previously been reported to interact with CB_2, included various transmembrane proteins; the highest-scored were a Rab GTPase-interacting protein putatively involved in membrane trafficking (YIPF1, Kranjc et al., 2017; two others from the same family were also hits), a cholesterol binding protein (translocator protein 2, TSPO2, Fan, Rone, & Papadopoulos, 2009), and a protein putatively involved in N-linked glycosylation and/or glycosylated protein membrane translocation (RFT1, Haeuptle et al., 2008). No prior-reported CB_2 interacting partners were detected as positive hits in the screen, though notably $G\alpha_i$ and arrestins 2 and 3 did not reach the confidence level for interaction with *any* GPCR (despite being included in the screen and producing positive hits with a handful of other signaling and trafficking proteins) (Luck et al., 2020; based on analysis via the IntAct database, Orchard et al., 2014). There was no overlap between the positive hits identified in this screen and the Sharaf et al. (2019) mass spectrometry screen discussed above (based on UniProt ID cross-reference; The Uniprot Consortium, 2019). While the novel putative interactors may be interesting to pursue, it is important to note that full length ORFs were utilized in this yeast two-hybrid screen and targeting integral transmembrane proteins to the yeast nucleus has been assumed to result in protein aggregation and presumably spurious hits (Bao, Redondo, Findlay, Walker, & Ponnambalam, 2009). A "LUMIER" ELISA-like screen designed to interrogate the chaperone interaction network indicated luciferase-tagged 26S proteasome non-ATPase regulatory subunit 2 (PSMD2), a component of the proteasome involved in degradation of ubiquitinated proteins, as interacting with a FLAG-tagged human CB_2 when expressed in HEK cells; no interacting proteins were identified for CB_1 (Taipale et al., 2014). Data from interaction studies are now being curated in readily-searchable databases (e.g., Orchard et al., 2014; Porras et al., 2020; Schweppe, Huttlin, Harper, & Gygi, 2018). With further efforts in screening, hopefully utilizing technologies appropriate for transmembrane proteins and in appropriate cellular context (e.g., Federspiel & Cristea, 2019; Snider et al., 2010), these tools provide useful resources to accelerate increased understanding of the GPCR interactome.

7 Perspectives and future vistas to expand the cannabinoid receptor interactome

There is widespread interest in targeting the cannabinoid system therapeutically. A handful drugs targeting CB_1 have been approved for clinical use, but the strong potential for unwanted psychotropic effects is a problematic and limiting factor. A CB_1 inverse agonist, SR141716A or Rimonabant, had been approved for clinical use (weight loss, smoking cessation) but was later withdrawn due to adverse effects relating to suicidal ideation (Sam, Salem, & Ghatei, 2011). Compounds targeting CB_2 are much less likely to induce psychoactive effects and a few have undergone or are currently in clinical trial. Results to date have been mixed, however, and no agents have yet reached the market. The cannabinoid system modulates a wide variety of physiological processes and there is huge scope for expanding our understanding from fundamental molecular control through to systems pharmacology and physiology. Protein-protein interactions with the cannabinoid receptors are a critical component of this puzzle, likely influencing all aspects of cannabinoid receptor-mediated physiologies.

At the most fundamental level, protein-protein interactions enable the cannabinoid receptors to induce signaling cascades and thereby influence cellular function. Some interacting proteins can allosterically alter receptor conformations and/or accessibility of receptor domains to other interacting partners, both of which can have critical effects on signaling. Interacting proteins that modulate receptor trafficking influence cellular responses by changing the population of receptors accessible to ligand and/or the local milieu of signaling mediators available to the receptor. Importantly, differences in interactome expression between cell types, and likely within cellular subdomains, across time, and in different states of physiology or pathology, will profoundly influence the ultimate outcome of a ligand interacting with a receptor in a particular cell at a particular time.

The vast majority of drug discovery efforts at the cannabinoid receptors, and indeed most other GPCRs, currently consider the monomeric receptor alone and focus on drugging the orthosteric or allosteric ligand binding sites. Complementing this approach with knowledge of protein-protein interactions with the receptors, and potentially even targeting these interactions, holds potential in improving the specificity of drug effects and expanding the potential for additional modulation not possible if targeting receptor monomers alone (Guidolin, Marcoli, Tortorella, Maura, & Agnati, 2019). The most obvious candidates for such targeting are GPCR dimers, wherein therapeutic agents (ligands, cell permeable peptides, antibodies) may be designed to target dual orthosteric sites and thereby refining drug effects both spatially and with regard to signaling responses. While exciting early development of cell permeable peptides with the potential to disrupt CB_1 dimers seems promising (Botta et al., 2019), to date the development of ligands targeting cannabinoid receptor-containing dimers has been less successful. While the likely mode of ligand entry to the cannabinoid receptors via the lipid bilayer complicates design considerably, and may preclude the effective development of bivalent ligands to dimer partners with water soluble ligands, the advent of improved structural knowledge for both CB_1 and CB_2, applied with modern modeling and dynamic simulation tools, may improve these prospects. Generally speaking, it seems as though interacting proteins responsible for modulating trafficking pathways will be less attractive therapeutic targets given these tend to have more redundancy between receptor types; for example, GASP1 and AP-3 interact with a variety of GPCRs. While competition strategies directed to the receptor might offer some specificity, there is some precedent for interacting proteins with more selectivity for the cannabinoid receptors, for example, CRIP1a, and these may be more amenable to successful therapeutic targeting while avoiding influence on other receptor systems.

Other than interactions with direct signaling mediators, our reading of the current literature indicates that only a few interacting partners for the cannabinoid receptors can be considered well established based on robust evidence gathered from multiple research groups, for example, CB_1 interactions with CRIP1a and D_2. We have discussed many other putative interacting proteins which have varying degrees of evidence associated to date. A particular area that is lacking for nearly all CB_1 and CB_2 interacting proteins is in evidence for existence and functional implications of the interactions *in vivo*. Indeed, the vast majority of characterization of cannabinoid receptor interacting proteins has been obtained from systems with heterologous or overexpression of one or both proteins. While these studies are necessary and important starting points, the majority of the techniques that are frequently utilized are vulnerable to false-positives due to bystander effects which are exacerbated by high protein expression levels. Although some degree of control and validation experiments are usually undertaken, convincing control experiments for techniques such as BRET and FRET are difficult to design and interpret (Bouvier, Heveker, Jockers, Marullo, & Milligan, 2007; James, Oliveira, Carmo, Iaboni, & Davis, 2006; Szalai et al., 2014). Furthermore, caution must be taken in interpreting data from modified proteins; for example, fluorophores and luminophores are often attached to receptor N- and C-termini which have strong potential to influence receptor subcellular distribution and/or ability to interact with partners—which could include promotion of non-physiological or disruption of physiological interactions. Some advances for reducing bystander effects such as mutations to BiFC fragments to reduce spontaneous assembly, and reversible systems with smaller tags like split NanoLuc may be useful in improving assay fidelity, but the latter assay has only been used once in CBR interaction studies to date (Dixon et al., 2016; Kodama & Hu, 2010). It is therefore important that the field continues to strive toward robust validation and contextualization, in particular, to study proteins at endogenous expression levels and stoichiometry, preferably in cells and tissues that express both proteins endogenously. Gene editing has recently been utilized to introduce bright and relatively small NanoLuc tags to GPCRs expressed under their endogenous promoters (Soave et al., 2021). Some CB_1 and CB_2 interactions have been studied via *in situ* PLA, but unfortunately this technique is one of the most permissive proximity techniques, with strong potential to report a positive result simply due to relatively high expression in the same region of a cell but not necessarily representing a physical interacting. Further, this technique relies on robust antibodies which are well-recognized to be difficult to obtain for GPCRs in general, but particularly for the cannabinoid receptors (e.g., Grimsey et al., 2008; Marchalant et al., 2014; Zhang et al., 2018).

A range of classical techniques have been applied to validate protein-protein interactions for other GPCRs and may also be useful to apply to cannabinoid receptor interactions. In particular, generation of receptor chimeras with swapped domains between putative interacting receptors can be illuminating when chimeric receptors are somehow defective when expressed alone, but have restored function when expressed in the putative interacting pair (e.g., Maggio, Vogel, & Wess, 1993). Some cannabinoid interactions have also been supported by peptide competition with putative interacting domains; it would be useful to follow up this work by investigating the effect of mutations of the putative interaction sites. Identification of these interaction points may then lend itself to generation of transgenic animals to understand the *in vivo* impact of the protein-protein interactions (Ferré et al., 2009).

Advances in fluorescence-based techniques utilizing live cell imaging and single particle tracking and/or recovery after photobleaching (FRAP) offer considerable improvements over population-based RET and other proximity assay screens in that these distinguish stable interaction from coincident collision and the subcellular location of the interaction is monitored inherently (e.g., Petersen et al., 2017). FRET between GPCR ligands had been utilized to demonstrate oligomers in native tissue and would constitute a welcome complement to antibody techniques (Albizu et al., 2010). Fluorescent ligands for CB_1 and CB_2 have been reported in recent years, though it remains to be seen whether any are suitable for *in situ* labeling of tissue and this may prove difficult due to the high lipophilicity of most cannabinoids (e.g., Bruno, Lembo, Novellino, Stornaiuolo, & Marinelli, 2014; Sarott et al., 2020; Singh, Oyagawa, et al., 2019). It would be interesting to explore whether a covalent attachment strategy for fluorescent GPCR ligands could assist in this regard (Stoddart et al., 2020).

The prior-discussed techniques are typically utilized in study already-detected or hypothesized interactions, but there are likely many more protein-protein interactions to discover in the cannabinoid receptor interactome. A few types of screening methodology have been applied to the cannabinoid receptors to date (e.g., yeast two-hybrid, affinity purification with mass spectroscopy); each has advantages and disadvantages, and though a good screen will yield a potential treasure trove of interactors, all have inherent false-positive rates and subsequent validation of potential hits constitutes a large task to distinguish genuine from noise. Of particular note, these screens likely only had the potential to detect constitutive interactions. Application of receptor stimuli and/or cellular models to represent physiological or pathological conditions may well reveal important interactions that may not have been present in screens to date. From the point of view of finding therapeutically useful protein-protein interactions, we might hope to discover further interaction partners with limited additional interactors and/or with interaction interfaces that are limited to one or a few targets—such as CRIP1a which to date has only one suggested interacting partner other than CB_1 (Mascia et al., 2017). To assist with identifying protein-protein interactions of relevance to a function of interest, and perhaps with specific CBR interactions, it may be useful to identify receptor mutations that modify said function and use these to screen for interaction vs the wild-type receptor (keeping in mind the importance of monitoring equivalent expression levels and subcellular distribution).

Additional future challenges for the field include consideration of alternative receptor forms and how these influence protein-protein interactions. This is particularly relevant for CB_2, for which multiple non-synonymous SNPs have been identified and considerable interspecies differences in receptor sequence and molecular pharmacology exist. Furthermore, as we look toward potential therapies targeted to or informed by cannabinoid receptor protein-protein interactions we will need to expand our understanding of the regulation and temporal nature of the interactions, as well as how any such interactions are modified in disease states. While we clearly have much to learn about the cannabinoid receptor interactome, the unprecedented availability of tools to study these interactions will support acceleration of this research and make this an exciting time for the field.

Conflict of interest statement

The authors have no conflict of interest to declare.

Acknowledgments

CO and NG were supported by funding from the Royal Society Te Apārangi Marsden Fund and the Health Research Council New Zealand (both to NG).

References

Agirregoitia, E., Carracedo, A., Subirán, N., Valdivia, A., Agirregoitia, N., Peralta, L., et al. (2010). The CB2 cannabinoid receptor regulates human sperm cell motility. *Fertility and Sterility*, *93*(5), 1378–1387. https://doi.org/10.1016/j.fertnstert.2009.01.153.

Ahmed, M. H., Kellogg, G. E., Selley, D. E., Safo, M. K., & Zhang, Y. (2014). Predicting the molecular interactions of CRIP1a-cannabinoid 1 receptor with integrated molecular modeling approaches. *Bioorganic and Medicinal Chemistry Letters*, *24*(4), 1158–1165. https://doi.org/10.1016/j.bmcl.2013.12.119.

Albizu, L., Cottet, M., Kralikova, M., Stoev, S., Seyer, R., Brabet, I., et al. (2010). Time-resolved FRET between GPCR ligands reveals oligomers in native tissues. *Nature Chemical Biology*, *6*(8), 587–594. https://doi.org/10.1038/nchembio.396.

Alsemarz, A., Lasko, P., & Fagotto, F. (2018). Limited significance of the in situ proximity ligation assay. *BioRxiv (Pre-Print)*, 411355. https://doi.org/10.1101/411355.

Angelats, E., Requesens, M., Aguinaga, D., Kreutz, M. R., Franco, R., & Navarro, G. (2018). Neuronal calcium and cAMP cross-talk mediated by cannabinoid CB1 receptor and EF-hand calcium sensor interactions. *Frontiers in Cell and Developmental Biology*, *6*(Jul), 67. https://doi.org/10.3389/fcell.2018.00067.

Appleton, K. M., Lee, M., Strungs, E. G., Nogueras-Ortiz, C., Gesty-Palmer, D., Yudowski, G. A., et al. (2019). Dissociation of β-arrestin-dependent desensitization and signaling using 'biased' parathyroid hormone receptor ligands. *Cell Press Sneak Peek (Pre-Print)*, *3422428*. https://doi.org/10.2139/ssrn.3422428.

Atwood, B. K., & MacKie, K. (2010). CB 2: A cannabinoid receptor with an identity crisis. *British Journal of Pharmacology*, *160*(3), 467–479. Wiley-Blackwell https://doi.org/10.1111/j.1476-5381.2010.00729.x.

Avet, C., Mancini, A., Breton, B., Le Gouill, C., Hauser, A. S., Normand, C., et al. (2020). Selectivity landscape of 100 therapeutically relevant GPCR profiled by an effector translocation-based BRET platform. *BioRxiv (Pre-Print)*, 2020.04.20.052027. https://doi.org/10.1101/2020.04.20.052027.

Baek, J. H., Darlington, C. L., Smith, P. F., & Ashton, J. C. (2013). Antibody testing for brain immunohistochemistry: Brain immunolabeling for the cannabinoid CB2 receptor. *Journal of Neuroscience Methods*, *216*(2), 87–95. https://doi.org/10.1016/j.jneumeth.2013.03.021.

Bagher, A. M., Laprairie, R. B., Kelly, M. E. M., & Denovan-Wright, E. M. (2016). Antagonism of dopamine receptor 2 long affects cannabinoid receptor 1 signaling in a cell culture model of striatal medium spiny projection neurons. *Molecular Pharmacology*, *89*(6), 652–666. https://doi.org/10.1124/mol.116.103465.

Bagher, A. M., Young, A. P., Laprairie, R. B., Toguri, J. T., Kelly, M. E. M., & Denovan-Wright, E. M. (2020). Heteromer formation between cannabinoid type 1 and dopamine type 2 receptors is altered by combination cannabinoid and antipsychotic treatments. *Journal of Neuroscience Research*, *98*(12), 2496–2509. https://doi.org/10.1002/jnr.24716.

Balenga, N. A., Martínez-Pinilla, E., Kargl, J., Schröder, R., Peinhaupt, M., Platzer, W., et al. (2014). Heteromerization of GPR55 and cannabinoid CB2 receptors modulates signalling. *British Journal of Pharmacology*, *171*(23), 5387–5406. https://doi.org/10.1111/bph.12850.

Bao, L., Redondo, C., Findlay, J. B. C., Walker, J. H., & Ponnambalam, S. (2009). Deciphering soluble and membrane protein function using yeast systems (review). *Molecular Membrane Biology*, *26*(3), 127–135. https://doi.org/10.1080/09687680802637652.

Bassé, F., Stout, J. G., Sims, P. J., & Wiedmer, T. (1996). Isolation of an erythrocyte membrane protein that mediates Ca2+- dependent transbilayer movement of phospholipid. *Journal of Biological Chemistry*, *271*(29), 17205–17210. https://doi.org/10.1074/jbc.271.29.17205.

Benito, C., Tolón, R. M., Pazos, M. R., Núñez, E., Castillo, A. I., & Romero, J. (2008). Cannabinoid CB 2 receptors in human brain inflammation. *British Journal of Pharmacology*, *153*(2), 277–285. https://doi.org/10.1038/sj.bjp.0707505.

Blasco-Benito, S., Moreno, E., Seijo-Vila, M., Tundidor, I., Andradas, C., Caffarel, M. M., et al. (2019). Therapeutic targeting of HER2-CB(2)R heteromers in HER2-positive breast cancer. *Proceedings of the National Academy of Sciences of the United States of America*, *116*(9), 3863–3872. https://doi.org/10.1073/pnas.1815034116.

Blume, L. C., Eldeeb, K., Bass, C. E., Selley, D. E., & Howlett, A. C. (2015). Cannabinoid receptor interacting protein (CRIP1a) attenuates CB1R signaling in neuronal cells. *Cellular Signalling*, *27*(3), 716–726. https://doi.org/10.1016/j.cellsig.2014.11.006.

Blume, L. C., Leone-Kabler, S., Luessen, D. J., Marrs, G. S., Lyons, E., Bass, C. E., et al. (2016). Cannabinoid receptor interacting protein suppresses agonist-driven CB1 receptor internalization and regulates receptor replenishment in an agonist-biased manner. *Journal of Neurochemistry*, *139*(3), 396–407. https://doi.org/10.1111/jnc.13767.

Blume, L. C., Patten, T., Eldeeb, K., Leone-Kabler, S., Ilyasov, A. A., Keegan, B. M., et al. (2017). Cannabinoid receptor interacting protein 1a competition with β-arrestin for CB1 receptor binding sites. *Molecular Pharmacology*, *91*(2), 75–86. https://doi.org/10.1124/mol.116.104638.

Bonaventura, J., Rico, A. J., Moreno, E., Sierra, S., Sánchez, M., Luquin, N., et al. (2014). L-DOPA-treatment in primates disrupts the expression of A2A adenosine-CB1 cannabinoid-D2 dopamine receptor heteromers in the caudate nucleus. *Neuropharmacology*, *79*, 90–100. https://doi.org/10.1016/j.neuropharm.2013.10.036.

Bonhaus, D. W., Chang, L. K., Kwan, J., & Martin, G. R. (1998). Dual activation and inhibition of adenylyl cyclase by cannabinoid receptor agonists: Evidence for agonist-specific trafficking of intracellular responses. *Journal of Pharmacology and Experimental Therapeutics*, *287*(3), 884–888. http://www.jpet.org.

Bornert, O., Møller, T. C., Boeuf, J., Candusso, M.-P., Wagner, R., Martinez, K. L., et al. (2013). Identification of a novel protein-protein interaction motif mediating interaction of GPCR-associated sorting proteins with G protein-coupled receptors. *PLoS One*, *8*(2), e56336. https://doi.org/10.1371/journal.pone.0056336.

Botta, J., Bibic, L., Killoran, P., McCormick, P. J., & Howell, L. A. (2019). Design and development of stapled transmembrane peptides that disrupt the activity of G-protein-coupled receptor oligomers. *Journal of Biological Chemistry*, *294*(45), 16587–16603. https://doi.org/10.1074/jbc.RA119.009160.

Bouaboula, M., Rinaldi, M., Carayon, P., Carillon, C., Delpech, B., Shire, D., et al. (1993). Cannabinoid-receptor expression in human leukocytes. *European Journal of Biochemistry*, *214*(1), 173–180. https://doi.org/10.1111/j.1432-1033.1993.tb17910.x.

Bouvier, M., Heveker, N., Jockers, R., Marullo, S., & Milligan, G. (2007). BRET analysis of GPCR oligomerization: Newer does not mean better. *Nature Methods*, *Vol. 4*(1), 3–4. Nature Publishing Group https://doi.org/10.1038/nmeth0107-3.

Brailoiu, G. C., Deliu, E., Marcu, J., Hoffman, N. E., Console-Bram, L., Zhao, P., et al. (2014). Differential activation of intracellular versus plasmalemmal CB2 cannabinoid receptors. *Biochemistry*, *53*(30), 4990–4999. https://doi.org/10.1021/bi500632a.

Brown, S. M., Wager-Miller, J., & Mackie, K. (2002). Cloning and molecular characterization of the rat CB2 cannabinoid receptor. *Biochimica et Biophysica Acta—Gene Structure and Expression*, *1576*(3), 255–264. https://doi.org/10.1016/S0167-4781(02)00341-X.

Bruno, A., Lembo, F., Novellino, E., Stornaiuolo, M., & Marinelli, L. (2014). Beyond radio-displacement techniques for identification of CB 1 ligands: The first application of a fluorescence-quenching assay. *Scientific Reports*, *4*(1), 1–9. https://doi.org/10.1038/srep03757.

Bushlin, I., Gupta, A., Stockton, S. D., Miller, L. K., & Devi, L. A. (2012). Dimerization with cannabinoid receptors allosterically modulates delta opioid receptor activity during neuropathic pain. *PLoS One*, *7*(12), e49789. https://doi.org/10.1371/journal.pone.0049789.

Bylund, D. B. (2007). Beta-2 adrenoceptor. In *xPharm: the comprehensive pharmacology reference* (pp. 1–12). Elsevier Inc. https://doi.org/10.1016/B978-008055232-3.60201-6.

Callén, L., Moreno, E., Barroso-Chinea, P., Moreno-Delgado, D., Cortés, A., Mallol, J., et al. (2012). Cannabinoid receptors CB1 and CB2 form functional heteromers in brain. *Journal of Biological Chemistry*, *287*(25), 20851–20865. https://doi.org/10.1074/jbc.M111.335273.

Cannaert, A., Storme, J., Franz, F., Auwärter, V., & Stove, C. P. (2016). Detection and activity profiling of synthetic cannabinoids and metabolites with a newly developed bio-assay. *Analytical Chemistry*, *88*(23). https://doi.org/10.1021/acs.analchem.6b02600.

Carlisle, S. J., Marciano-Cabral, F., Staab, A., Ludwick, C., & Cabral, G. A. (2002). Differential expression of the CB2 cannabinoid receptor by rodent macrophages and macrophage-like cells in relation to cell activation. *International Immunopharmacology*, *2*(1), 69–82. https://doi.org/10.1016/S1567-5769(01)00147-3.

Carriba, P., Navarro, G., Ciruela, F., Ferré, S., Casadó, V., Agnati, L., et al. (2008). Detection of heteromerization of more than two proteins by sequential BRET-FRET. *Nature Methods*, *5*(8), 727–733. https://doi.org/10.1038/nmeth.1229.

Carriba, P., Ortiz, O., Patkar, K., Justinova, Z., Stroik, J., Themann, A., et al. (2007). Striatal adenosine A2A and cannabinoid CB1 receptors form functional heteromeric complexes that mediate the motor effects of cannabinoids. *Neuropsychopharmacology*, *32*(11), 2249–2259. https://doi.org/10.1038/sj.npp.1301375.

Castaneda, J. T., Harui, A., Kiertscher, S. M., Roth, J. D., & Roth, M. D. (2013). Differential expression of intracellular and extracellular CB2 cannabinoid receptor protein by human peripheral blood leukocytes. *Journal of Neuroimmune Pharmacology*, *8*(1), 323–332. https://doi.org/10.1007/s11481-012-9430-8.

Castillo, P. E., Younts, T. J., Chávez, A. E., & Hashimotodani, Y. (2012). Endocannabinoid signaling and synaptic function. *Neuron*, *76*(1), 70–81. https://doi.org/10.1016/j.neuron.2012.09.020.

Catani, M. V., Gasperi, V., Catanzaro, G., Baldassarri, S., Bertoni, A., Sinigaglia, F., et al. (2010). Human platelets express authentic CB1 and CB2 receptors. *Current Neurovascular Research*, *7*(4), 311–318. https://doi.org/10.2174/156720210793180774.

Chabre, M., Deterre, P., & Antonny, B. (2009). The apparent cooperativity of some GPCRs does not necessarily imply dimerization. *Trends in Pharmacological Sciences*, *30*(4), 182–187. https://doi.org/10.1016/j.tips.2009.01.003.

Christie, M. J. (2006). Opioid and cannabinoid receptors: Friends with benefits or just close friends? *British Journal of Pharmacology*, *148*(4), 385–386. https://doi.org/10.1038/sj.bjp.0706756.

Coke, C. J., Scarlett, K. A., Chetram, M. A., Jones, K. J., Sandifer, B. J., Davis, A. S., et al. (2016). Simultaneous activation of induced heterodimerization between CXCR4 chemokine receptor and cannabinoid receptor 2 (CB2) reveals a mechanism for regulation of tumor progression. *Journal of Biological Chemistry*, *291*(19), 9991–10005. https://doi.org/10.1074/jbc.M115.712661.

Console-Bram, L., Brailoiu, E., Brailoiu, G. C., Sharir, H., & Abood, M. E. (2014). Activation of GPR18 by cannabinoid compounds: A tale of biased agonism. *British Journal of Pharmacology*, *171*(16), 3908–3917. https://doi.org/10.1111/bph.12746.

Cruz, S. L., Sánchez-Miranda, E., Castillo-Arellano, J. I., Cervantes-Villagrana, R. D., Ibarra-Sánchez, A., & González-Espinosa, C. (2018). Anandamide inhibits FcεRI-dependent degranulation and cytokine synthesis in mast cells through CB2 and GPR55 receptor activation. Possible involvement of CB2-GPR55 heteromers. *International Immunopharmacology*, *64*, 298–307. https://doi.org/10.1016/j.intimp.2018.09.006.

Daigle, T. L., Kwok, M. L., & Mackie, K. (2008). Regulation of CB $_1$ cannabinoid receptor internalization by a promiscuous phosphorylation-dependent mechanism. *Journal of Neurochemistry*, *106*(1), 70–82. https://doi.org/10.1111/j.1471-4159.2008.05336.x.

Dalton, G. D., Carney, S. T., Marshburn, J. D., Norford, D. C., & Howlett, A. C. (2020). CB1 cannabinoid receptors stimulate Gβγ-GRK2-mediated FAK phosphorylation at tyrosine 925 to regulate ERK activation involving neuronal focal adhesions. *Frontiers in Cellular Neuroscience*, *14*, 176. https://doi.org/10.3389/fncel.2020.00176. In this issue.

De Jesús, M. L., Sallés, J., Meana, J. J., & Callado, L. F. (2006). Characterization of CB1 cannabinoid receptor immunoreactivity in postmortem human brain homogenates. *Neuroscience*, *140*(2), 635–643. https://doi.org/10.1016/j.neuroscience.2006.02.024.

Delgado-Peraza, F., Ahn, K. H., Nogueras-Ortiz, C., Mungrue, I. N., Mackie, K., Kendall, D. A., et al. (2016). Mechanisms of biased β-arrestin-mediated signaling downstream from the cannabinoid 1 receptor. *Molecular Pharmacology*, *89*(6), 618–629. https://doi.org/10.1124/mol.115.103176.

Dhopeshwarkar, A., & Mackie, K. (2016). Functional selectivity of CB2 cannabinoid receptor ligands at a canonical and noncanonical pathway. *The Journal of Pharmacology and Experimental Therapeutics*, *358*(2), 342–351. https://doi.org/10.1124/jpet.116.232561.

Diez-Alarcia, R., Ibarra-Lecue, I., Lopez-Cardona, Á. P., Meana, J., Gutierrez-Adán, A., Callado, L. F., et al. (2016). Biased agonism of three different cannabinoid receptor agonists in mouse brain cortex. *Frontiers in Pharmacology*, *7*(Nov), 415. https://doi.org/10.3389/fphar.2016.00415.

Dixon, A. S., Schwinn, M. K., Hall, M. P., Zimmerman, K., Otto, P., Lubben, T. H., et al. (2016). NanoLuc complementation reporter optimized for accurate measurement of protein interactions in cells. *ACS Chemical Biology*, *11*, 24. https://doi.org/10.1021/acschembio.5b00753.

Domenici, M. R., Azad, S. C., Marsicano, G., Schierloh, A., Wotjak, C. T., Dodt, H. U., et al. (2006). Cannabinoid receptor type 1 located on presynaptic terminals of principal neurons in the forebrain controls glutamatergic synaptic transmission. *Journal of Neuroscience*, *26*(21), 5794–5799. https://doi.org/10.1523/JNEUROSCI.0372-06.2006.

Donthamsetti, P., Quejada, J. R., Javitch, J. A., Gurevich, V. V., & Lambert, N. A. (2015). Using bioluminescence resonance energy transfer (BRET) to characterize agonist-induced arrestin recruitment to modified and unmodified G protein-coupled receptors. *Current Protocols in Pharmacology*, *70*(1), 2.14.1–2.14.14. https://doi.org/10.1002/0471141755.ph0214s70.

Dvorakova, M., Kubik-Zahorodna, A., Straiker, A., Sedlacek, R., Hajkova, A., Mackie, K., et al. (2021). SGIP1 is involved in regulation of emotionality, mood, and nociception and tunes in vivo signaling of cannabinoid receptor 1. *British Journal of Pharmacology*, *178*(7), 1588–1604. https://doi.org/10.1111/bph.15383.

Ehrhart, J., Obregon, D., Mori, T., Hou, H., Sun, N., Bai, Y., et al. (2005). Stimulation of cannabinoid receptor 2 (CB2) suppresses microglial activation. *Journal of Neuroinflammation*, *2*(1), 29. https://doi.org/10.1186/1742-2094-2-29.

Ellis, J., Pediani, J. D., Canals, M., Milasta, S., & Milligan, G. (2006). Orexin-1 receptor-cannabinoid CB1 receptor heterodimerization results in both ligand-dependent and -independent coordinated alterations of receptor localization and function. *Journal of Biological Chemistry*, *281*(50), 38812–38824. https://doi.org/10.1074/jbc.M602494200.

Esteban, P. F., Garcia-Ovejero, D., Paniagua-Torija, B., Moreno-Luna, R., Arredondo, L. F., Zimmer, A., et al. (2020). Revisiting CB1 cannabinoid receptor detection and the exploration of its interacting partners. *Journal of Neuroscience Methods*, *337*, 108680. https://doi.org/10.1016/j.jneumeth.2020.108680.

Evenseth, L. S. M., Gabrielsen, M., & Sylte, I. (2020). The GABAB receptor—Structure, ligand binding and drug development. *Molecules*, *25*(13), 3093. https://doi.org/10.3390/molecules25133093.

Fan, J., Rone, M. B., & Papadopoulos, V. (2009). Translocator protein 2 is involved in cholesterol redistribution during erythropoiesis. *Journal of Biological Chemistry*, *284*(44), 30484–30497. https://doi.org/10.1074/jbc.M109.029876.

Federspiel, J. D., & Cristea, I. M. (2019). Considerations for identifying endogenous protein complexes from tissue via immunoaffinity purification and quantitative mass spectrometry. In C. Evans, P. Wright, & J. Noirel (Eds.), *Vol. 1977. Mass spectrometry of proteins. Methods in molecular biology* (pp. 115–143). New York, NY: Humana Press. https://doi.org/10.1007/978-1-4939-9232-4_9.

Fernández-Fernández, C., Decara, J., Bermúdez-Silva, F. J., Sánchez, E., Morales, P., Gómez-Cañas, M., et al. (2013). Description of a bivalent cannabinoid ligand with hypophagic properties. *Archiv der Pharmazie*, *346*(3), 171–179. https://doi.org/10.1002/ardp.201200392.

Fernández-Ruiz, J., & González, S. (2005). Cannabinoid control of motor function at the basal ganglia. In *Vol. 168. Cannabinoids* (pp. 479–507). Springer-Verlag. https://doi.org/10.1007/3-540-26573-2_16.

Ferré, S., Baler, R., Bouvier, M., Caron, M. G., Devi, L. A., Durroux, T., et al. (2009). Building a new conceptual framework for receptor heteromers. *Nature Chemical Biology*, *5*(3), 131–134. Nature Publishing Group https://doi.org/10.1038/nchembio0309-131.

Ferré, S., Casadó, V., Devi, L. A., Filizola, M., Jockers, R., Lohse, M. J., et al. (2014). G protein–coupled receptor oligomerization revisited: Functional and pharmacological perspectives. *Pharmacological Reviews*, *66*(2), 413–434. https://doi.org/10.1124/pr.113.008052.

Ferreira, S. G., Gonçalves, F. Q., Marques, J. M., Tomé, R., Rodrigues, R. J., Nunes-Correia, I., et al. (2015). Presynaptic adenosine a 2A receptors dampen cannabinoid CB 1 receptor-mediated inhibition of corticostriatal glutamatergic transmission. *British Journal of Pharmacology*, *172*(4), 1074–1086. https://doi.org/10.1111/bph.12970.

Filppula, S., Yaddanapudi, S., Mercier, R., Xu, W., Pavlopoulos, S., & Makriyannis, A. (2004). Purification and mass spectroscopic analysis of human CB2 cannabinoid receptor expressed in the baculovirus system. *The Journal of Peptide Research*, *64*(6), 225–236. https://doi.org/10.1111/j.1399-3011.2004.00188.x.

Finlay, D. B., Cawston, E. E., Grimsey, N. L., Hunter, M. R., Korde, A., Vemuri, V. K., et al. (2017). Gαs signalling of the CB1 receptor and the influence of receptor number. *British Journal of Pharmacology*, *174*(15), 2545–2562. https://doi.org/10.1111/bph.13866.

Finlay, D. B., Joseph, W. R., Grimsey, N. L., & Glass, M. (2016). GPR18 undergoes a high degree of constitutive trafficking but is unresponsive to N-arachidonoyl glycine. *PeerJ*, *4*(3), e1835. https://doi.org/10.7717/peerj.1835.

Fraguas-Sánchez, A. I., Martín-Sabroso, C., & Torres-Suárez, A. I. (2018). Insights into the effects of the endocannabinoid system in cancer: A review. *British Journal of Pharmacology*, *175*(13), 2566–2580. https://doi.org/10.1111/bph.14331.

Franco, R., Reyes-Resina, I., Aguinaga, D., Lillo, A., Jiménez, J., Raïch, I., et al. (2019). Potentiation of cannabinoid signaling in microglia by adenosine A2A receptor antagonists. *Glia*, *67*(12), 2410–2423. https://doi.org/10.1002/glia.23694.

Franklin, J. M., & Carrasco, G. A. (2013). G-protein receptor kinase 5 regulates the cannabinoid receptor 2-induced up-regulation of serotonin 2A receptors. *Journal of Biological Chemistry*, *288*(22), 15712–15724. https://doi.org/10.1074/jbc.M113.454843.

Fujita, K. I., Maeda, D., Xiao, Q., & Srinivasula, S. M. (2011). Nrf2-mediated induction of p62 controls toll-like receptor-4-driven aggresome-like induced structure formation and autophagic degradation. *Proceedings of the National Academy of Sciences of the United States of America*, *108*(4), 1427–1432. https://doi.org/10.1073/pnas.1014156108.

Fuxe, K., Marcellino, D., Borroto-Escuela, D. O., Guescini, M., Fernández-Dueñas, V., Tanganelli, S., et al. (2010). Adenosine-dopamine interactions in the pathophysiology and treatment of CNS disorders. *CNS Neuroscience and Therapeutics*, *16*(3), e18. https://doi.org/10.1111/j.1755-5949.2009.00126.x.

Galiègue, S., Mary, S., Marchand, J., Dussossoy, D., Carrière, D., Carayon, P., et al. (1995). Expression of central and peripheral cannabinoid receptors in human immune tissues and leukocyte subpopulations. *European Journal of Biochemistry*, *232*(1), 54–61. https://doi.org/10.1111/j.1432-1033.1995.tb20780.x.

Gantz, I., Muraoka, A., Yang, Y. K., Samuelson, L. C., Zimmerman, E. M., Cook, H., et al. (1997). Cloning and chromosomal localization of a gene (GPR18) encoding a novel seven transmembrane receptor highly expressed in spleen and testis. *Genomics*, *42*(3), 462–466. https://doi.org/10.1006/geno.1997.4752.

García-Gutiérrez, M. S., Navarrete, F., Navarro, G., Reyes-Resina, I., Franco, R., Lanciego, J. L., et al. (2018). Alterations in gene and protein expression of cannabinoid CB 2 and GPR55 receptors in the dorsolateral prefrontal cortex of suicide victims. *Neurotherapeutics*, *15*(3), 796–806. https://doi.org/10.1007/s13311-018-0610-y.

Ghosh, S., Preet, A., Groopman, J. E., & Ganju, R. K. (2006). Cannabinoid receptor CB2 modulates the CXCL12/CXCR4-mediated chemotaxis of T lymphocytes. *Molecular Immunology*, *43*(14), 2169–2179. https://doi.org/10.1016/j.molimm.2006.01.005.

Gillis, A., Kliewer, A., Kelly, E., Henderson, G., Christie, M. J., Schulz, S., et al. (2020). Critical assessment of G protein-biased agonism at the μ-opioid receptor. *Trends in Pharmacological Sciences*, *41*(12), 947–959. https://doi.org/10.1016/j.tips.2020.09.009.

Glass, M., Dragunow, M., & Faull, R. L. M. (2000). The pattern of neurodegeneration in Huntington's disease: A comparative study of cannabinoid, dopamine, adenosine and GABA(A) receptor alterations in the human basal ganglia in Huntington's disease. *Neuroscience*, *97*(3), 505–519. https://doi.org/10.1016/S0306-4522(00)00008-7.

Glass, M., & Felder, C. C. (1997). Concurrent stimulation of cannabinoid CB1 and dopamine D2 receptors augments cAMP accumulation in striatal neurons: Evidence for a Gs linkage to the CB1 receptor. *The Journal of Neuroscience*, *17*(14), 5327. https://doi.org/10.1523/jneurosci.17-14-05327.1997.

Glass, M., Govindpani, K., Furkert, D. P., Hurst, D. P., Reggio, P. H., & Flanagan, J. U. (2016). One for the price of two…are bivalent ligands targeting cannabinoid receptor dimers capable of simultaneously binding to both receptors? *Trends in Pharmacological Sciences*, *37*(5), 353–363. https://doi.org/10.1016/j.tips.2016.01.010.

Glass, M., & Northup, J. K. (1999). Agonist selective regulation of G proteins by cannabinoid CB1 and CB2 receptors. *Molecular Pharmacology*, *56*(6), 1362–1369. https://doi.org/10.1124/mol.56.6.1362.

González, S., Cebeira, M., & Fernández-Ruiz, J. (2005). Cannabinoid tolerance and dependence: A review of studies in laboratory animals. *Pharmacology Biochemistry and Behavior*, *81*(2 Spec. Iss), 300–318. https://doi.org/10.1016/j.pbb.2005.01.028.

Gowran, A., Murphy, C. E., & Campbell, V. A. (2009). Δ^9-Tetrahydrocannabinol regulates the p53 post-translational modifiers murine double minute 2 and the small ubiquitin MOdifier protein in the rat brain. *FEBS Letters*, *583*(21), 3412–3418. https://doi.org/10.1016/j.febslet.2009.09.056.

Grimsey, N. L., Goodfellow, C. E., Dragunow, M., & Glass, M. (2011). Cannabinoid receptor 2 undergoes Rab5-mediated internalization and recycles via a Rab11-dependent pathway. *Biochimica et Biophysica Acta*, *1813*(8), 1554–1560. https://doi.org/10.1016/j.bbamcr.2011.05.010.

Grimsey, N. L., Goodfellow, C. E., Scotter, E. L., Dowie, M. J., Glass, M., & Graham, E. S. (2008). Specific detection of CB1 receptors; cannabinoid CB1 receptor antibodies are not all created equal! *Journal of Neuroscience Methods*, *171*(1), 78–86. https://doi.org/10.1016/j.jneumeth.2008.02.014.

Grimsey, N. L., Graham, E. S., Dragunow, M., & Glass, M. (2010). Cannabinoid receptor 1 trafficking and the role of the intracellular pool: Implications for therapeutics. *Biochemical Pharmacology*, *80*(7), 1050–1062. https://doi.org/10.1016/j.bcp.2010.06.007.

Guggenhuber, S., Alpar, A., Chen, R., Schmitz, N., Wickert, M., Mattheus, T., et al. (2016). Cannabinoid receptor-interacting protein Crip1a modulates CB1 receptor signaling in mouse hippocampus. *Brain Structure and Function*, *221*(4), 2061–2074. https://doi.org/10.1007/s00429-015-1027-6.

Guidolin, D., Marcoli, M., Tortorella, C., Maura, G., & Agnati, L. F. (2019). Receptor-receptor interactions as a widespread phenomenon: Novel targets for drug development? *Frontiers in Endocrinology*, *10*(Feb), 53. Frontiers Media S.A https://doi.org/10.3389/fendo.2019.00053.

Gyombolai, P., Pap, D., Turu, G., Catt, K. J., Bagdy, G., & Hunyady, L. (2012). Regulation of endocannabinoid release by G proteins: A paracrine mechanism of G protein-coupled receptor action. *Molecular and Cellular Endocrinology*, *353*(1–2), 29–36. https://doi.org/10.1016/j.mce.2011.10.011.

Haeuptle, M. A., Pujol, F. M., Neupert, C., Winchester, B., Kastaniotis, A. J. J., Aebi, M., et al. (2008). Human RFT1 deficiency leads to a disorder of N-linked glycosylation. *American Journal of Human Genetics*, *82*(3), 600–606. https://doi.org/10.1016/j.ajhg.2007.12.021.

Hájková, A., Techlovská, Š., Dvořáková, M., Chambers, J. N., Kumpošt, J., Hubálková, P., et al. (2016). SGIP1 alters internalization and modulates signaling of activated cannabinoid receptor 1 in a biased manner. *Neuropharmacology*, *107*, 201–214. https://doi.org/10.1016/j.neuropharm.2016.03.008.

Hay, D. L., Garelja, M. L., Poyner, D. R., & Walker, C. S. (2018). Update on the pharmacology of calcitonin/CGRP family of peptides: IUPHAR review 25. *British Journal of Pharmacology*, *175*(1), 3–17. https://doi.org/10.1111/bph.14075.

He, F., Kumar, A., & Song, Z.-H. H. (2012). Heat shock protein 90 is an essential molecular chaperone for CB2 cannabinoid receptor-mediated signaling in trabecular meshwork cells. *Molecular Vision*, *18*, 2839–2846. https://pubmed.ncbi.nlm.nih.gov/23233786.

He, F., Qiao, Z. H., Cai, J., Pierce, W., He, D. C., & Song, Z. H. (2007). Involvement of the 90-kDa heat shock protein (Hsp-90) in CB2 cannabinoid receptor-mediated cell migration: A new role of Hsp-90 in migration signaling of a G protein-coupled receptor. *Molecular Pharmacology*, *72*(5), 1289–1300. https://doi.org/10.1124/mol.107.036566.

Hebert, T. E., Moffett, S., Morello, J. P., Loisel, T. P., Bichet, D. G., Barret, C., et al. (1996). A peptide derived from a β 2-adrenergic receptor transmembrane domain inhibits both receptor dimerization and activation. *Journal of Biological Chemistry*, *271*(27), 16384–16392. https://doi.org/10.1074/jbc.271.27.16384.

Heydorn, A., Søndergaard, B. P., Ersbøll, B., Holst, B., Nielsen, F. C., Haft, C. R., et al. (2004). A library of 7TM receptor C-terminal tails: Interactions with the proposed post-endocytic sorting proteins ERM-binding phosphoprotein 50 (EBP50), N-ethylmaleimide-sensitive factor (NSF), sorting nexin 1 (SNX1), and G protein-coupled receptor-associated sorting protein (GASP). *Journal of Biological Chemistry*, *279*(52), 54291–54303. https://doi.org/10.1074/jbc.M406169200.

Hojo, M., Sudo, Y., Ando, Y., Minami, K., Takada, M., Matsubara, T., et al. (2008). μ-Opioid receptor forms a functional heterodimer with cannabinoid CB1 receptor: Electrophysiological and FRET assay analysis. *Journal of Pharmacological Sciences*, *108*(3), 308–319. https://doi.org/10.1254/jphs.08244FP.

Hua, T., Li, X., Wu, L., Iliopoulos-Tsoutsouvas, C., Wang, Y., Wu, M., et al. (2020). Activation and signaling mechanism revealed by cannabinoid receptor-Gi complex structures. *Cell*, *180*(4), 655–665.e18. https://doi.org/10.1016/j.cell.2020.01.008.

Huang, G., Pemp, D., Stadtmüller, P., Nimczick, M., Heilmann, J., & Decker, M. (2014). Design, synthesis and in vitro evaluation of novel uni- and bivalent ligands for the cannabinoid receptor type 1 with variation of spacer length and structure. *Bioorganic and Medicinal Chemistry Letters*, *24*(17), 4209–4214. https://doi.org/10.1016/j.bmcl.2014.07.038.

Hudson, B. D., Hébert, T. E., & Kelly, M. E. (2010). Physical and functional interaction between CB1 cannabinoid receptors and β2-adrenoceptors. *British Journal of Pharmacology*, *160*(3), 627–642. https://doi.org/10.1111/j.1476-5381.2010.00681.x.

Ibsen, M. S., Finlay, D. B., Patel, M., Javitch, J. A., Glass, M., & Grimsey, N. L. (2019). Cannabinoid CB1 and CB2 receptor-mediated arrestin translocation: Species, subtype, and agonist-dependence. *Frontiers in Pharmacology*, *10*, 350. https://doi.org/10.3389/fphar.2019.00350.

Inoue, A., Raimondi, F., Kadji, F. M. N., Singh, G., Kishi, T., Uwamizu, A., et al. (2019). Illuminating G-protein-coupling selectivity of GPCRs. *Cell*, *177*(7), 1933–1947.e25. https://doi.org/10.1016/j.cell.2019.04.044.

Irving, A. J., Coutts, A. A., Harvey, J., Rae, M. G., Mackie, K., Bewick, G. S., et al. (2000). Functional expression of cell surface cannabinoid CB1 receptors on presynaptic inhibitory terminals in cultured rat hippocampal neurons. *Neuroscience*, *98*(2), 253–262. https://doi.org/10.1016/S0306-4522(00)00120-2.

James, J. R., Oliveira, M. I., Carmo, A. M., Iaboni, A., & Davis, S. J. (2006). A rigorous experimental framework for detecting protein oligomerization using bioluminescence resonance energy transfer. *Nature Methods*, *3*(12), 1001–1006. https://doi.org/10.1038/nmeth978.

Jäntti, M. H., Mandrika, I., & Kukkonen, J. P. (2014). Human orexin/hypocretin receptors form constitutive homo- and heteromeric complexes with each other and with human CB1 cannabinoid receptors. *Biochemical and Biophysical Research Communications*, *445*(2), 486–490. https://doi.org/10.1016/j.bbrc.2014.02.026.

Jin, W., Brown, S., Roche, J. P., Hsieh, C., Celver, J. P., Kovoor, A., et al. (1999). Distinct domains of the CB1 cannabinoid receptor mediate desensitization and internalization. *Journal of Neuroscience*, *19*(10), 3773–3780. https://doi.org/10.1523/jneurosci.19-10-03773.1999.

Kaplan, B. L. F. (2013). The role of CB1 in immune modulation by cannabinoids. *Pharmacology and Therapeutics*, *137*(3), 365–374. https://doi.org/10.1016/j.pharmthera.2012.12.004.

Kargl, J., Balenga, N., Parzmair, G. P., Brown, A. J., Heinemann, A., & Waldhoer, M. (2012). The cannabinoid receptor CB1 modulates the signaling properties of the lysophosphatidylinositol receptor GPR55. *Journal of Biological Chemistry*, *287*(53), 44234–44248. https://doi.org/10.1074/jbc.M112.364109.

Kearn, C. S., Blake-Palmer, K., Daniel, E., Mackie, K., & Glass, M. (2005). Concurrent stimulation of cannabinoid CB1 and dopamine D2 receptors enhances heterodimer formation: A mechanism for receptor cross-talk? *Molecular Pharmacology*, *67*(5), 1697–1704. https://doi.org/10.1124/mol.104.006882.

Khajehali, E., Malone, D. T., Glass, M., Sexton, P. M., Christopoulos, A., & Leach, K. (2015). Biased agonism and biased allosteric modulation at the CB1 cannabinoid receptors. *Molecular Pharmacology*, *88*(2), 368–379. https://doi.org/10.1124/mol.115.099192.

Khan, S. S., & Lee, F. J. S. (2014). Delineation of domains within the cannabinoid CB1 and dopamine D2 receptors that mediate the formation of the heterodimer complex. *Journal of Molecular Neuroscience*, *53*(1), 10–21. https://doi.org/10.1007/s12031-013-0181-7.

Kim, K., & Chung, K. Y. (2020). Many faces of the GPCR-arrestin interaction. *Archives of Pharmacal Research*, *43*(9), 890–899. https://doi.org/10.1007/s12272-020-01263-w.

Kim, J. Y., & Ozato, K. (2009). The sequestosome 1/p62 attenuates cytokine gene expression in activated macrophages by inhibiting IFN regulatory factor 8 and TNF receptor-associated factor 6/NF-κB activity. *The Journal of Immunology*, *182*(4), 2131–2140. https://doi.org/10.4049/jimmunol.0802755.

Kleyer, J., Nicolussi, S., Taylor, P., Simonelli, D., Furger, E., Anderle, P., et al. (2012). Cannabinoid receptor trafficking in peripheral cells is dynamically regulated by a binary biochemical switch. *Biochemical Pharmacology*, *83*(10), 1393–1412. https://doi.org/10.1016/J.BCP.2012.02.014.

Kodama, Y., & Hu, C.-D. (2010). An improved bimolecular fluorescence complementation assay with a high signal-to-noise ratio. *BioTechniques*, *49*(5), 793–805. https://doi.org/10.2144/000113519.

Kodama, Y., & Hu, C.-D. D. (2012). Bimolecular fluorescence complementation (BiFC): A 5-year update and future perspectives. *BioTechniques*, *53*(5), 285–298. https://doi.org/10.2144/000113943.

Köfalvi, A., Moreno, E., Cordomí, A., Cai, N. S., Fernández-Dueñas, V., Ferreira, S. G., et al. (2020). Control of glutamate release by complexes of adenosine and cannabinoid receptors. *BMC Biology*, *18*(1), 9. https://doi.org/10.1186/s12915-020-0739-0.

Kouznetsova, M., Kelley, B., Shen, M., & Thayer, S. A. (2002). Desensitization of cannabinoid-mediated presynaptic inhibition of neurotransmission between rat hippocampal neurons in culture. *Molecular Pharmacology*, *61*(3), 477–485. https://doi.org/10.1124/mol.61.3.477.

Kranjc, T., Dempsey, E., Cagney, G., Nakamura, N., Shields, D. C., & Simpson, J. C. (2017). Functional characterisation of the YIPF protein family in mammalian cells. *Histochemistry and Cell Biology*, *147*(4), 439–451. https://doi.org/10.1007/s00418-016-1527-3.

Krishna Kumar, K., Shalev-Benami, M., Robertson, M. J., Hu, H., Banister, S. D., Hollingsworth, S. A., et al. (2019). Structure of a signaling cannabinoid receptor 1-G protein complex. *Cell*, *176*(3), 448–458.e12. https://doi.org/10.1016/j.cell.2018.11.040.

Kwon, D. H., Park, O. H., Kim, L., Jung, Y. O., Park, Y., Jeong, H., et al. (2018). Insights into degradation mechanism of N-end rule substrates by p62/SQSTM1 autophagy adapter. *Nature Communications*, *9*(1), 1–13. https://doi.org/10.1038/s41467-018-05825-x.

Laezza, C., Pagano, C., Navarra, G., Pastorino, O., Proto, M. C., Fiore, D., et al. (2020). The endocannabinoid system: A target for cancer treatment. *International Journal of Molecular Sciences*, *21*(3), 747. https://doi.org/10.3390/ijms21030747.

Lambert, N. A., & Javitch, J. A. (2014). CrossTalk opposing view: Weighing the evidence for class A GPCR dimers, the jury is still out. *The Journal of Physiology*, *592*(12), 2443–2445. https://doi.org/10.1113/jphysiol.2014.272997.

Lan, T.-H., Liu, Q., Li, C., Wu, G., & Lambert, N. A. (2012). Sensitive and high resolution localization and tracking of membrane proteins in live cells with BRET. *Traffic*, *13*(11), 1450–1456. https://doi.org/10.1111/j.1600-0854.2012.01401.x.

Lan, T. H., Liu, Q., Li, C., Wu, G., Steyaert, J., & Lambert, N. A. (2015). BRET evidence that β2 adrenergic receptors do not oligomerize in cells. *Scientific Reports*, *5*, 10166. https://doi.org/10.1038/srep10166.

Laprairie, R. B., Bagher, A. M., Kelly, M. E. M., & Denovan-Wright, E. M. (2016). Biased type 1 cannabinoid receptor signaling influences neuronal viability in a cell culture model of Huntington diseases. *Molecular Pharmacology*, *89*(3), 364–375. https://doi.org/10.1124/mol.115.101980.

Lauckner, J. E., Hille, B., & Mackie, K. (2005). The cannabinoid agonist WIN55,212-2 increases intracellular calcium via CB1 receptor coupling to Gq/11 G proteins. *Proceedings of the National Academy of Sciences of the United States of America*, *102*(52), 19144–19149. https://doi.org/10.1073/pnas.0509588102.

Le Naour, M., Akgün, E., Yekkirala, A., Lunzer, M. M., Powers, M. D., Kalyuzhny, A. E., et al. (2013). Bivalent ligands that target μ opioid (MOP) and cannabinoid1 (CB 1) receptors are potent analgesics devoid of tolerance. *Journal of Medicinal Chemistry*, *56*(13), 5505–5513. https://doi.org/10.1021/jm4005219.

Lee, M. H., Appleton, K. M., Strungs, E. G., Kwon, J. Y., Morinelli, T. A., Peterson, Y. K., et al. (2016). The conformational signature of β-arrestin2 predicts its trafficking and signalling functions. *Nature*, *531*(7596), 665–668. https://doi.org/10.1038/nature17154.

Lee, S. P., O'Dowd, B. F., Ng, G. Y. K., Varghese, G., Akil, H., Mansour, A., et al. (2000). Inhibition of cell surface expression by mutant receptors demonstrates that D2 dopamine receptors exist as oligomers in the cell. *Molecular Pharmacology*, *58*(1), 120–128. https://doi.org/10.1124/mol.58.1.120.

Leterrier, C., Bonnard, D., Carrel, D., Rossier, J., & Lenkei, Z. (2004). Constitutive endocytic cycle of the CB1 cannabinoid receptor. *Journal of Biological Chemistry*, *279*(34), 36013–36021. https://doi.org/10.1074/jbc.M403990200.

Leterrier, C., Lainé, J., Darmon, M., Boudin, H., Rossier, J., & Lenkei, Z. (2006). Constitutive activation drives compartment-selective endocytosis and axonal targeting of type 1 cannabinoid receptors. *Journal of Neuroscience*, *26*(12), 3141–3153. https://doi.org/10.1523/JNEUROSCI.5437-05.2006.

Li, Y., & Kim, J. (2015). Neuronal expression of CB2 cannabinoid receptor mRNAs in the mouse hippocampus. *Neuroscience*, *311*, 253–267. https://doi.org/10.1016/j.neuroscience.2015.10.041.

Lin, X., Li, S., Zhao, Y., Ma, X., Zhang, K., He, X., et al. (2013). Interaction domains of p62: A bridge between p62 and selective autophagy. *DNA and Cell Biology*, *32*(5), 220–227. https://doi.org/10.1089/dna.2012.1915.

Liu, B., Song, S., Jones, P. M., & Persaud, S. J. (2015). GPR55: From orphan to metabolic regulator? *Pharmacology and Therapeutics*, *145*, 35–42. https://doi.org/10.1016/j.pharmthera.2014.06.007.

Lu, H. C., & Mackie, K. (2016). An introduction to the endogenous cannabinoid system. *Biological Psychiatry*, *79*(7), 516–525. https://doi.org/10.1016/j.biopsych.2015.07.028.

Lu, V. B., Puhl, H. L., & Ikeda, S. R. (2013). N-arachidonyl glycine does not activate G protein-coupled receptor 18 signaling via canonical pathways. *Molecular Pharmacology*, *83*(1), 267–282. https://doi.org/10.1124/mol.112.081182.

Lu, C., Shi, L., Sun, B., Zhang, Y., Hou, B., Sun, Y., et al. (2017). A single intrathecal or intraperitoneal injection of CB2 receptor agonist attenuates bone cancer pain and induces a time-dependent modification of GRK2. *Cellular and Molecular Neurobiology*, *37*(1), 101–109. https://doi.org/10.1007/s10571-016-0349-0.

Luck, K., Kim, D. K., Lambourne, L., Spirohn, K., Begg, B. E., Bian, W., et al. (2020). A reference map of the human binary protein interactome. *Nature*, *580*(7803), 402–408. https://doi.org/10.1038/s41586-020-2188-x.

Luu, W., Hart-Smith, G., Sharpe, L. J., & Brown, A. J. (2015). The terminal enzymes of cholesterol synthesis, DHCR24 and DHCR7, interact physically and functionally. *Journal of Lipid Research*, *56*(4), 888–897. https://doi.org/10.1194/jlr.M056986.

Maggio, R., Vogel, Z., & Wess, J. (1993). Coexpression studies with mutant muscarinic/adrenergic receptors provide evidence for intermolecular "cross-talk" between G-protein-linked receptors. *Proceedings of the National Academy of Sciences of the United States of America*, *90*(7), 3103–3107. https://doi.org/10.1073/pnas.90.7.3103.

Manzanares, J., Julian, M., & Carrascosa, A. (2006). Role of the cannabinoid system in pain control and therapeutic implications for the management of acute and chronic pain episodes. *Current Neuropharmacology*, *4*(3), 239–257. https://doi.org/10.2174/157015906778019527.

Marcellino, D., Carriba, P., Filip, M., Borgkvist, A., Frankowska, M., Bellido, I., et al. (2008). Antagonistic cannabinoid CB1/dopamine D2 receptor interactions in striatal CB1/D2 heteromers. A combined neurochemical and behavioral analysis. *Neuropharmacology*, *54*(5), 815–823. https://doi.org/10.1016/j.neuropharm.2007.12.011.

Marchalant, Y., Brownjohn, P. W., Bonnet, A., Kleffmann, T., & Ashton, J. C. (2014). Validating antibodies to the cannabinoid CB2 receptor: Antibody sensitivity is not evidence of antibody specificity. *Journal of Histochemistry & Cytochemistry*, *62*(6), 395–404. https://doi.org/10.1369/0022155414530995.

Martínez-Pinilla, E., Reyes-Resina, I., Oñatibia-Astibia, A., Zamarbide, M., Ricobaraza, A., Navarro, G., et al. (2014). CB1 and GPR55 receptors are co-expressed and form heteromers in rat and monkey striatum. *Experimental Neurology*, *261*, 44–52. https://doi.org/10.1016/j.expneurol.2014.06.017.

Martínez-Pinilla, E., Rico, A. J., Rivas-Santisteban, R., Lillo, J., Roda, E., Navarro, G., et al. (2020). Expression of GPR55 and either cannabinoid CB1 or CB2 heteroreceptor complexes in the caudate, putamen, and accumbens nuclei of control, parkinsonian, and dyskinetic non-human primates. *Brain Structure and Function*, *225*(7), 2153–2164. https://doi.org/10.1007/s00429-020-02116-4.

Martini, L., Thompson, D., Kharazia, V., & Whistler, J. L. (2010). Differential regulation of behavioral tolerance to WIN55,212-2 by GASP1. *Neuropsychopharmacology*, *35*(6), 1363–1373. https://doi.org/10.1038/npp.2010.6.

Martini, L., Waldhoer, M., Pusch, M., Kharazia, V., Fong, J., Lee, J. H., et al. (2007). Ligand-induced down-regulation of the cannabinoid 1 receptor is mediated by the G-protein-coupled receptor-associated sorting protein GASP1. *The FASEB Journal*, *21*(3), 802–811. https://doi.org/10.1096/fj.06-7132com.

Mascia, F., Klotz, L., Lerch, J., Ahmed, M. H., Zhang, Y., & Enz, R. (2017). CRIP1a inhibits endocytosis of G-protein coupled receptors activated by endocannabinoids and glutamate by a common molecular mechanism. *Journal of Neurochemistry*, *141*(4), 577–591. https://doi.org/10.1111/jnc.14021.

Maurice, P., Kamal, M., & Jockers, R. (2011). Asymmetry of GPCR oligomers supports their functional relevance. *Trends in Pharmacological Sciences*, *32*(9), 514–520. https://doi.org/10.1016/j.tips.2011.05.006.

McDonald, N. A., Henstridge, C. M., Connolly, C. N., & Irving, A. J. (2007). An essential role for constitutive endocytosis, but not activity, in the axonal targeting of the CB1 cannabinoid receptor. *Molecular Pharmacology*, *71*(4), 976–984. https://doi.org/10.1124/mol.106.029348.

McHugh, D., Page, J., Dunn, E., & Bradshaw, H. B. (2012). Δ9-tetrahydrocannabinol and N-arachidonyl glycine are full agonists at GPR18 receptors and induce migration in human endometrial HEC-1B cells. *British Journal of Pharmacology*, *165*(8), 2414–2424. https://doi.org/10.1111/j.1476-5381.2011.01497.x.

Mielnik, C. A., Lam, V. M., & Ross, R. A. (2021). CB1 allosteric modulators and their therapeutic potential in CNS disorders. *Progress in Neuro-Psychopharmacology and Biological Psychiatry*, *106*, 110163. https://doi.org/10.1016/j.pnpbp.2020.110163.

Miller, L. K., & Devi, L. A. (2011). The highs and lows of cannabinoid receptor expression in disease: Mechanisms and their therapeutic implications. *Pharmacological Reviews*, *63*(3), 461–470. https://doi.org/10.1124/pr.110.003491.

Milligan, G. (2013). The prevalence, maintenance, and relevance of g protein-coupled receptor oligomerization. *Molecular Pharmacology*, *84*(1), 158–169. https://doi.org/10.1124/mol.113.084780.

Mnpotra, J. S., Qiao, Z., Cai, J., Lynch, D. L., Grossfield, A., Leioatts, N., et al. (2014). Structural basis of G protein-coupled receptor-Gi protein interaction: Formation of the cannabinoid CB2 receptor-Gi protein complex. *Journal of Biological Chemistry*, *289*(29), 20259–20272. https://doi.org/10.1074/jbc.M113.539916.

Morena, M., & Campolongo, P. (2014). The endocannabinoid system: An emotional buffer in the modulation of memory function. *Neurobiology of Learning and Memory*, *112*, 30–43. https://doi.org/10.1016/j.nlm.2013.12.010.

Moreno, E., Andradas, C., Medrano, M., Caffarel, M. M., Pérez-Gómez, E., Blasco-Benito, S., et al. (2014). Targeting CB2-GPR55 receptor heteromers modulates cancer cell signaling. *Journal of Biological Chemistry*, *289*(32), 21960–21972. https://doi.org/10.1074/jbc.M114.561761.

Moreno, E., Chiarlone, A., Medrano, M., Puigdellívol, M., Bibic, L., Howell, L. A., et al. (2018). Singular location and signaling profile of adenosine A2A-cannabinoid CB1 receptor heteromers in the dorsal striatum. *Neuropsychopharmacology*, *43*(5), 964–977. https://doi.org/10.1038/npp.2017.12.

Mukherjee, S., Adams, M., Whiteaker, K., Daza, A., Kage, K., Cassar, S., et al. (2004). Species comparison and pharmacological characterization of rat and human CB 2 cannabinoid receptors. *European Journal of Pharmacology*, *505*(1–3), 1–9. https://doi.org/10.1016/j.ejphar.2004.09.058.

Mukhopadhyay, S., & Howlett, A. C. (2001). CB1 receptor-G protein association: Subtype selectivity is determined by distinct intracellular domains. *European Journal of Biochemistry*, *268*(3), 499–505. https://doi.org/10.1046/j.1432-1327.2001.01810.x.

Mukhopadhyay, S., McIntosh, H. H., Houston, D. B., & Howlett, A. C. (2000). The CB1 cannabinoid receptor juxtamembrane C-terminal peptide confers activation to specific G proteins in brain. *Molecular Pharmacology*, *57*(1), 162–170. http://www.molpharm.org.

Nasser, M. W., Qamri, Z., Deol, Y. S., Smith, D., Shilo, K., Zou, X., et al. (2011). Crosstalk between chemokine receptor CXCR4 and cannabinoid receptor CB2 in modulating breast cancer growth and invasion. *PLoS One*, *6*(9), e23901. https://doi.org/10.1371/journal.pone.0023901.

Navarro, G., Borroto-Escuela, D., Angelats, E., Etayo, Í., Reyes-Resina, I., Pulido-Salgado, M., et al. (2018). Receptor-heteromer mediated regulation of endocannabinoid signaling in activated microglia. Role of CB1 and CB2 receptors and relevance for Alzheimer's disease and levodopa-induced dyskinesia. *Brain, Behavior, and Immunity*, *67*, 139–151. https://doi.org/10.1016/j.bbi.2017.08.015.

Navarro, G., Ferré, S., Cordomi, A., Moreno, E., Mallol, J., Casadó, V., et al. (2010). Interactions between intracellular domains as key determinants of the quaternary structure and function of receptor heteromers. *Journal of Biological Chemistry*, *285*(35), 27346–27359. https://doi.org/10.1074/jbc.M110.115634.

Nealon, C. M., Henderson-Redmond, A. N., Hale, D. E., & Morgan, D. J. (2019). Tolerance to WIN55,212-2 is delayed in desensitization-resistant S426A/S430A mice. *Neuropharmacology*, *148*, 151–159. https://doi.org/10.1016/j.neuropharm.2018.12.026.

Nguyen, P. T., Schmid, C. L., Raehal, K. M., Selley, D. E., Bohn, L. M., & Sim-Selley, L. J. (2012). β-Arrestin2 regulates cannabinoid CB 1 receptor signaling and adaptation in a central nervous system region-dependent manner. *Biological Psychiatry*, *71*(8), 714–724. https://doi.org/10.1016/j.biopsych.2011.11.027.

Niehaus, J. L., Liu, Y., Wallis, K. T., Egertová, M., Bhartur, S. G., Mukhopadhyay, S., et al. (2007). CB1 cannabinoid receptor activity is modulated by the cannabinoid receptor interacting protein CRIP 1a. *Molecular Pharmacology*, *72*(6), 1557–1566. https://doi.org/10.1124/mol.107.039263.

Nilsson, O., Fowler, C. J., & Jacobsson, S. O. P. (2006). The cannabinoid agonist WIN 55,212-2 inhibits TNF-α-induced neutrophil transmigration across ECV304 cells. *European Journal of Pharmacology*, *547*(1), 165–173. https://doi.org/10.1016/j.ejphar.2006.07.016.

Nogueras-Ortiz, C., Roman-Vendrell, C., Mateo-Semidey, G. E., Liao, Y.-H., Kendall, D. A., & Yudowski, G. A. (2017). Retromer stops beta-arrestin 1–mediated signaling from internalized cannabinoid 2 receptors. *Molecular Biology of the Cell*, *28*(24), 3554–3561. https://doi.org/10.1091/mbc.e17-03-0198.

Ochel, H. J., Eichhorn, K., & Gademann, G. (2001). Geldanamycin: The prototype of a class of antitumor drugs targeting the heat shock protein 90 family of molecular chaperones. *Cell Stress and Chaperones*, *6*(2), 105–112. https://www.ncbi.nlm.nih.gov/pmc/articles/PMC434387/.

Oddi, S., Dainese, E., Sandiford, S., Fezza, F., Lanuti, M., Chiurchiù, V., et al. (2012). Effects of palmitoylation of Cys 415 in helix 8 of the CB 1 cannabinoid receptor on membrane localization and signalling. *British Journal of Pharmacology*, *165*(8), 2635–2651. https://doi.org/10.1111/j.1476-5381.2011.01658.x.

Oliver, E. E., Hughes, E. K., Puckett, M. K., Chen, R., Lowther, W. T., & Howlett, A. C. (2020). Cannabinoid receptor interacting protein 1a (CRIP1a) in health and disease. *Biomolecules*, *10*(12), 1–22. https://doi.org/10.3390/biom10121609.

Olsen, R. H. J., DiBerto, J. F., English, J. G., Glaudin, A. M., Krumm, B. E., Slocum, S. T., et al. (2020). TRUPATH, an open-source biosensor platform for interrogating the GPCR transducerome. *Nature Chemical Biology*, *16*(8), 841–849. https://doi.org/10.1038/s41589-020-0535-8.

Orchard, S., Ammari, M., Aranda, B., Breuza, L., Briganti, L., Broackes-Carter, F., et al. (2014). The MIntAct project—IntAct as a common curation platform for 11 molecular interaction databases. *Nucleic Acids Research*, *42*(D1), D358–D363. https://doi.org/10.1093/nar/gkt1115. Database https://www.ebi.ac.uk/intact/.

Orr, A. G., Orr, A. L., Li, X. J., Gross, R. E., & Traynelis, S. F. (2009). Adenosine A2A receptor mediates microglial process retraction. *Nature Neuroscience*, *12*(7), 872–878. https://doi.org/10.1038/nn.2341.

Oyagawa, C. R. M., de la Harpe, S. M., Saroz, Y., Glass, M., Vernall, A. J., & Grimsey, N. L. (2018). Cannabinoid receptor 2 signalling bias elicited by 2,4,6-trisubstituted 1,3,5-triazines. *Frontiers in Pharmacology*, *9*, 1202. https://www.frontiersin.org/article/10.3389/fphar.2018.01202.

Pandey, P., Roy, K. K., & Doerksen, R. J. (2020). Negative allosteric modulators of cannabinoid receptor 2: Protein modeling, binding site identification and molecular dynamics simulations in the presence of an orthosteric agonist. *Journal of Biomolecular Structure and Dynamics*, *38*(1), 32–47. https://doi.org/10.1080/07391102.2019.1567384.

Park, S. Y., & Guo, X. (2014). Adaptor protein complexes and intracellular transport. *Bioscience Reports*, *34*(4), 381–390. https://doi.org/10.1042/BSR20140069.

Parker, S. L., Parker, M. S., Estes, A. M., Wong, Y. Y., Sah, R., Sweatman, T., et al. (2008). The neuropeptide Y (NPY) Y2 receptors are largely dimeric in the kidney, but monomeric in the forebrain. *Journal of Receptors and Signal Transduction*, *28*(3), 245–263. https://doi.org/10.1080/10799890802084341.

Pérez-Benito, L., Henry, A., Matsoukas, M. T., Lopez, L., Pulido, D., Royo, M., et al. (2018). The size matters? A computational tool to design bivalent ligands. *Bioinformatics*, *34*(22), 3857–3863. https://doi.org/10.1093/bioinformatics/bty422.

Pérez-Gómez, E., Andradas, C., Blasco-Benito, S., Caffarel, M. M., García-Taboada, E., Villa-Morales, M., et al. (2015). Role of cannabinoid receptor CB2 in HER2 pro-oncogenic signaling in breast cancer. *Journal of the National Cancer Institute*, *107*(6), djv077. https://doi.org/10.1093/jnci/djv077.

Persidsky, Y., Fan, S., Dykstra, H., Reichenbach, N. L., Rom, S., & Ramirez, S. H. (2015). Activation of cannabinoid type two receptors (CB2) diminish inflammatory responses in macrophages and brain endothelium. *Journal of Neuroimmune Pharmacology*, *10*(2), 302–308. https://doi.org/10.1007/s11481-015-9591-3.

Petersen, J., Wright, S. C., Rodríguez, D., Matricon, P., Lahav, N., Vromen, A., et al. (2017). Agonist-induced dimer dissociation as a macromolecular step in G protein-coupled receptor signaling. *Nature Communications*, *8*(1), 1–15. https://doi.org/10.1038/s41467-017-00253-9.

Pickel, V. M., Chan, J., Kearn, C. S., & Mackie, K. (2006). Targeting dopamine D2 and cannabinoid-1 (CB1) receptors in rat nucleus accumbens. *Journal of Comparative Neurology*, *495*(3), 299–313. https://doi.org/10.1002/cne.20881.

Pin, J. P., Kniazeff, J., Prézeau, L., Liu, J. F., & Rondard, P. (2019). GPCR interaction as a possible way for allosteric control between receptors. *Molecular and Cellular Endocrinology*, *486*, 89–95. https://doi.org/10.1016/j.mce.2019.02.019.

Plouffe, B., Thomsen, A. R. B., & Irannejad, R. (2020). Emerging role of compartmentalized G protein-coupled receptor signaling in the cardiovascular field. *ACS Pharmacology and Translational Science*, *3*(2), 221–236. https://doi.org/10.1021/acsptsci.0c00006.

Porras, P., Barrera, E., Bridge, A., Del-Toro, N., Cesareni, G., Duesbury, M., et al. (2020). Towards a unified open access dataset of molecular interactions. *Nature Communications*, *11*(1), 1–12. https://doi.org/10.1038/s41467-020-19942-z. Database http://www.imexconsortium.org/.

Przybyla, J. A., & Watts, V. J. (2010). Ligand-induced regulation and localization of cannabinoid CB1 and dopamine D2L receptor heterodimers. *The Journal of Pharmacology and Experimental Therapeutics*, *332*(3), 710–719. https://doi.org/10.1124/jpet.109.162701.

Reggio, P. H., & Shore, D. M. (2015). The therapeutic potential of orphan GPCRs, GPR35 and GPR55. *Frontiers in Pharmacology*, *6*(Mar), 69. https://doi.org/10.3389/fphar.2015.00069.

Reyes-Resina, I., Navarro, G., Aguinaga, D., Canela, E. I., Schoeder, C. T., Załuski, M., et al. (2018). Molecular and functional interaction between GPR18 and cannabinoid CB2 G-protein-coupled receptors. Relevance in neurodegenerative diseases. *Biochemical Pharmacology*, *157*, 169–179. https://doi.org/10.1016/j.bcp.2018.06.001.

Rios, C., Gomes, I., & Devi, L. A. (2006). μ opioid and CB1 cannabinoid receptor interactions: Reciprocal inhibition of receptor signaling and neuritogenesis. *British Journal of Pharmacology*, *148*(4), 387–395. https://doi.org/10.1038/sj.bjp.0706757.

Rozenfeld, R., Bushlin, I., Gomes, I., Tzavaras, N., Gupta, A., Neves, S., et al. (2012). Receptor heteromerization expands the repertoire of cannabinoid signaling in rodent neurons. *PLoS One*, *7*(1), e29239. https://doi.org/10.1371/journal.pone.0029239.

Rozenfeld, R., & Devi, L. A. (2008). Regulation of CB1 cannabinoid receptor trafficking by the adaptor protein AP-3. *The FASEB Journal*, *22*(7), 2311. https://doi.org/10.1096/FJ.07-102731.

Rozenfeld, R., Gupta, A., Gagnidze, K., Lim, M. P., Gomes, I., Lee-Ramos, D., et al. (2011). AT1R-CB1 R heteromerization reveals a new mechanism for the pathogenic properties of angiotensin II. *EMBO Journal*, *30*(12), 2350–2363. https://doi.org/10.1038/emboj.2011.139.

Sam, A. H., Salem, V., & Ghatei, M. A. (2011). Rimonabant: From RIO to ban. *Journal of Obesity*, *2011*, 432607. https://doi.org/10.1155/2011/432607.

Sanz, L., Sanchez, P., Lallena, M. J., Diaz-Meco, M. T., & Moscat, J. (1999). The interaction of p62 with RIP links the atypical PKCs to NF-κB activation. *EMBO Journal*, *18*(11), 3044–3053. https://doi.org/10.1093/emboj/18.11.3044.

Sarott, R. C., Westphal, M. V., Pfaff, P., Korn, C., Sykes, D. A., Gazzi, T., et al. (2020). Development of high-specificity fluorescent probes to enable cannabinoid type 2 receptor studies in living cells. *Journal of the American Chemical Society*, *142*(40), 16953–16964. https://doi.org/10.1021/jacs.0c05587.

Saroz, Y., Kho, D. T., Glass, M., Graham, E. S., & Grimsey, N. L. (2019). Cannabinoid receptor 2 (CB2) signals via G-alpha-s and induces IL-6 and IL-10 cytokine secretion in human primary leukocytes. *ACS Pharmacology & Translational Science*, *2*(6), 414–428. https://doi.org/10.1021/acsptsci.9b00049.

Scarlett, K. A., White, E. S. Z., Coke, C. J., Carter, J. R., Bryant, L. K., & Hinton, C. V. (2018). Agonist-induced CXCR4 and CB2 heterodimerization inhibits Ga13/RhoA-mediated migration. *Molecular Cancer Research*, *16*(4), 728–739. https://doi.org/10.1158/1541-7786.MCR-16-0481.

Schopf, F. H., Biebl, M. M., & Buchner, J. (2017). The HSP90 chaperone machinery. *Nature Reviews Molecular Cell Biology*, *18*(6), 345–360. https://doi.org/10.1038/nrm.2017.20.

Schweppe, D. K., Huttlin, E. L., Harper, J. W., & Gygi, S. P. (2018). BioPlex display: An interactive suite for large-scale AP-MS protein-protein interaction data. *Journal of Proteome Research*, *17*(1), 722–726. https://doi.org/10.1021/acs.jproteome.7b00572. Database https://bioplex.hms.harvard.edu/.

Sharaf, A., Mensching, L., Keller, C., Rading, S., Scheffold, M., Palkowitsch, L., et al. (2019). Systematic affinity purification coupled to mass spectrometry identified p62 as part of the cannabinoid receptor CB2 interactome. *Frontiers in Molecular Neuroscience*, *12*, 224. https://doi.org/10.3389/fnmol.2019.00224.

Shim, J. Y., Ahn, K. H., & Kendall, D. A. (2013). Molecular basis of cannabinoid CB1 receptor coupling to the G protein heterotrimer Gαiβγ; identification of key CB1 contacts with the C-terminal helix α5 of Gαi. *Journal of Biological Chemistry*, *288*(45), 32449–32465. https://doi.org/10.1074/jbc.M113.489153.

Sierra, S., Gomes, I., & Devi, L. A. (2017). Class A GPCRs: Cannabinoid and opioid receptor heteromers. In K. Herrick-Davis, G. Milligan, & G. Di Giovanni (Eds.), *Vol. 33. G-protein-coupled receptor dimers. Receptors* (pp. 173–206). Cham: Humana Press. https://doi.org/10.1007/978-3-319-60174-8_7.

Simonin, F., Karcher, P., Boeuf, J. J.-M., Matifas, A., & Kieffer, B. L. (2004). Identification of a novel family of G protein-coupled receptor associated sorting proteins. *Journal of Neurochemistry*, *89*(3), 766–775. https://doi.org/10.1111/j.1471-4159.2004.02411.x.

Singh, P., Ganjiwale, A., Howlett, A. C., & Cowsik, S. M. (2017). In silico interaction analysis of cannabinoid receptor interacting protein 1b (CRIP1b)—CB1 cannabinoid receptor. *Journal of Molecular Graphics and Modelling*, *77*, 311–321. https://doi.org/10.1016/j.jmgm.2017.09.006.

Singh, P., Ganjiwale, A., Howlett, A. C., & Cowsik, S. M. (2019). Molecular interaction between distal C-terminal domain of the CB1 cannabinoid receptor and cannabinoid receptor interacting proteins (CRIP1a/CRIP1b). *Journal of Chemical Information and Modeling*, *59*(12), 5294–5303. https://doi.org/10.1021/acs.jcim.9b00948.

Singh, S., Oyagawa, C. R. M., Macdonald, C., Grimsey, N. L., Glass, M., & Vernall, A. J. (2019). Chromenopyrazole-based high affinity, selective fluorescent ligands for cannabinoid type 2 receptor. *ACS Medicinal Chemistry Letters*, *10*(2), 209–214. https://doi.org/10.1021/acsmedchemlett.8b00597.

Śledziński, P., Zeyland, J., Słomski, R., & Nowak, A. (2018). The current state and future perspectives of cannabinoids in cancer biology. *Cancer Medicine*, *7*(3), 765–775. https://doi.org/10.1002/cam4.1312.

Smith, T. H., Blume, L. C., Straiker, A., Cox, J. O., David, B. G., Secor McVoy, J. R., et al. (2015). Cannabinoid receptor-interacting protein 1a modulates CB1 receptor signaling and regulation. *Molecular Pharmacology*, *87*(4), 747–765. https://doi.org/10.1124/mol.114.096495.

Smith, T. H., Sim-Selley, L. J., & Selley, D. E. (2010). Cannabinoid CB 1 receptor-interacting proteins: Novel targets for central nervous system drug discovery? *British Journal of Pharmacology*, *160*(3), 454–466. https://doi.org/10.1111/j.1476-5381.2010.00777.x.

Snider, J., Kittanakom, S., Damjanovic, D., Curak, J., Wong, V., & Stagljar, I. (2010). Detecting interactions with membrane proteins using a membrane two-hybrid assay in yeast. *Nature Protocols*, *5*(7), 1281–1293. https://doi.org/10.1038/nprot.2010.83.

Soave, M., Stoddart, L. A., White, C. W., Kilpatrick, L. E., Goulding, J., Briddon, S. J., et al. (2021). Detection of genome-edited and endogenously expressed G protein-coupled receptors. *The FEBS Journal*, *288*(8), 2585–2601. https://doi.org/10.1111/febs.15729.

Soethoudt, M., Grether, U., Fingerle, J., Grim, T. W., Fezza, F., de Petrocellis, L., et al. (2017). Cannabinoid CB(2) receptor ligand profiling reveals biased signalling and off-target activity. *Nature Communications*, *8*, 13958. https://doi.org/10.1038/ncomms13958.

Song, C., & Howlett, A. C. (1995). Rat brain cannabinoid receptors are N-linked glycosylated proteins. *Life Sciences*, *56*(23–24), 1983–1989. https://doi.org/10.1016/0024-3205(95)00179-A.

Stampanoni Bassi, M., Sancesario, A., Morace, R., Centonze, D., & Iezzi, E. (2017). Cannabinoids in Parkinson's disease. *Cannabis and Cannabinoid Research*, *2*(1), 21–29. https://doi.org/10.1089/can.2017.0002.

Stoddart, L. A., Kindon, N. D., Otun, O., Harwood, C. R., Patera, F., Veprintsev, D. B., et al. (2020). Ligand-directed covalent labelling of a GPCR with a fluorescent tag in live cells. *Communications Biology*, *3*(1), 1–9. https://doi.org/10.1038/s42003-020-01451-w.

Szabo, B., & Schlicker, E. (2005). Effects of cannabinoids on neurotransmission. In *Cannabinoids* (pp. 327–365). Springer-Verlag. Issue 168 https://doi.org/10.1007/3-540-26573-2_11.

Szalai, B., Hoffmann, P., Prokop, S., Erdélyi, L., Várnai, P., & Hunyady, L. (2014). Improved methodical approach for quantitative BRET analysis of G protein coupled receptor dimerization. *PLoS One*, *9*(10), e109503. https://doi.org/10.1371/journal.pone.0109503.

Szidonya, L., Cserzo, M., & Hunyady, L. (2008). Dimerization and oligomerization of G-protein-coupled receptors: Debated structures with established and emerging functions. *Journal of Endocrinology*, *196*(3), 435–453. BioScientifica https://doi.org/10.1677/JOE-07-0573.

Taipale, M., Tucker, G., Peng, J., Krykbaeva, I., Lin, Z. Y., Larsen, B., et al. (2014). A quantitative chaperone interaction network reveals the architecture of cellular protein homeostasis pathways. *Cell*, *158*(2), 434–448. https://doi.org/10.1016/j.cell.2014.05.039.

Tanveer, R., McGuinness, N., Daniel, S., Gowran, A., & Campbell, V. A. (2012). Cannabinoid receptors and neurodegenerative diseases. *Wiley Interdisciplinary Reviews: Membrane Transport and Signaling*, *1*(5), 633–639. https://doi.org/10.1002/wmts.64.

Tappe-Theodor, A., Agarwal, N., Katona, I., Rubino, T., Martini, L., Swiercz, J., et al. (2007). A molecular basis of analgesic tolerance to cannabinoids. *Journal of Neuroscience*, *27*(15), 4165–4177. https://doi.org/10.1523/JNEUROSCI.5648-06.2007.

Tebano, M. T., Martire, A., & Popoli, P. (2012). Adenosine A2A-cannabinoid CB1 receptor interaction: An integrative mechanism in striatal glutamatergic neurotransmission. In *Vol. 1476. Brain research* (pp. 108–118). Elsevier. https://doi.org/10.1016/j.brainres.2012.04.051.

Teixeira-Clerc, F., Julien, B., Grenard, P., Van Nhieu, J. T., Deveaux, V., Li, L., et al. (2006). CB1 cannabinoid receptor antagonism: A new strategy for the treatment of liver fibrosis. *Nature Medicine*, *12*(6), 671–676. https://doi.org/10.1038/nm1421.

Terrillon, S., & Bouvier, M. (2004). Roles of G-protein-coupled receptor dimerization. *EMBO Reports*, *5*(1), 30–34. https://doi.org/10.1038/sj.embor.7400052.

The Uniprot Consortium. (2019). UniProt: A worldwide hub of protein knowledge. *Nucleic Acids Research*, *47*(D1), D506–D515. https://doi.org/10.1093/nar/gky1049. Database https://www.uniprot.org/.

Thomsen, A. R. B., Jensen, D. D., Hicks, G. A., & Bunnett, N. W. (2018). Therapeutic targeting of endosomal G-protein-coupled receptors. *Trends in Pharmacological Sciences*, *39*(10), 879–891. https://doi.org/10.1016/j.tips.2018.08.003.

Tiulpakov, A., White, C. W., Abhayawardana, R. S., See, H. B., Chan, A. S., Seeber, R. M., et al. (2016). Mutations of vasopressin receptor 2 including novel L312S have differential effects on trafficking. *Molecular Endocrinology*, *30*(8), 889–904. https://doi.org/10.1210/me.2016-1002.

Trevaskis, J., Walder, K., Foletta, V., Kerr-Bayles, L., McMillan, J., Cooper, A., et al. (2005). Src homology 3-domain growth factor receptor-bound 2-like (endophilin) interacting protein 1, a novel neuronal protein that regulates energy balance. *Endocrinology*, *146*(9), 3757–3764. https://doi.org/10.1210/en.2005-0282.

Turcotte, C., Blanchet, M. R., Laviolette, M., & Flamand, N. (2016). The CB2 receptor and its role as a regulator of inflammation. *Cellular and Molecular Life Sciences*, *73*(23), 4449–4470. https://doi.org/10.1007/s00018-016-2300-4.

Udoh, M., Santiago, M., Devenish, S., McGregor, I. S., & Connor, M. (2019). Cannabichromene is a cannabinoid CB2 receptor agonist. *British Journal of Pharmacology*, *176*(23), 4537–4547. https://doi.org/10.1111/bph.14815.

Uezu, A., Horiuchi, A., Kanda, K., Kikuchi, N., Umeda, K., Tsujita, K., et al. (2007). SGIP1α is an endocytic protein that directly interacts with phospholipids and Eps15. *Journal of Biological Chemistry*, *282*(36), 26481–26489. https://doi.org/10.1074/jbc.M703815200.

Viñals, X., Moreno, E., Lanfumey, L., Cordomí, A., Pastor, A., de La Torre, R., et al. (2015). Cognitive impairment induced by delta9-tetrahydrocannabinol occurs through Heteromers between cannabinoid CB1 and serotonin 5-HT2A receptors. *PLoS Biology*, *13*(7), e1002194. https://doi.org/10.1371/journal.pbio.1002194.

Wager-Miller, J., Westenbroek, R., & Mackie, K. (2002). Dimerization of G protein-coupled receptors: CB1 cannabinoid receptors as an example. *Chemistry and Physics of Lipids*, *121*(1–2), 83–89. https://doi.org/10.1016/S0009-3084(02)00151-2.

Ward, R. J., Pediani, J. D., & Milligan, G. (2011). Heteromultimerization of cannabinoid CB 1 receptor and orexin OX 1 receptor generates a unique complex in which both protomers are regulated by orexin A. *Journal of Biological Chemistry*, *286*(43), 37414–37428. https://doi.org/10.1074/jbc.M111.287649.

Whistler, J. L., Enquist, J., Marley, A., Fong, J., Gladher, F., Tsuruda, P., et al. (2002). Modulation of postendocytic sorting of G protein-coupled receptors. *Science*, *297*(5581), 615–620. https://doi.org/10.1126/science.1073308.

Witkin, J. M., Tzavara, E. T., & Nomikos, G. G. (2005). A role for cannabinoid CB1 receptors in mood and anxiety disorders. *Behavioural Pharmacology*, *16*(5–6), 315–331. https://doi.org/10.1097/00008877-200509000-00005.

Wright, K. L., Duncan, M., & Sharkey, K. A. (2008). Cannabinoid CB2 receptors in the gastrointestinal tract: A regulatory system in states of inflammation. *British Journal of Pharmacology*, *153*(2), 263–270. https://doi.org/10.1038/sj.bjp.0707486.

Wu, D.-F., Yang, L.-Q., Goschke, A., Stumm, R., Brandenburg, L.-O., Liang, Y.-J., et al. (2008). Role of receptor internalization in the agonist-induced desensitization of cannabinoid type 1 receptors. *Journal of Neurochemistry*, *104*(4), 1132–1143. https://doi.org/10.1111/j.1471-4159.2007.05063.x.

Xing, C., Zhuang, Y., Xu, T. H., Feng, Z., Zhou, X. E., Chen, M., et al. (2020). Cryo-EM structure of the human cannabinoid receptor CB2-Gi signaling complex. *Cell*, *180*(4), 645–654.e13. https://doi.org/10.1016/j.cell.2020.01.007.

Xu, W., Filppula, S. A., Mercier, R., Yaddanapudi, S., Pavlopoulos, S., Cai, J., et al. (2005). Purification and mass spectroscopic analysis of human CB1 cannabinoid receptor functionally expressed using the baculovirus system. *Journal of Peptide Research*, *66*(3), 138–150. https://doi.org/10.1111/j.1399-3011.2005.00283.x.

Youker, R. T., & Voet, D. (2020). Fluorescence fluctuation techniques for the investigation of structure-function relationships of G-protein-coupled receptors. In *Fluorescence methods for investigation of living cells and microorganisms* IntechOpen. https://doi.org/10.5772/intechopen.93229.

Zhang, Y., Gilliam, A., Maitra, R., Damaj, M. I., Tajuba, J. M., Seltzman, H. H., et al. (2010). Synthesis and biological evaluation of bivalent ligands for the cannabinoid 1 receptor. *Journal of Medicinal Chemistry*, *53*(19), 7048–7060. https://doi.org/10.1021/jm1006676.

Zhang, R., Kim, T. K., Qiao, Z. H., Cai, J., Pierce, W. M., & Song, Z. H. (2007). Biochemical and mass spectrometric characterization of the human CB2 cannabinoid receptor expressed in Pichia pastoris-importance of correct processing of the N-terminus. *Protein Expression and Purification*, *55*(2), 225–235. https://doi.org/10.1016/j.pep.2007.03.018.

Zhang, H.-Y., Shen, H., Jordan, C. J., Liu, Q.-R., Gardner, E. L., Bonci, A., et al. (2018). CB 2 receptor antibody signal specificity: Correlations with the use of partial CB 2-knockout mice and anti-rat CB 2 receptor antibodies. *Acta Pharmacologica Sinica*, *40*, 398–409. https://doi.org/10.1038/s41401-018-0037-3.

Zhao, H., Guo, L., Zhao, H., Zhao, J., Weng, H., & Zhao, B. (2015). CXCR4 over-expression and survival in cancer: A system review and meta-analysis. *Oncotarget*, *6*(7), 5022–5040. https://doi.org/10.18632/oncotarget.3217.

Zheng, C., Chen, L., Chen, X., He, X., Yang, J., Shi, Y., et al. (2013). The second intracellular loop of the human cannabinoid CB2 receptor governs G protein coupling in coordination with the carboxyl terminal domain. *PLoS One*, *8*(5), e63262. https://doi.org/10.1371/journal.pone.0063262.

Zhou, Q., Zhao, J., Stout, J. G., Luhm, R. A., Wiedmer, T., & Sims, P. J. (1997). Molecular cloning of human plasma membrane phospholipid scramblase. *Journal of Biological Chemistry*, *272*(29), 18240–18244. https://doi.org/10.1074/jbc.272.29.18240.

Zou, S., & Kumar, U. (2018). Cannabinoid receptors and the endocannabinoid system: Signaling and function in the central nervous system. *International Journal of Molecular Sciences*, *19*(3), 833. https://doi.org/10.3390/ijms19030833.

Zou, S., & Kumar, U. (2019). Somatostatin and cannabinoid receptors crosstalk in protection of huntingtin knock-in striatal neuronal cells in response to quinolinic acid. *Neurochemistry International*, *129*, 104518. https://doi.org/10.1016/j.neuint.2019.104518.

Zou, S., Somvanshi, R. K., & Kumar, U. (2017). Somatostatin receptor 5 is a prominent regulator of signaling pathways in cells with coexpression of cannabinoid receptors 1. *Neuroscience*, *340*, 218–231. https://doi.org/10.1016/j.neuroscience.2016.10.056.

Zvonok, N., Yaddanapudi, S., Williams, J., Dai, S., Dong, K., Rejtar, T., et al. (2007). Comprehensive proteomic mass spectrometric characterization of human cannabinoid CB2 receptor. *Journal of Proteome Research*, *6*(6), 2068–2079. https://doi.org/10.1021/pr060671h.

CHAPTER

6

Purinergic GPCR transmembrane residues involved in ligand recognition and dimerization

Veronica Salmaso, Shanu Jain, and Kenneth A. Jacobson*

Molecular Recognition Section, Laboratory of Bioorganic Chemistry, National Institute of Diabetes and Digestive and Kidney Diseases, National Institutes of Health, Bethesda, MD, United States

**Corresponding author: e-mail address: kennethj@niddk.nih.gov*

Chapter outline

Abstract

We compare the GPCR-ligand interactions and highlight important residues for recognition in purinergic receptors—from both X-ray crystallographic and cryo-EM structures. These include A_1 and A_{2A} adenosine receptors, and P2Y_1 and P2Y_{12} receptors that respond to ADP

Methods in Cell Biology, Volume 166, ISSN 0091-679X, https://doi.org/10.1016/bs.mcb.2021.06.001

and other nucleotides. These receptors are important drug discovery targets for immune, metabolic and nervous system disorders. In most cases, orthosteric ligands are represented, except for one allosteric $P2Y_1$ antagonist. This review catalogs the residues and regions that engage in contacts with ligands or with other GPCR protomers in dimeric forms. Residues that are in proximity to bound ligands within purinergic GPCR families are correlated. There is extensive conservation of recognition motifs between adenosine receptors, but the $P2Y_1$ and $P2Y_{12}$ receptors are each structurally distinct in their ligand recognition. Identifying common interaction features for ligand recognition within a receptor class that has multiple structures available can aid in the drug discovery process.

1 Introduction

GPCR structural biology has accelerated the discovery of new drug molecules, for example, enabling the *de novo* design of A_{2A} adenosine receptor (AR) antagonist AZD4635 (Fig. 1), which is in clinical trials for cancer (Borodovsky et al., 2020; Congreve, de Graaf, Swain, & Tate, 2020). X-ray crystallography has been used to determine most of these structures (Qu, Wang, & Wu, 2020; Wacker, Stevens, & Roth, 2017). In addition, cryogenic electron microscopy (cryo-EM) is increasingly used and has the advantage of more readily including protein partners such as the associated G proteins and detecting multiple conformations of the same protein (Safdari, Pandey, Shukla, & Dutta, 2018). GPCRs contain seven transmembrane helical domains (TMs), but curiously not all 7TM proteins are GPCRs (Vasiliauskaite-Brooks, Healey, & Granier, 2019). In most of the known structures of GPCR-ligand complexes, the recognition of small molecules involves amino acid residues in the TM region facing the centroid within the heptahelical bundle. The canonical, orthosteric ligand binding site of Class A (rhodopsin-like) GPCRs has largely conserved features (Ballesteros and Weinstein, 1995; van Rhee & Jacobson, 1996; Venkatakrishnan et al., 2013). This conserved structure-function aids in predicting the modes of interaction of competitive small molecule ligands, both for cases in which a protein structure is lacking (Jung et al., 2020) and when a known receptor structure is applied to predicting the interactions of different ligand classes (Rodriguez et al., 2016). The correspondence of key TM residues between receptor subtypes and their shared functionality in ligand binding and activation has been analyzed (Lebon et al., 2011; Wacker et al., 2017; Warne, Edwards, Dore, Leslie, & Tate, 2019; Xu et al., 2011) for Class A GPCRs. There is structural information available, as well, for non-class A GPCRs, in which the degree of homology with Class A is low, for example, a recently determined Class D Ste2 receptor dimer (Velazhahan et al., 2021). Key water molecules, H-bonding networks and molecular (micro)switches for activation within the TM region that are shared among various Class A GPCRs have been defined (Filipek, 2019; Venkatakrishnan et al., 2019; White et al., 2018; Zhou et al., 2019). GPCRmd (http://gpcrmd.org/) is an open source online platform for molecular dynamics (MD) simulations and visualization of dynamic ligand and interhelical interactions, conserved structural motifs, water molecules, H-bonding and other interactions in GPCRs (Rodriguez-Espigares et al., 2020).

Allosteric small molecule modulators of GPCRs have been shown to bind in nearly all regions of the general structure (Wold, Chen, Cunningham, & Zhou, 2019). Nevertheless, many of the allosteric binding regions that are distal or distinct from the canonical binding site still may involve TM residues. This review does not cover the GPCR intracellular loops (ILs), which interact with the C-terminus of the alpha subunit of the G protein or with β-arrestin (Carpenter & Tate, 2017; Ciancetta, Rubio, Lieberman, & Jacobson, 2019), but it emphasizes the role of TM residues.

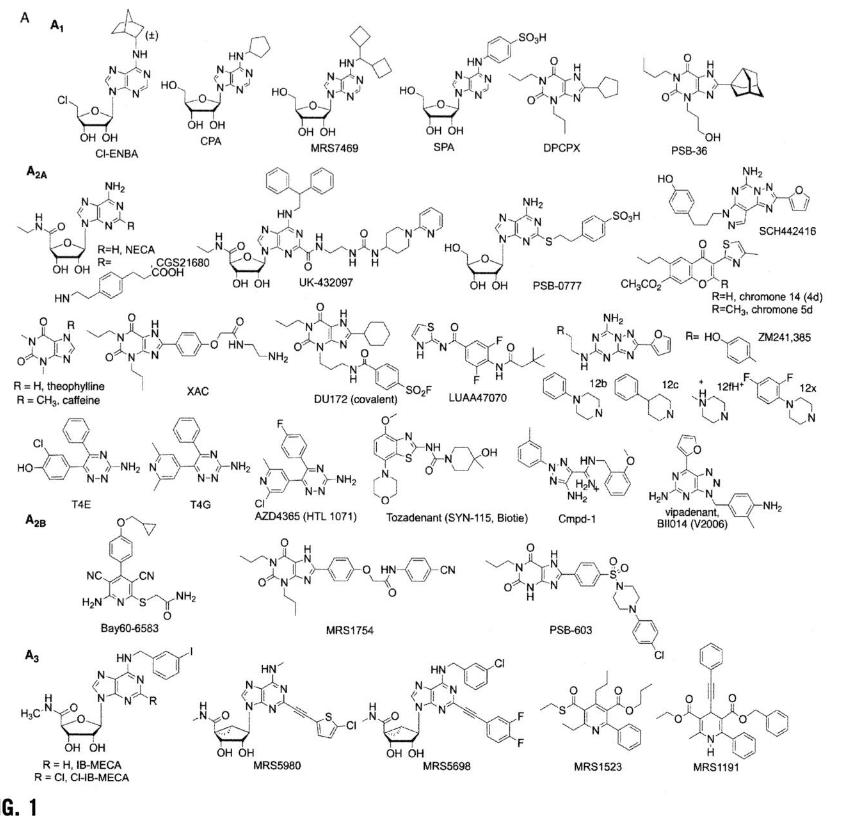

FIG. 1

Structures of pharmacological probes for ARs (A) and P2YRs (B) shown in Table 1, including ligands present in reported structures. The structures of adenosine, ADP and ATP are not shown. Theophylline and caffeine (shown under A_{2A}AR) are nonselective antagonists.

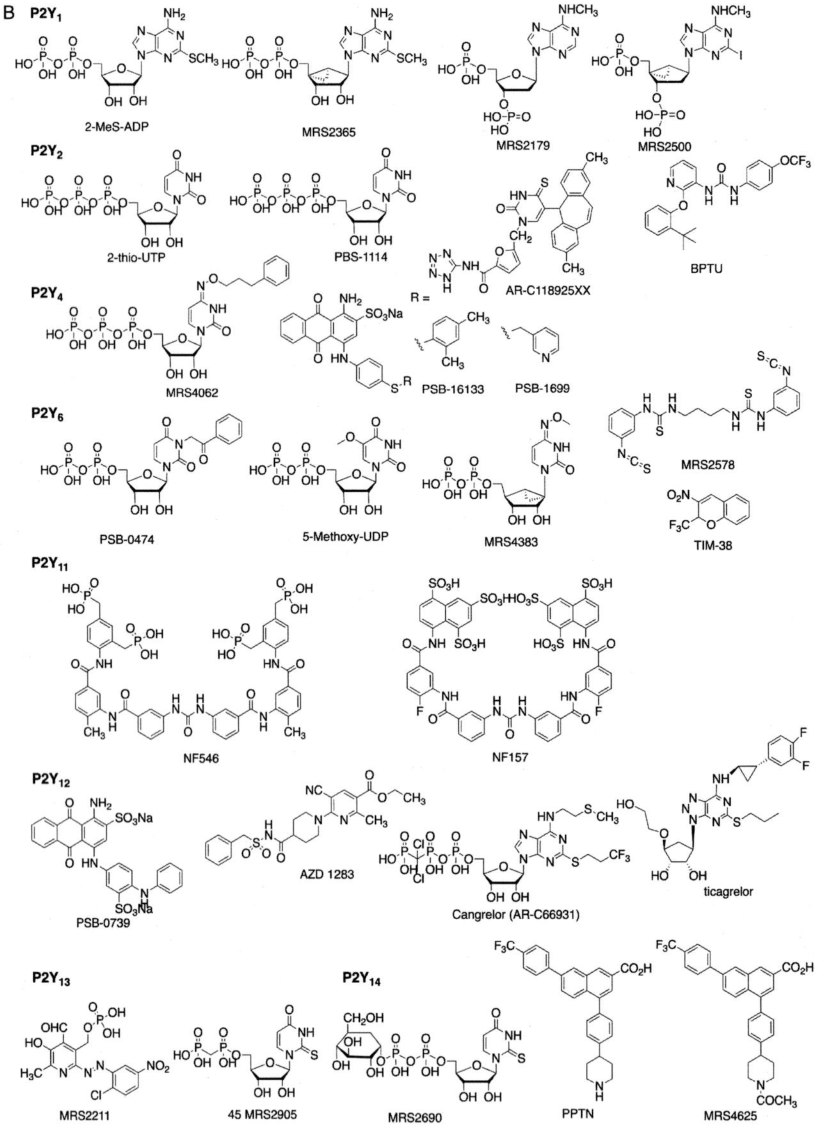
B
$P2Y_1$
2-MeS-ADP
MRS2365
MRS2179
MRS2500
$P2Y_2$
2-thio-UTP
PBS-1114
AR-C118925XX
BPTU
$P2Y_4$
MRS4062
R =
PSB-16133
PSB-1699
$P2Y_6$
PSB-0474
5-Methoxy-UDP
MRS4383
MRS2578
TIM-38
$P2Y_{11}$
NF546
NF157
$P2Y_{12}$
PSB-0739
AZD 1283
Cangrelor (AR-C66931)
ticagrelor
$P2Y_{13}$
$P2Y_{14}$
MRS2211
45 MRS2905
MRS2690
PPTN
MRS4625

FIG. 1—CONT'D

The extracellular loops (ELs) that have their own effects on GPCR ligand binding and activation (Cao et al., 2018; Marullo et al., 2020), and data for some EL residues are included in this review. EL2, in particular is typically in direct contact with the orthosteric ligands of Class A GPCRs.

GPCR dimerization is also now recognized as being important biologically (García-Recio et al., 2020; Pin, Kniazeff, Prezeau, Liu, & Rondard, 2019). Class C GPCRs form obligatory functional dimers, but there is increasing evidence for functional dimerization and its pharmacological consequences within Class A receptors, as well. It appears that the monomers and dimers may exist in an equilibrium. Thus, we discuss information about the TM regions that are engaged in putative contact during dimerization with other GPCR protomers.

2 AR and P2YR ligands and structures

In this review, we focus on residues of purinergic GPCRs, i.e., ARs and PY receptors (P2YRs), that are involved in ligand and protein recognition. These represent 12 Class A GPCRs (4 ARs and 8 P2YRs), which couple to various G proteins: $G_i/_o$, G_s or G_q (Table 1). Representative members of these receptor families have been determined structurally by X-ray or cryo-EM methods (Table 2): specifically, A_1AR, A_{2A}AR, P2Y_1R and P2Y_{12}R (all human). In particular, the A_{2A}AR has served as a model system for structural probing in many studies. An abundance of ligands has been discovered for these receptors, which have important drug discovery potential (Jacobson et al., 2021). These receptors are important drug discovery targets for immune, inflammatory, cardiovascular, metabolic and nervous system disorders. Some of these orthosteric ligands have been tested in clinical trials or are being developed for mainly chronic, but also some acute conditions, such as stroke (Pedata et al., 2014). Thus, it is valuable for future drug discovery efforts to analyze the commonality between ARs and within the two subfamilies of the P2YRs (Attah et al., 2020; Ciancetta & Jacobson, 2018; Gutierrez-de-Teran, Sallander, & Sotelo, 2017; Jespers et al., 2017; Jung et al., 2020).

Numerous selective agonist and antagonist ligands (Jacobson et al., 2021) have been reported for the ARs (Fig. 1A) and P2YRs (Fig. 1B), and receptor complexes have been determined structurally for some of these agonists/antagonists, including both subtype selective and nonselective ligands (Table 2). In the case of ARs, three-dimensional structures have been solved for two receptor subtypes with either agonists or antagonists: A_1AR and A_{2A}AR. Ligand binding in the orthosteric binding site has many parallels between the A_1 and A_{2A}ARs structures and with the theoretical models of A_{2B} and A_3ARs (Kose et al., 2018; Tosh et al., 2020; Tosh et al., 2020). There are also many ligand-interacting amino acid residues in common between nucleoside agonists and diverse, non-nucleoside antagonists of the ARs. Although most AR agonists are adenosine derivatives, pyridine-3,5-dicarbonitrile agonists and partial agonists have been extensively explored (Dal Ben et al., 2019). Structural information on P2YRs is also available for two receptor subtypes, P2Y_1R and P2Y_{12}R,

Table 1 G protein coupling of ARs and P2YRs and their ligands used as pharmacological probes (Jacobson et al., 2020; Jacobson, IJzerman, & Müller, 2021).

Receptor family	Subtype (agonist)		G protein	Synthetic agonists; antagonists (K_i in nM at human receptor, unless noted)
ARs	A_1		G_i, G_o	Cl-ENBA (0.51), CPA (2.3), MRS7469 (2.14), SPA (7.92); DPCPX (3.9), PSB-36 (0.7)
	A_{2A}		G_s, G_{olf}	CGS21680 (27), UK432097 (4), PSB-0777 (360); SCH442416 (4.1)
	A_{2B}		G_s, G_q	Bay60-6583 (114); MRS1754 (1.97), PSB-603 (0.55)
	A_3		G_i	IB-MECA (1.8), Cl-IB-MECA (1.4), MRS5980 (0.70), MRS5698 (3.5); MRS1523 (19), MRS1191 (31)
P2YRs	$P2Y_1$-like	$P2Y_1$ (ADP)	G_q	MRS2365 (0.40); 2-MeS-ADP (2510)[a]; MRS2179 (331), MRS2500 (0.79), BPTU (6.0)
		$P2Y_2$ (ATP, UTP)	G_q, G_i	2-thio-UTP (35), PBS-1114 (135); AR-C118925XX (58)
		$P2Y_4$ (UTP)	G_q, G_i	MRS4062 (23); PSB-16133 (234), PSB-1699 (407)
		$P2Y_6$ (UDP)	G_q	MRS4383 (493)[b], PSB-0474 (71), 5-OMe-UDP (79); MRS2578 (372)[b], TIM-38 (4300)[b]
		$P2Y_{11}$ (ATP)	G_q, G_s	NF546 (537); NF157 (45)[c]
	$P2Y_{12}$-like	$P2Y_{12}$ (ADP)	G_i	2-MeS-ADP (5)[a]; cangrelor (0.40), ticagrelor (13), AZD1283 (31.6), PSB-0739 (0.16)
		$P2Y_{13}$ (ADP)	G_i	2-MeS-ADP (19)[a]; MRS2211 (1620)
		$P2Y_{14}$ (UDPG, UDP)	G_i	MRS2905 (0.92), MRS2690 (49); PPTN (0.44), MRS4625 (27.6)[b]

Adenosine is the principal native agonist for the AR subtypes, while the more diverse endogenous P2YR ligands are indicated above. P2YR affinities are found in Jacobson et al. (2020). Structures of AR ligands are shown in Fig. 1.
[a]Activates $P2Y_1$, $P2Y_{12}$ and $P2Y_{13}$ receptors.
[b]IC_{50} value.
[c]Antagonist of $P2Y_1$ and $P2Y_{11}$ receptors.

with only antagonists in the former example, but agonists and antagonists are represented in the latter. The parallelism that can be observed in the case of ARs, is missing in the case of P2YRs, where more variability has been observed among three-dimensional structures.

Table 2 Reported AR and P2YR experimental structures (in the Protein Data Bank (PDB)) and their characteristics.

Receptor	Activation state	Ligand	PDB ID (References)
ARs			
A_1AR	Inactive	PSB36	5N2S (Cheng et al., 2017)
		DU172 (covalent)	5UEN (Glukhova et al., 2017)
	Active (G_i)	Adenosine[a]	6D9H (Draper-Joyce et al., 2018)
$A_{2A}AR$	Inactive	ZM241384	3EML (Jaakola et al., 2008), 3PWH (Doré et al., 2011), 3VG9 & 3VGA (Hino et al., 2012), 5UVI (Martin-Garcia et al., 2017), 5JTB (Melnikov et al., 2017), 6AQF (Eddy et al., 2018), 6MH8 (Martin-Garcia et al., 2019), 6S0L & 6S0Q (Nass et al., 2020)
		ZM241384, Na^+ allosteric mod	4EIY (Liu et al., 2012), 5IU4 (Segala et al., 2016), 5K2A & 5K2B & 5K2C & 5K2D (Batyuk et al., 2016), 5NLX & 5NM2 & 5NM4 (Weinert et al., 2017), 5VRA (Broecker et al., 2018), 5OLG (Rucktooa et al., 2018), 6JZH (Shimazu et al., 2019), 6LPJ & 6LPK & 6LPL (Ihara et al., 2020), 6PS7 (Ishchenko et al., 2019), 6WQA (Lee et al., 2020)
		12b, Na^+ allosteric mod	5IUA (Segala et al., 2016)
		12x, Na^+ allosteric mod	5IUB (Segala et al., 2016)
		12c, Na^+ allosteric mod	5IU7 (Segala et al., 2016)
		12f, Na^+ allosteric mod	5IU8 (Segala et al., 2016)
		XAC	3REY (Doré et al., 2011)
		Caffeine	3RFM (Doré et al., 2011)
		Caffeine, Na^+ allosteric mod	5MZP (Cheng et al., 2017)
		Theophylline, Na^+ allosteric mod	5MZJ (Cheng et al., 2017)
		PSB36, Na^+ allosteric mod	5N2R (Cheng et al., 2017)
		T4G	3UZA (Congreve et al., 2012)
		T4E	3UZC (Congreve et al., 2012)
		T4E, Na^+ allosteric mod	5OLZ & 5OM1 & 5OM4 (Rucktooa et al., 2018)

Continued

Table 2 Reported AR and P2YR experimental structures (in the Protein Data Bank (PDB)) and their characteristics.—cont'd

Receptor	Activation state	Ligand	PDB ID (References)
		AZD4635, Na^+ allosteric mod	6GT3 (Borodovsky et al., 2020)
		Tozadenant, Na^+ allosteric mod	5OLO (Rucktooa et al., 2018)
		Cmpd-1	5UIG (Sun et al., 2017)
		LUAA47070, Na^+ allosteric mod	5OLV (Rucktooa et al., 2018)
		Vipadenant, Na^+ allosteric mod	5OLH (Rucktooa et al., 2018)
		Chromone 4d, Na^+ allosteric mod	6ZDR (Jespers et al., 2020)
		Chromone 5d, Na^+ allosteric mod	6ZDR (Jespers et al., 2020)
	Intermediate	UK432097	3QAK (Xu et al., 2011), 5WF5 & 5WF6 (White et al., 2018)
		Adenosine	2YDO (Lebon et al., 2011)
		NECA	2YDV (Lebon et al., 2011)
		CGS21680	4UG2 & 4UHR (Lebon, Edwards, Leslie, & Tate, 2015)
	Active (bound to mini-G_s)	NECA	5G53 (Carpenter, Nehmé, Warne, Leslie, & Tate, 2016)
		NECA[a]	6GDG (García-Nafría, Lee, Bai, Carpenter, & Tate, 2018)
P2YRs			
$P2Y_1R$	Inactive	MRS2500	4XNV (Zhang et al., 2015)
		BPTU	4XNW (Zhang et al., 2015)
$P2Y_{12}R$	Inactive	AZD1283	4NTJ (Zhang et al., 2014)
	Intermediate	2-MeSADP	4PXZ (Zhang et al., 2014)
		2-MeSATP	4PY0 (Zhang, Zhang, Gao, Zhang, et al., 2014)

The main technique used to solve the structures was by X-ray crystallography, except where noted.
[a]Solved by cryo-EM.

We have categorized GPCR-ligand interactions, i.e., identified important residues needed to coordinate the ligands—either from X-ray crystallographic or cryo-EM structures—for $G_{i/o}$-coupled A_1 and G_s-coupled A_{2A} receptors (Glukhova et al., 2017; Jaakola et al., 2008; Lebon et al., 2011; Xu et al., 2011), G_q-coupled $P2Y_1$ and G_i-coupled $P2Y_{12}$ receptors (Zhang et al., 2015; Zhang, Zhang, Gao,

Paoletta, et al., 2014; Zhang, Zhang, Gao, Zhang, et al., 2014). Often, there is site-directed mutagenesis (SDM) data gathered in multiple studies to support these structures. We have investigated the tridimensional structures deposited in the Protein Data Bank (PDB) (rcsb.org; Berman et al., 2000), analyzed the contacts between ligands and receptor residues and collected the results in bar plots (Fig. 2) enabling the highlighting of most interacting residues. An automatic procedure was employed using a Schrödinger (Schrödinger Release 2020–4: Maestro, Schrödinger, LLC, New York, NY, 2020) command line script to detect ligand-receptor interactions on the PDB complexes, previously prepared with the Protein Preparation Wizard of the Schrödinger suite. Contacts defined as "good contacts" were taken into account, described by the distance of two atoms divided by the sum of their van der Waals radii within the range of 1.30 and 0.89. Furthermore, H-bonds, water mediated H-bonds, π–π interactions (face-to-face or edge-to-face), salt bridges and halogen bonds were examined, and these interaction patterns are discussed in the text. Chain A or R of the PDB files and the alternate state with higher occupancy were employed. TM residues are highlighted using the Ballesteros-Weinstein notation (Ballesteros & Weinstein, 1995) in addition to their number in the protein sequence (x·yz, where x is the helix number and yz is the amino acid residue number relative to the most conserved residue in that helix set as 50).

3 Analysis of small molecule recognition by TM residues of adenosine receptors

3.1 AR structures and models

X-ray and cryo-EM structures have been reported for two AR subtypes, specifically A_1AR and $A_{2A}AR$, for a total of 3 structures in the first case and 57 in the second (Table 2). The structures of the two subtypes show high similarity, with major differences confined to the loops. In particular, EL2 has a completely different conformation in the two receptors, and even the same receptor ($A_{2A}AR$) with different bound ligands can show a large reorganization of EL2 (Xu et al., 2011). Human ARs have high sequence identity, ranging from a maximum of 46% for the A_1AR-A_3AR pair to a minimum of 30% for the $A_{2A}AR$-A_3AR pair (sequence identity data retrieved from GPCRdb; https://gpcrdb.org/similaritymatrix/render; Kooistra et al., 2021). The similarities among the TM structures of the two subtypes, together with their high sequence identity among ARs, makes structural information for A_1AR and $A_{2A}AR$ useful also to build homology models for $A_{2B}AR$ and A_3AR.

Experimental structures and models have been used extensively in medicinal chemistry to design new ligands of the same chemical family and to discover novel chemotypes for a given receptor by virtual screening. For example, the discovery of a sub-micromolar $A_{2A}AR$ antagonist by virtually screening around 40,000 lead-like compounds in the dark chemical matter was accomplished by Carlsson et al. (Ballante et al., 2020). The use of purinergic structure for rational drug design has been recently reviewed (Salmaso & Jacobson, 2020b).

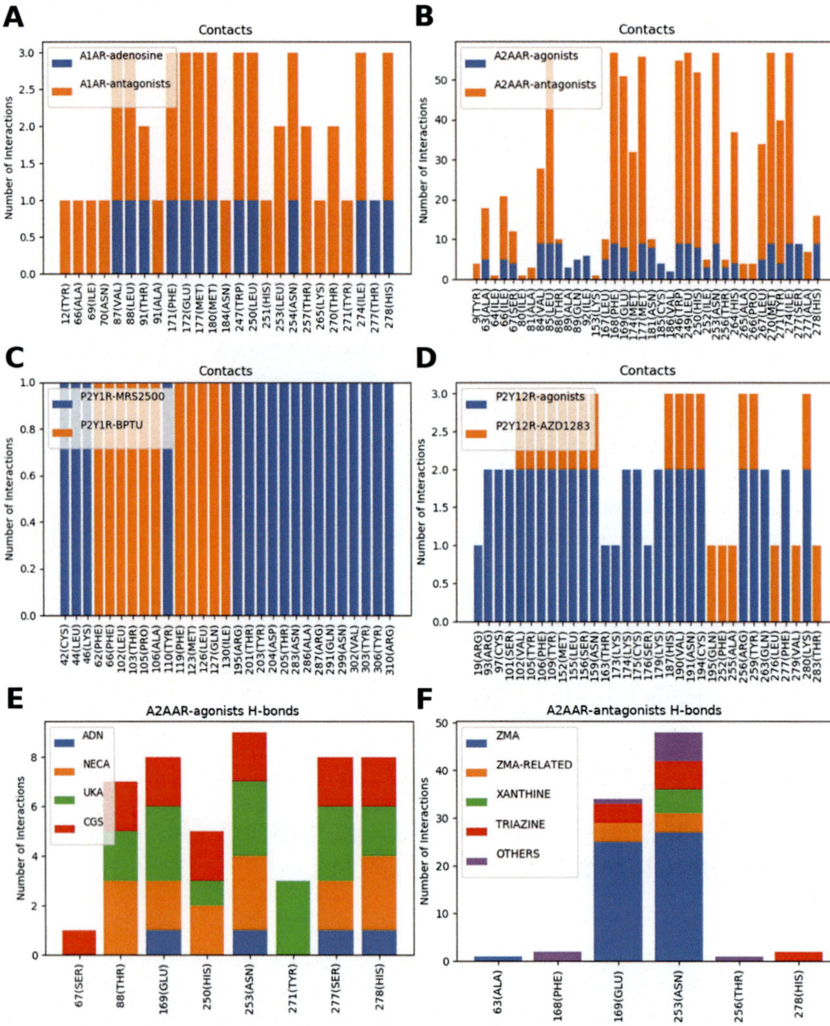

FIG. 2

The purinergic GPCR structures deposited in the Protein Data Bank (PDB) so far were surveyed to compare the interaction patterns of different ligands bound to ARs and P2YRs. Maestro by Schrödinger (Schrödinger Release 2020–4: Maestro, Schrödinger, LLC, New York, NY, 2020) was employed to prepare (Protein Preparation Wizard) and analyze the PDB structures, and the data were then plotted with matplotlib using python scripts. Contacts defined as "good contacts" in Maestro were taken into account (distance of two atoms divided by the sum of their van der Waals radii within the range of 1.30 and 0.89). Panels (A–D) show residues in contact with the ligand in the complexes deposited in the PDB. The height of the bar plots represents the number of structures where the contact is observed, with a possible maximum height, corresponding to the number of deposited structures, equal to 3, 57, 2, 3 in the case of A_1AR, $A_{2A}AR$, $P2Y_1R$ and $P2Y_{12}R$, respectively (panels A–D). Different colors highlight structures with agonists and antagonists in the case of A_1AR, $A_{2A}AR$ and $P2Y_{12}R$ (panels A, B and D), while discriminate the orthosteric antagonist MRS2500 from the allosteric BPTU in the case of $P2Y_1R$ (panel C). A focus on the H-bonds between $A_{2A}AR$ residues and agonists (panel E) or antagonists (panel F) is further provided. Ligand abbreviations: *ADN*, adenosine; *UKA*, UK432097; *CGS*, CGS21680; *ZMA*, ZM241385.

3.1.1 A_1AR structures

Three structures have been reported for A_1AR, one in the agonist-bound active state (PDB ID: 6D9H, bound to adenosine) and two in the antagonist-bound inactive state (PDB IDs: 5N2S and 5UEN, respectively bound to PSB36 and the covalent antagonist DU172), where both antagonists have a xanthine structure (Fig. 3A and C). A total of 25 residues have been found in contact with a ligand in at least one of the three tridimensional structures, and among them 12 residues are common to all the three ligands (Fig. 2A): $V87^{3.32}$, $L88^{3.33}$, $T91^{3.36}$ (mutated into A91 in structure 5N2S), $F171^{EL2}$, $E172^{EL2}$, $M177^{5.35}$, $M180^{5.38}$, $W247^{6.48}$, $L250^{6.51}$, $N254^{6.55}$, $I274^{7.39}$ and $H278^{7.43}$. A contact between adenosine and $T277^{7.42}$ distinguishes this agonist from the two antagonists, which instead make additional contacts with $L253^{6.54}$, $T257^{6.58}$, $T270^{7.35}$ (and with $I69^{2.64}$, $N184^{5.42}$, $H251^{6.52}$ and $K265^{EL3}$ in the case of PSB36 in structure 5N2S, and with $Y12^{1.35}$, $A66^{2.61}$, $N70^{2.65}$ and $Y271^{7.36}$ in the case of DU172 in structure 5UEN).

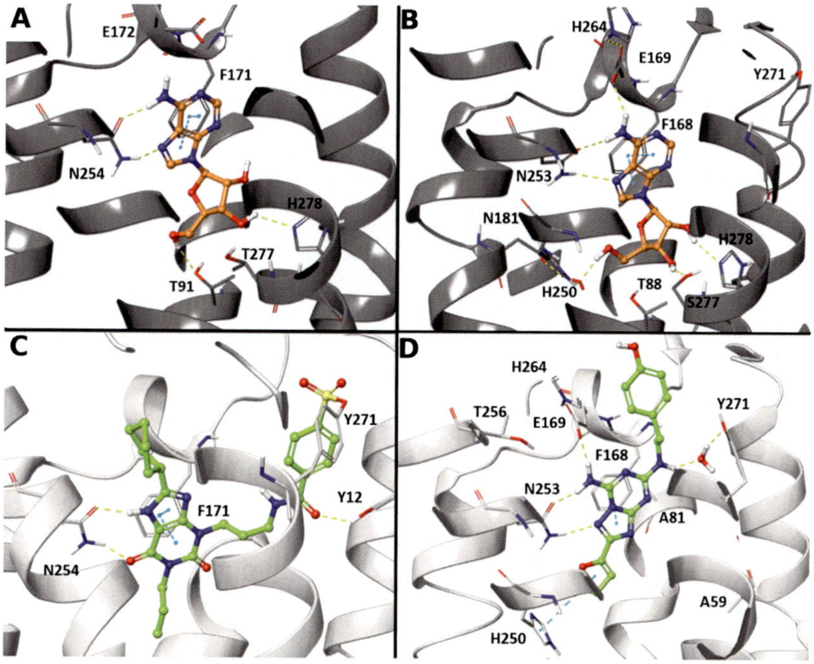

FIG. 3

(A) and (C) Experimental binding mode of the agonist adenosine (orange sticks) and of the covalent antagonist DU172 (green) to A_1AR (dark and light gray when bound to agonist and antagonist, in structures 6D9H and 5UEN, respectively). (B) and (D) Experimental binding mode of the agonist adenosine (orange sticks) and of the antagonist ZM241385 (green) to $A_{2A}AR$ (dark and light gray when bound to agonist and antagonist, in structures 2YDO and 4EIY, respectively). Relevant receptor residues discussed in the text are shown by sticks.

AR-conserved N254$^{6.55}$ is engaged in a bidentate H-bond with the exocyclic amine at position 6, N^{6}H, and with the position 7 of the adenine scaffold in the case of adenosine (agonist), and with the carbonyl at position 6 and the NH at position 7 of the xanthine scaffold of the antagonists. The aromatic base scaffolds are furthermore involved in a π–π stacking interaction with F171^{EL2}. Adenosine is involved in additional H-bonds with T91$^{3.36}$ through the 5′-hydroxy group and H278$^{7.43}$ through the 3′-hydroxy group, and is in contact with T277$^{7.42}$, whose equivalent position in A_{2A}AR, i.e., S277$^{7.42}$, interacts through H-bonds with agonists. The antagonists are engaged in fewer H-bonds compared to agonists, with only the additional involvement of Y12$^{1.35}$ in the case of the covalently bound xanthine DU172, but the cyclohexyl (DU172) and 3-tricyclo[3.3.1.0^{3,7}]nonanyl (of xanthine derivative PSB36) substitution at position 8 face the extracellular portion of the receptor making hydrophobic contacts with F171^{EL2}, E172^{EL2}, M177$^{5.35}$, L250$^{6.51}$, L253$^{6.54}$, T257$^{6.58}$ and T270$^{7.35}$ (and K265^{EL3} in structure 5N2S). Moreover, the long propyl-carbamoyl-benzene-1-sulfonyl substituent at position 3 of DU172, covalently bound to Y271$^{7.36}$, makes further interactions with TM1 (residue Y12$^{1.35}$) and TM2 (residues A66$^{2.61}$ and N70$^{2.65}$). However, the shorter 3-hydroxypropyl group of PSB36 engages TM2 only through I69$^{2.64}$. Differently, PSB36 presents a butyl substituent at position 1 which is bulkier than the propyl substituent of DU172, and probes the orthosteric pocket more deeply, entering in contact with N184$^{5.42}$ and H251$^{6.52}$.

3.1.2 A_{2A}AR structures

In the case of A_{2A}AR, 57 experimental structures have been deposited in the PDB so far. An ensemble of 38 residues appears in contact with a ligand in at least one PDB structure, and among them, 10 residues are common to all (i.e., F168^{EL2}, L249$^{6.51}$, N253$^{6.55}$, M270$^{7.35}$, I274$^{7.39}$) or nearly all (~90% or more) (i.e., L85$^{3.33}$, E169^{EL2}, M177$^{5.38}$, W246$^{6.48}$, H250$^{6.52}$) structures. Thus, these 10 residues define the orthosteric binding pocket of A_{2A}AR regardless of the agonist or antagonist character of the ligand (Fig. 2B). Moreover, seven residues are only in proximity to antagonists (i.e., Y9$^{1.35}$, I64$^{2.62}$, I80$^{3.28}$, A81$^{3.29}$, K153^{EL2}, A265^{EL3}, P266$^{7.31}$). However, they interact in a small fraction of all the PDB structures (less than 10%), so they cannot be considered a fingerprint for this category of compounds. Four residues are only in proximity to agonists (i.e., Q89$^{3.37}$, which is mutated into A89 in some structures, I92$^{3.40}$, C185$^{5.46}$, and V186$^{5.47}$). These residues define a pocket between TM3 and TM5 that is the deepest region of the orthosteric site, which typically hosts the ethyl-amido group at position 5′ of most of the agonists, but it is not explored by any of the antagonists.

In 9 PDB structures, the A_{2A}AR is bound to an agonist, in particular to adenosine (PDB ID: 2YDO), NECA (PDB IDs: 2YDV, 5G53 and 6GDG), CGS21680 (PDB IDs: 4UG2, 4UHR) and UK432097 (PDB IDs: 3QAK, 5WF5, 5WF6). Among these structures, 5G53 and 6GDG are in the active state, with the receptor bound to an engineered G protein, mini-Gs (truncated $G\alpha_s$) alone or in a heterotrimeric state, respectively. The agonists share a similar binding mode (Fig. 3B), where the ligands are surrounded by V84$^{3.32}$, L85$^{3.33}$, T88$^{3.36}$, F168^{EL2}, M177$^{5.38}$, W246$^{6.48}$, L249$^{6.51}$,

N253$^{6.55}$, M270$^{7.35}$, I274$^{7.39}$, S277$^{7.42}$, H278$^{7.43}$ in all the structures, and by E169^{EL2}, Q89$^{3.37}$ (mutated into A89 in some structures), N181$^{5.42}$, H250$^{6.52}$ in nearly all (~90%) of the structures (Fig. 2B). Particularly, agonists are involved in a π–π stacking interaction between F168^{EL2} and the adenine base, in a bidentate H-bond between N253$^{6.55}$ and the endocyclic N7 position of the adenine base and the exocyclic N^{6}H, which interacts also with E169^{EL2} (missing only in structure 6GDG), in a H-bond between S277$^{7.42}$ and the 3′-hydroxyl group and between H278$^{7.43}$ and the 2′-hydroxyl group (Fig. 2E).

Few differences can be observed among the agonists: T88$^{3.36}$ makes a H-bond with all the ligands' 5′-amido substituent, while adenosine's 5′-hydroxyl group is involved in a water mediated H-bond with N181$^{5.42}$ and H250$^{6.52}$ (PDB ID: 2YDO), with H250$^{6.52}$ interacting instead with the carbonyl of the 5′-amido substituent of NECA, CGS21680 and UK432097 in some of the reported structures. E169^{EL2}, involved in a salt bridge with H264^{EL3}, interacts through a H-bond with the unsubstituted N^{6}H moiety of adenosine, NECA (not in 6GDG) and CGS21680. However, in the case of UK432097, the E169^{EL2}–H264^{EL3} salt bridge is broken, and E169^{EL2} makes a bidentate H-bond with the ureidic group of the substituent at position 2, interacting also with Y271$^{7.36}$. The ethylamide substituent at position 5′ distinguishes the other agonists from adenosine, and enters the binding pocket more deeply, making contacts with Q89$^{3.37}$ (mutated into A89 in some structures), I92$^{3.40}$, C185$^{5.46}$, and V186$^{5.47}$ in different structures. As compared to adenosine and NECA, agonists CGS21680 and UK432097 bear a long substituent at position 2, which fills the orthosteric binding pocket entering in contact with residues of TM2 (A63$^{2.61}$, I66$^{2.64}$ and S67$^{2.65}$), EL2 (L167^{EL2}), EL3 (H264^{EL3}) and TM7 (L267$^{7.32}$ and Y271$^{7.36}$) in some of the PDB structures. In addition, UK432097's bulky *N*^{6}-2, 2-diphenylethyl substituent also engages the tip of TM6 (I252$^{6.54}$ and T256$^{6.58}$). Several conserved water molecules are noted to surround the AR agonists. For example, the stabilizing contribution of the conserved water bridging the adenine nitrogen at position 3 and the 2′-hydroxyl has been explored using a combination of free energy perturbation (FEP) and chemical synthesis (Matricon et al., 2021). The penalty for driving this water out of the binding site is a dramatic reduction of affinity at all AR subtypes.

Forty-eight structures of antagonist-bound A_{2A}ARs are present in the PDB, with [1,2,4]triazolo[2,3-*a*][1,3,5]triazine derivative ZM241385 in 27 cases (PDB IDs: 3EML, 3PWH, 3VG9, 3VGA, 4EIY, 5IU4, 5JTB, 5K2A, 5K2B, 5K2C, 5K2D, 5NLX, 5NM2, 5NM4, 5OLG, 5UVI, 5VRA, 6AQF, 6JZH, 6LPJ, 6LPK, 6LPL, 6MH8, 6PS7, 6S0L, 6S0Q, 6WQA, a ZM241385-related compound in 4 cases (PDB IDs: 5IU7, 5IU8, 5IUA, 5IUB), a xanthine derivative in 5 cases (PDB IDs: 3REY, 3RFM, 5MZJ, 5MZP, 5N2R), a 1,2,4-triazin-3-amine derivative, with an aryl substituent at position 5 and 6, in 6 cases (PDB IDs: 3UZA, 3UZC, 5OLZ, 5OM1, 5OM4, 6GT3), vipadenant (PDB ID: 5OLH), tozadenant (PDB ID: 5OLO), LUAA47070 (PDB ID: 5OLV), a triazole-carboximidamide, Compound-1 (PDB ID: 5UIG), chromone 14 (4d) (PDB ID: 6ZDR), chromone 5d (PDB ID: 6ZDV) in one case each. Nearly all the antagonists are in contact (in at least 90% of the

PDB structures) with the pattern of 10 residues cited at the beginning of the paragraph, comprising: L85$^{3.33}$, F168^{EL2}, E169^{EL2}, M177$^{5.38}$, W246$^{6.48}$, L249$^{6.51}$, H250$^{6.52}$, N253$^{6.55}$, M270$^{7.35}$ and I274$^{7.39}$ (Fig. 2B). N253$^{6.55}$ is involved in a direct H-bond with the ligand in all the structures, and a π–π stacking interaction with F168^{EL2} is often observed as well (Fig. 3D). A π–π edge-to-face interaction with H250$^{6.52}$ is made by several ligands (ZM241385 and ZM241385-related compounds, 1,2,4-triazin-3-amine derivatives, vipadenant and LUAA47070), and a H-bond with E169^{EL2} is encountered in most complexes with ZM241385, ZM241385-related compounds and 1,2,4-triazin-3-amine derivatives, and in the case of vipadenant. However, some ligand-specific interactions are observed for these antagonists.

In the case of ZM241385 and ZM241385-related compounds, the [1,2,4]triazolo [1,5-*a*][1,3,5]triazine aromatic scaffold participates in the π–π face-to-face interaction with F168^{EL2}, and N1 (or sometimes the furan oxygen atom) and the 7-amino group interacts with N253$^{6.55}$ through bidentate H-bonding (even if with a deviation from ideal geometry in the case of structures 3VG9 and 3VGA) (Fig. 2F). In addition, a π–π edge-to-face interaction is observed between the furan ring and H250$^{6.52}$, and a H-bond is formed between the 7-amino group and E169^{EL2} (missing only in structures 3PWH, 3VG9, 3VGA). A water molecule bridges a H-bond between Y271$^{7.36}$ and the 5-amino group in the 75% of the PDB structures, and often co-crystalized water molecules interact also with nitrogen atoms at positions 3 and 4 of the ligand. Moreover, in the case of ZM241385-related compounds bearing a piperidine or piperazine moiety in the 5-substituent, a salt bridge with E169^{EL2} can be observed.

X-ray structures for A_{2A}AR in complex with four different xanthine antagonists have been reported so far, in particular XAC (PDB ID: 3REY), caffeine (PDB IDs: 3RFM, 5MZP), theophylline (PDB ID: 5MZJ), PSB36 (PDB ID: 5N2R). Caffeine and theophylline are naturally-occurring prototypical AR antagonists, and antagonist XAC was introduced in 1985 as a high affinity, functionalized congener for coupling to carrier or reporter moieties (Jacobson, 2009). The extended position 8 ethylamino chain of XAC was later confirmed in an A_{2A}AR structure to be reaching toward the extracellular region, thus explaining its ability to be acylated without preventing receptor binding. Xanthine bases are involved in the π–π face-to-face interaction with F168^{EL2}, while the H-bond interaction with N253$^{6.55}$ is fulfilled by the carbonyl at position 6 (maintained in both the alternate states of caffeine observed in structure 5MZP), accompanied by the free NH at position 7 in theophylline and PSB36. In order to maintain a H-bond between N253$^{6.55}$ and the carbonyl of XAC in structure 3REY, N253$^{6.55}$ is flipped as compared to its common conformation. Moreover, a water molecule mediates an interaction between carbonyl at position 2 of theophylline, caffeine and PSB36 with A81$^{3.29}$ in structures 5MZJ, 5MZP and 5N2R, respectively.

In the case of the 1,2,4-triazin-3-amine A_{2A}AR antagonists such as T4E (PDB ID: 3UZC, 5OLZ, 5OM1, 5OM4), T4G (PDB ID: 3UZA) and AZD4635 (PDB ID: 6GT3), the triazine moiety participates in the π–π interaction with F168^{EL2}. This interaction is face-to-face in most cases, apart from structure 3UZA where it is closer to an edge-to-face interaction, and in structure 3UZC where the conformation of

F168^{EL2} does not permit an ideal π–π interaction. The exocyclic 3-amino group and the endocyclic N4 are involved in the typical bidentate H-bond with N253$^{6.55}$ (conserved across all ARs), and the exocyclic 3-amino group also interacts with E169^{EL2} in all structures except 3UZA and 3UZC, where the salt bridge between E169^{EL2} and H264^{EL3} is not formed (Fig. 2F). The 5-phenyl group makes a π–π edge-to-face interaction with H250$^{6.52}$, while the 6-aryl group, consisting in a phenol in the case of compound T4E (ODB IDs: 3UZC, 5OLZ, 5OM1, 5OM4), interacts though its p-hydroxyl group with H278$^{7.43}$, and with A59$^{2.57}$ through a water mediated H-bond (not present in structure 3UZC).

Triazolo[4,5-d]pyrimidin-5-amine derivative vipadenant (PDB ID: 5OLH) interacts through its [1,2,3]triazolo[4,5-d]pyrimidine structure with F168^{EL2} and through its 5-amine and N^{6}H with N253$^{2.55}$, making the typical π–π stacking interaction and bidentate H-bond. The furan ring attached to the position 7 makes a π–π edge-to-face interaction with H250$^{6.52}$ and an H-bond with A81$^{3.29}$.

In the case of tozadenant (PDB ID: 5OLO), the typical π–π interaction with F168^{EL2} is granted by the benzothiazole structure, which contributes through its nitrogen atom to the bidentate hydrogen with N253$^{6.55}$, together with the exocyclic nitrogen attached to position 2 of the scaffold. T256$^{6.58}$ is engaged in an additional H-bond with the hydroxyl group at the position 4 of the piperidine moiety.

Thiazole derivative LUAA47070 (PDB ID: 5OLV) lacks the π–π stacking with F168^{EL2}, even if its benzamide structure assumes a position not far from enabling that interaction. The amide nitrogen makes the H-bond with N253$^{6.55}$, accompanied by the nitrogen of the attached thiazole group, which is also engaged in the π–π edge-to-face interaction with H250$^{6.52}$.

In the case of structure 5UIG (Sun et al., 2017), the interactions of antagonist Compound-1 with F168^{EL2} and N253$^{6.55}$ are present with the 5-amino-1,2,3-triazole scaffold; while in the case of chromones 4d and 5d in structures 6ZDR and 6ZDV, the π–π face-to-face interaction with F168^{EL2} is accomplished by the chromone group, while N253$^{6.55}$ makes a single H-bond with the nitrogen of the thiazole ring (Jespers et al., 2020).

3.1.3 Examples of AR models

The overall sequence identity between A_3AR and A_1AR is 46%, and it goes up to 56% focusing on TM regions; while the overall sequence identity between A_3AR and $A_{2A}AR$ is 30%, with 47% value for the TM regions (sequence identity data retrieved from GPCRdb; https://gpcrdb.org/similaritymatrix/render; Kooistra et al., 2021). The high sequence similarity, especially of the TM region, among ARs makes A_1AR and $A_{2A}AR$ good templates for homology modeling of the other AR subtypes. For instance, a hybrid model of A_3AR has been recently built using an $A_{2A}AR$ structure as template for the TM helices and an A_1AR structure for TM2 (Tosh, Salmaso, Rao, Bitant, et al., 2020; Tosh, Salmaso, Rao, Campbell, et al., 2020). The upper part of TM2 is observed to move outward from the TM bundle in the case of antagonist-bound A_1AR, and this is in agreement with an established hypothesis for A_3AR, where the outwardly displaced TM2 would allow binding of agonists with bulky substituents at position 2. A similar model was previously built using the TM2 of opsin, outward displaced as well (Tosh et al., 2012). An A_3AR homology model enabled understanding of agonism by NECA-like amides vs 5′-CH_2OH derivatives (Dal Ben et al., 2014). These models enable docking of agonists in a conformation analogous to that experimentally assumed by agonists at A_1AR and $A_{2A}AR$ binding sites and described in the previous paragraphs. In particular, the H-bonds of hydroxyl groups 2′ and 3′, and of the hydroxyl or amidic moiety at 5′, the π–π interaction of the aromatic base, the bidentate H-bonds of the adenine 6-amino group and N7 might occur in the A_3AR with residues at equivalent positions to the other ARs, in particular with $H272^{7.43}$, $S271^{7.42}$, $T94^{3.36}$, $F168^{EL2}$ and $N250^{6.55}$. An $A_{2B}AR$ homology model based on a ZM241385-$A_{2A}AR$ complex (PDB ID: 3EML) was reported (Kose et al., 2018), which features a close similarity to other AR subtypes. The proposed binding mode of a fluorescent $A_{2B}AR$ antagonist derived from a 3-propylxanthine bore a close resemblance to the xanthine pose in the XAC-$A_{2A}AR$ complex (PDB ID: 3REY).

4 Analysis of small molecule recognition by TM residues of P2Y receptors

4.1 P2YR structures and models

There are two subfamilies of P2YRs, based on both sequence identity and on G protein coupling. The eight P2Y receptors have much more sequence diversity than the four AR subtypes, with sequence identity values below 20% between human $P2Y_1$-like and $P2Y_{12}$-like P2YRs. The X-ray structures of two P2Y receptors (Table 2) have been reported ($P2Y_1$ and $P2Y_{12}$), representative of each of the two subfamilies (Zhang et al., 2015; Zhang, Zhang, Gao, Paoletta, et al., 2014; Zhang, Zhang, Gao, Zhang, et al., 2014). These two P2YR subtypes are much less similar to each other than the intrafamily similarity of ARs, with a sequence identity of 19% (sequence identity data retrieved from GPCRdb; https://gpcrdb.org/similaritymatrix/render; Kooistra et al., 2021). Ligand binding in the orthosteric binding sites have identified common residues in the recognition of nucleotide agonists, which are completely different between $P2Y_1$ and $P2Y_{12}$Rs. However, several features are in common between the $P2Y_1$ and $P2Y_{12}$Rs. There is a predominance of positively charged Lys and Arg residues in the EL regions, which coordinate the negatively charged phosphate moieties of the nucleotide ligands. Also, an unusual disulfide bridge is formed between the N-terminus and EL3 that is common to all P2YRs. This disulfide bridge was first discovered by SDM and led to early modeling suggesting that the ELs are important in the path taken by nucleotide ligands (Moro, Hoffmann, & Jacobson, 1999). Modeling of other P2YR subtypes based on the X-ray structures has been reported (Attah et al., 2020; Jung et al., 2020; Toti et al., 2017).

4.1.1 $P2Y_1R$ structures

Two X-ray structures have been deposited in the PDB so far for $P2Y_1R$. The receptor is co-crystallized with an orthosteric antagonist, MRS2500 (PDB ID: 4XNW), in one structure, and to one allosteric antagonist, BPTU (PDB ID: 4XNV), in the other, and the two ligands bind to two different binding pockets, with no common residues (Fig. 4A and C). MRS2500 is a nucleotide derivative, but its ribose ring has been sterically constrained by a bicyclic ring system (methanocarba) in a receptor-preferred North conformation, unlike native ribose which is freely twisting between North and South conformations. 18 residues in $P2Y_1R$ appear in contact only with the nucleotide antagonist, while 11 residues are in proximity only to the allosteric antagonist, bound on the outer surface in contact with the phospholipid bilayer (Fig. 2C).

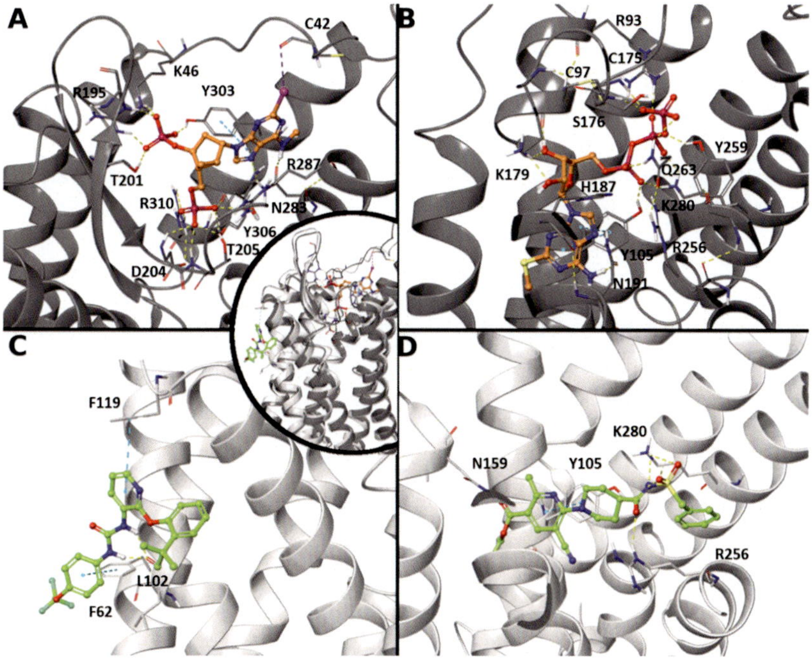

FIG. 4

(A) and (C) Experimental binding mode of the orthosteric antagonist MRS2500 (orange sticks) and of the negative allosteric modulator BPTU (green) to $P2Y_1R$ (dark and light gray when bound to MRS2500 and BPTU, in structure 4XNW and 4XNV, respectively). (B) and (D) Experimental binding mode of the agonist 2MeSATP (orange sticks) and of the antagonist AZD1283 (green) to $P2Y_{12}R$ (dark and light gray when bound to agonist and antagonist, in structures 4PY0 and 4NTJ, respectively). Relevant receptor residues discussed in the text are shown by sticks.

Bicyclic nucleotide derivative MRS2500 binds the receptor close to the extracellular region, at the interface among the N-terminus, extracellular loops and tip of TM6 and TM7. This antagonist is surrounded by $C42^{N\text{-ter}}$, $L44^{N\text{-ter}}$ and $K46^{N\text{-ter}}$ of the N-terminus, $Y110^{2.63}$, $R195^{EL2}$, $T201^{EL2}$, $Y203^{EL2}$, $D204^{EL2}$ and $T205^{EL2}$ of EL2, $N283^{6.58}$, $A286^{6.61}$, $R287^{6.62}$, $Q291^{EL3}$ of EL3, $N299^{7.28}$, $V302^{7.31}$, $Y303^{7.32}$, $Y306^{7.35}$, $R310^{7.39}$ (Fig. 2C). Among these residues, $N283^{6.58}$ interacts though a bidentate H-bond with the endocyclic N7 and exocyclic N^6H positions of the adenine scaffold, $D204^{EL2}$, $T205^{EL2}$, $Y306^{7.35}$ and $R310^{7.39}$ are involved in H-bonds with the 5′-phosphate group, which interacts also with $R287^{6.62}$ though a water mediated H-bond, $K46^{N\text{-ter}}$, $R195^{EL2}$, $T201^{EL2}$ and $Y303^{7.32}$ are involved in H-bonds with the 3′-phosphate group, with $K46^{N\text{-ter}}$ and $R195^{EL2}$ contributing also through salt bridges (Fig. 4A). In addition, $Y303^{7.32}$ makes an edge-to-face π–π interaction with the adenine aromatic scaffold, and the $C42^{N\text{-ter}}$ carbonyl is involved in a halogen bond (at a ~180 degree angle) with the iodine atom at position 2 of the base. Differently, BPTU binds at the interface between TM1, TM2 and TM3 on the membrane side, out of the TM bundle, in a binding site defined by $F62^{1.43}$, $F66^{1.47}$, $L102^{2.55}$, $T103^{2.56}$, $P105^{2.58}$, $A106^{2.59}$, $F119^{EL2}$, $M123^{3.24}$, $L126^{3.27}$, $Q127^{3.28}$, $I130^{3.31}$ (Fig. 2C), with the backbone carbonyl of $L102^{2.55}$ interacting through a H-bond with the ureidic potion of the ligand, and $F62^{1.43}$ and $F119^{EL2}$ through an edge-to-face π–π interaction with the phenyl and pyridine substituents of the urea group of BPTU (Fig. 4C).

4.1.2 $P2Y_{12}R$ structures

In the case of the $P2Y_{12}R$, two agonist-bound (2MeSATP and 2MeSADP, in PDB ID: 4PY0 and 4PXZ, respectively) and one non-nucleotide antagonist bound (AZD1283, PDB ID: 4NTJ) X-ray structures are reported. The three ligands bind the receptor in a similar region, sharing the contacts with residues $V102^{3.30}$, $Y105^{3.33}$, $F106^{3.34}$, $Y109^{3.37}$, $M152^{4.53}$, $L155^{4.56}$, $S156^{4.57}$, $N159^{4.60}$, $H187^{5.36}$, $V190^{5.39}$, $N191^{5.40}$, $C194^{5.43}$, $R256^{6.55}$, $Y259^{6.58}$, and $K280^{7.35}$. Thus, 15 residues in $P2Y_{12}R$ appear in proximity of both agonist and antagonist ligands, while 6 residues are in contact only with the antagonist, and 12 are in contact only with agonists (Fig. 2D). However, it is to be noted that the ELs in the $P2Y_{12}R$ structures are mostly undefined in the antagonist-bound structure but well-defined with nucleotide agonists bound.

$Y105^{3.33}$ is involved in a π–π stacking interaction in all the structures, interacting with the adenine base in the case of the agonists and with the pyridine in the case of the antagonists; $R256^{6.55}$ and $K280^{7.35}$ contribute through H-bonds with the 5′-phosphate in the case of the agonists and with the sulfonyl-carbamoyl substituent in the case of AZD1283. However, ligand binding of the non-nucleotide antagonist AZD1283 is unlike other purinergic receptors and has no similar structure among other GPCRs. The antagonist-bound structure consistently deviates from the agonist-bound $P2Y_{12}R$ structures: the tips of TM5, TM6 and TM7 bent toward the membrane environment, and the disulfide bond between $C175^{EL2}$ and $C97^{3.25}$, linking EL2 to TM3, is broken. This is unusual for a GPCR structure, as this same disulfide is considered a conserved structural feature in other Class A receptors.

AZD1283 in fact spans the TM bundle, going from a cleft defined by TM3, TM4 and TM5, where the ethyl ester engages N159$^{4.60}$ in a H-bond and occupies the same hydrophobic pocket that hosts the 2-methylsulfanyl substituent of 2MeSADP and 2MeSATP in the agonist-bound structures, to the interface between TM7 and TM6, where it is surrounded by T283$^{7.38}$, K280$^{7.35}$, V279$^{7.34}$, L276$^{7.31}$, Y259$^{6.58}$, R256$^{6.55}$, A255$^{6.54}$, F252$^{6.51}$, stabilizing the ligand's benzylic group through an edge-to-face π–π interaction (Fig. 4D). Differently, the TM bundle of the agonist-bound structures is more compact, and the two agonists share the same binding mode where the adenine base, involved in the π–π stacking interaction with Y105$^{3.33}$, directs the exocyclic amino group toward the bottom of the pocket, where it is engaged in a H-bond with N191$^{5.40}$, and the ribose ring toward the extracellular region, where it makes H-bonds with H187$^{5.36}$ through the 2′-hydroxyl group and K179^{EL2} and C97$^{3.25}$ through the 3′-hydroxyl group (Fig. 4B). The α-phosphate interacts through H-bonds with Y105$^{3.33}$, Q263$^{6.62}$, in addition to the aforementioned R256$^{6.55}$ and K280$^{7.35}$, which have also a salt bridge character. The β-phosphate interacts via H-bonds with Y259$^{6.58}$, K280$^{7.35}$ and R93$^{3.21}$ (also salt bridge in the case of the last two), while the γ-phosphate of 2MeSATP interacts with R93$^{3.21}$, C175^{EL2}, S176^{EL2}.

4.1.3 Example of P2YR models

As an example of modeling for P2YRs, we report P2Y$_{14}$R subtype, whose antagonists may have therapeutic interest in treating inflammatory diseases, asthma, acute kidney injury, etc. P2Y$_{14}$R share with P2Y$_{12}$R 44% sequence (44% overall sequence identity, 52% sequence identity of TM regions) (sequence identity data retrieved from GPCRdb; https://gpcrdb.org/similaritymatrix/render; Kooistra et al., 2021), so X-ray structures of the second can be used as templates for homology modeling. Specifically, a model was built using the 2-MeSADP-bound P2Y$_{12}$R structure as a template and then refined using MD (Salmaso & Jacobson, 2020a) simulations after docking of the known biaryl antagonist PPTN (Junker et al., 2016). This model has been used to drive the design and synthesis of PPTN analogs as P2Y$_{14}$R antagonists. The suggested binding mode of PPTN and related compounds involves a salt bridge between the ligand's carboxylate and K77$^{2.60}$ and K277$^{7.35}$, which behave also as H-bonds donor together with Y102$^{3.33}$ (Jung et al., 2020; Yu et al., 2018). Cation-π interactions stabilize the aromatic moieties of the ligands, involving R253$^{6.55}$ and R274$^{7.32}$. Although the pose of P2Y$_{14}$R antagonists does not resemble the experimental binding mode of P2Y$_{12}$R ligands, and even though PPTN is selective for P2Y$_{14}$R subtype, some of the residues involved in binding are equivalent to those enumerated for P2Y$_{12}$R in the previous paragraph, i.e., Y102$^{3.33}$, K277$^{7.35}$ and R253$^{6.55}$.

5 Receptor domains involved in dimerization

There is still some controversy about the stoichiometric nature of class A GPCRs (Felce et al., 2017). While class C GPCRs are known to dimerize, there is no consensus in the case of class A receptors. Evidence suggests that various ARs and

P2YRs can form putative dimers, either homo- or heterodimers, with other receptor protomers (Borroto-Escuela et al., 2020; Hill, May, Kellam, & Woolard, 2014; Hinz et al., 2018; Navarro et al., 2016; Neumann, Müller, & Namasivayam, 2020). The contact regions for this dimerization have been determined experimentally, in limited cases, and through modeling (Al-Shar'i & Al-Balas, 2019). In some cases, the association between protomers is facilitated by loop regions of the receptors, for example, intracellular loops of the A_{2A}AR (Fernandez-Duenas et al., 2012), but also a critical role of TM residues is proposed (Johnston, Wang, Provasi, & Filizola, 2012; Townsend-Nicholson, Altwaijry, Potterton, Morao, & Heifetz, 2019). A tool for exploring the location of residues involved in GPCR dimerization based on reported structures is available on the web (DIMERBOW, http://lmc.uab.es/dimerbow/; García-Recio et al., 2020).

An early modeling study of potential A_3AR homodimerization pointed to a likely TM4-TM5 interface (Kim & Jacobson, 2006). A_3AR homodimerization was later demonstrated pharmacologically (May, Bridge, Stoddart, Briddon, & Hill, 2011), and the TM4-TM5 interface was confirmed as an interface also for A_{2A}AR-D_2R heterodimerization, by means of peptide segments from the sequences of A_{2A}AR's TM4 and TM5 which interfered with heterodimerization (Borroto-Escuela et al., 2018). The same interface, which was experimentally observed in the X-ray structure of the oligomeric β_1-adrenergic receptor (Huang, Chen, Zhang, & Huang, 2013), was proposed for homodimerization of A_1AR and A_{2A}AR, while the same work suggested the TM5-TM6 interface for a A_1AR-A_{2A}AR heterotetramer (Navarro et al., 2016). Moreover, A_1AR crystallized in dimeric form in one of the X-ray structures (PDB ID: 5UEN), resulting in the burial of hydrophobic areas of TM4 and TM5 (Glukhova et al., 2017). The role of TM5 in A_{2A}AR dimerization was previously anticipated in a study exploring the self-assembly of TM5 in oligomeric structures, which also found that a mutation of M193 (5.54) alters the dimerization capability of the whole length A_{2A}AR (Thevenin, Lazarova, Roberts, & Robinson, 2005).

6 Summary

We compare the GPCR-ligand interactions and highlight important residues for recognition—from both X-ray crystallographic or cryo-EM structures for purinergic receptors. These include A_1 and A_{2A}ARs and P2Y_1 and P2Y_{12}Rs that respond to nucleotides. Molecular modeling, including MD simulation, has been successful in predicting ligand interactions across most of this family of 12 purinergic GPCRs (Salmaso & Jacobson, 2020a). This review catalogs the residues of the binding pocket, with specific interest in transmembrane helical (TM) regions, that engage in contacts with ligands or with other GPCR protomers in dimeric forms. In most cases, orthosteric ligands are represented, except for one allosteric P2Y_1 antagonist. There is extensive conservation of recognition motifs within the adenosine receptor, but the P2Y_1 and P2Y_{12} receptors are each structurally distinct in their ligand recognition.

Acknowledgment

We thank the NIDDK Intramural Research Program for support (ZIADK-031126).

References

Al-Shar'i, N. A., & Al-Balas, Q. A. (2019). Molecular dynamics simulations of adenosine receptors: Advances, applications and trends. *Current Pharmaceutical Design*, *25*, 783–816.

Attah, I. Y., Neumann, A., Al-Hroub, H., Rafehi, M., Baqi, Y., Namasivayam, V., et al. (2020). Ligand binding and activation of UTP-activated G protein-coupled P2Y2 and P2Y4 receptors elucidated by mutagenesis, pharmacological and computational studies. *Biochimica et Biophysica Acta—General Subjects*, *1864*(3), 129501. https://doi.org/10.1016/j.bbagen.2019.129501.

Ballante, F., Rudling, A., Zeifman, A., Luttens, A., Vo, D. D., Irwin, J. J., et al. (2020). Docking finds GPCR ligands in dark chemical matter. *Journal of Medicinal Chemistry*, *63*(2), 613–620. https://doi.org/10.1021/acs.jmedchem.9b01560.

Ballesteros, J. A., & Weinstein, H. (1995). Integrated methods for the construction of three dimensional models and computational probing of structure function relations in G protein-coupled receptors. *Methods in Neurosciences*, *25*, 366–428.

Batyuk, A., Galli, L., Ishchenko, A., Han, G. W., Gati, C., Popov, P. A., et al. (2016). Native phasing of x-ray free-electron laser data for a G protein-coupled receptor. *Science Advances*, *2*, e1600292. https://doi.org/10.1126/sciadv.1600292.

Berman, H. M., Westbrook, J., Feng, Z., Gilliland, G., Bhat, T. N., Weissig, H., et al. (2000). The Protein Data Bank. *Nucleic Acids Research*, *28*(1), 235–242. https://doi.org/10.1093/nar/28.1.235.

Borodovsky, A., Barbon, C. M., Wang, Y., Ye, M., Prickett, L., Chandra, D., et al. (2020). Small molecule AZD4635 inhibitor of A2AR signaling rescues immune cell function including CD103(+) dendritic cells enhancing anti-tumor immunity. *Journal for Immunotherapy of Cancer*, *8*(2). https://doi.org/10.1136/jitc-2019-000417.

Borroto-Escuela, D. O., Ferraro, L., Narvaez, M., Tanganelli, S., Beggiato, S., Liu, F., et al. (2020). Multiple adenosine-dopamine (A2A-D2 like) heteroreceptor complexes in the brain and their role in schizophrenia. *Cell*, *9*(5). https://doi.org/10.3390/cells9051077.

Borroto-Escuela, D. O., Rodriguez, D., Romero-Fernandez, W., Kapla, J., Jaiteh, M., Ranganathan, A., et al. (2018). Mapping the Interface of a GPCR dimer: A structural model of the A2A adenosine and D2 dopamine receptor heteromer. *Frontiers in Pharmacology*, *9*, 829. https://doi.org/10.3389/fphar.2018.00829.

Broecker, J., Morizumi, T., Ou, W.-L., Klingel, V., Kuo, A., Kissick, D. J., et al. (2018). High-throughput in situ X-ray screening of and data collection from protein crystals at room temperature and under cryogenic conditions. *Nature Protocols*, *13*, 260–292. https://doi.org/10.1038/nprot.2017.135.

Cao, R., Giorgetti, A., Bauer, A., Neumaier, B., Rossetti, G., & Carloni, P. (2018). Role of extracellular loops and membrane lipids for ligand recognition in the neuronal adenosine receptor type 2A: An enhanced sampling simulation study. *Molecules*, *23*(10). https://doi.org/10.3390/molecules23102616.

Carpenter, B., Nehmé, R., Warne, T., Leslie, A. G. W., & Tate, C. G. (2016). Structure of the adenosine A(2A) receptor bound to an engineered G protein. *Nature*, *536*, 104–107. https://doi.org/10.1038/nature18966.

Carpenter, B., & Tate, C. G. (2017). Active state structures of G protein-coupled receptors highlight the similarities and differences in the G protein and arrestin coupling interfaces. *Current Opinion in Structural Biology*, *45*, 124–132. https://doi.org/10.1016/j.sbi.2017.04.010.

Cheng, R. K. Y., Segala, E., Robertson, N., Deflorian, F., Doré, A. S., Errey, J. C., et al. (2017). Structures of human A1 and A2A adenosine receptors with xanthines reveal determinants of selectivity. *Structure*, *25*, 1275–1285.e4. https://doi.org/10.1016/j.str.2017.06.012.

Ciancetta, A., & Jacobson, K. A. (2018). Breakthrough in GPCR crystallography and its impact on computer-aided drug design. *Methods in Molecular Biology*, *1705*, 45–72. https://doi.org/10.1007/978-1-4939-7465-8_3.

Ciancetta, A., Rubio, P., Lieberman, D. I., & Jacobson, K. A. (2019). A3 adenosine receptor activation mechanisms: Molecular dynamics analysis of inactive, active, and fully active states. *Journal of Computer-Aided Molecular Design*, *33*(11), 983–996. https://doi.org/10.1007/s10822-019-00246-4.

Congreve, M., Andrews, S. P., Doré, A. S., Hollenstein, K., Hurrell, E., Langmead, C. J., et al. (2012). Discovery of 1,2,4-triazine derivatives as adenosine A(2A) antagonists using structure based drug design. *Journal of Medicinal Chemistry*, *55*, 1898–1903. https://doi.org/10.1021/jm201376w.

Congreve, M., de Graaf, C., Swain, N. A., & Tate, C. G. (2020). Impact of GPCR structures on drug discovery. *Cell*, *181*(1), 81–91. https://doi.org/10.1016/j.cell.2020.03.003.

Dal Ben, D., Buccioni, M., Lambertucci, C., Kachler, S., Falgner, N., Marucci, G., et al. (2014). Different efficacy of adenosine and NECA derivatives at the human A3 adenosine receptor: Insight into the receptor activation switch. *Biochemical Pharmacology*, *87*(2), 321–331.

Dal Ben, D., Lambertucci, C., Buccioni, M., Martí Navia, A., Marucci, G., Spinaci, A., et al. (2019). Non-nucleoside agonists of the adenosine receptors: An overview. *Pharmaceuticals (Basel, Switzerland)*, *12*(4), 150. https://doi.org/10.3390/ph12040150.

Doré, A. S., Robertson, N., Errey, J. C., Ng, I., Hollenstein, K., Tehan, B., et al. (2011). Structure of the adenosine A(2A) receptor in complex with ZM241385 and the xanthines XAC and caffeine. *Structure*, *19*, 1283–1293. https://doi.org/10.1016/j.str.2011.06.014.

Draper-Joyce, C. J., Khoshouei, M., Thal, D. M., Liang, Y.-L., Nguyen, A. T. N., Furness, S. G. B., et al. (2018). Structure of the adenosine-bound human adenosine A1 receptor-Gi complex. *Nature*, *558*, 559–563. https://doi.org/10.1038/s41586-018-0236-6.

Eddy, M. T., Lee, M.-Y., Gao, Z.-G., White, K. L., Didenko, T., Horst, R., et al. (2018). Allosteric coupling of drug binding and intracellular signaling in the A2A adenosine receptor. *Cell*, *172*, 68–80.e12. https://doi.org/10.1016/j.cell.2017.12.004.

Felce, J. H., Latty, S. L., Knox, R. G., Mattick, S. R., Lui, Y., Lee, S. F., et al. (2017). Receptor quaternary organization explains G protein-coupled receptor family structure. *Cell Reports*, *20*(11), 2654–2665. https://doi.org/10.1016/j.celrep.2017.08.072.

Fernandez-Duenas, V., Gomez-Soler, M., Jacobson, K. A., Kumar, S. T., Fuxe, K., Borroto-Escuela, D. O., et al. (2012). Molecular determinants of A2AR-D2R allosterism: Role of the intracellular loop 3 of the D2R. *Journal of Neurochemistry*, *123*(3), 373–384. https://doi.org/10.1111/j.1471-4159.2012.07956.x.

Filipek, S. (2019). Molecular switches in GPCRs. *Current Opinion in Structural Biology*, *55*, 114–120. https://doi.org/10.1016/j.sbi.2019.03.017.

García-Nafría, J., Lee, Y., Bai, X., Carpenter, B., & Tate, C. G. (2018). Cryo-EM structure of the adenosine A2A receptor coupled to an engineered heterotrimeric G protein. *elife*, *7*. https://doi.org/10.7554/eLife.35946.

García-Recio, A., Navarro, G., Franco, R., Olivella, M., Guixà-González, R., & Cordomí, A. (2020). DIMERBOW: Exploring possible GPCR dimer interfaces. *Bioinformatics*, *36*, 3271–3272.

Glukhova, A., Thal, D. M., Nguyen, A. T., Vecchio, E. A., Jorg, M., Scammells, P. J., et al. (2017). Structure of the adenosine A1 receptor reveals the basis for subtype selectivity. *Cell*, *168*(5), 867–877.e813. https://doi.org/10.1016/j.cell.2017.01.042.

Gutierrez-de-Teran, H., Sallander, J., & Sotelo, E. (2017). Structure-based rational design of adenosine receptor ligands. *Current Topics in Medicinal Chemistry*, *17*(1), 40–58. https://doi.org/10.2174/1568026616666160719164207.

Hill, S. J., May, L. T., Kellam, B., & Woolard, J. (2014). Allosteric interactions at adenosine A (1) and A(3) receptors: New insights into the role of small molecules and receptor dimerization. *British Journal of Pharmacology*, *171*(5), 1102–1113. https://doi.org/10.1111/bph.12345.

Hino, T., Arakawa, T., Iwanari, H., Yurugi-Kobayashi, T., Ikeda-Suno, C., Nakada-Nakura, Y., et al. (2012). G-protein-coupled receptor inactivation by an allosteric inverse-agonist antibody. *Nature*, *482*, 237–240. https://doi.org/10.1038/nature10750.

Hinz, S., Navarro, G., Borroto-Escuela, D., Seibt, B. F., Ammon, Y. C., de Filippo, E., et al. (2018). Adenosine A2A receptor ligand recognition and signaling is blocked by A2B receptors. *Oncotarget*, *9*(17), 13593–13611. https://doi.org/10.18632/oncotarget.24423.

Huang, J., Chen, S., Zhang, J. J., & Huang, X. Y. (2013). Crystal structure of oligomeric beta1-adrenergic G protein-coupled receptors in ligand-free basal state. *Nature Structural & Molecular Biology*, *20*(4), 419–425. https://doi.org/10.1038/nsmb.2504.

Ihara, K., Hato, M., Nakane, T., Yamashita, K., Kimura-Someya, T., Hosaka, T., et al. (2020). Isoprenoid-chained lipid EROCOC17+4: A new matrix for membrane protein crystallization and a crystal delivery medium in serial femtosecond crystallography. *Scientific Reports*, *10*, 19305. https://doi.org/10.1038/s41598-020-76277-x.

Ishchenko, A., Stauch, B., Han, G. W., Batyuk, A., Shiriaeva, A., Li, C., et al. (2019). Toward G protein-coupled receptor structure-based drug design using X-ray lasers. *IUCrJ*, *6*, 1106–1119. https://doi.org/10.1107/S2052252519013137.

Jaakola, V. P., Griffith, M. T., Hanson, M. A., Cherezov, V., Chien, E. Y., Lane, J. R., et al. (2008). The 2.6 angstrom crystal structure of a human A2A adenosine receptor bound to an antagonist. *Science*, *322*(5905), 1211–1217. https://doi.org/10.1126/science.1164772.

Jacobson, K. A. (2009). Functionalized congener approach to the design of ligands for G protein–coupled receptors (GPCRs). *Bioconjugate Chemistry*, *20*, 1816–1835.

Jacobson, K. A., Delicado, E. G., Gachet, C., Kennedy, C., von Kügelgen, I., Li, B., et al. (2020). Update of P2Y receptor pharmacology: IUPHAR Review 27. *British Journal of Pharmacology*, *177*, 2413–2433. https://doi.org/10.1111/bph.15005.

Jacobson, K. A., IJzerman, A. P., & Müller, C. E. (2021). Medicinal chemistry of P2 and adenosine receptors: Common scaffolds adapted for multiple targets. *Biochemical Pharmacology*, *187*, 114311. https://doi.org/10.1016/j.bcp.2020.114311.

Jespers, W., Oliveira, A., Prieto-Diaz, R., Majellaro, M., Aqvist, J., Sotelo, E., et al. (2017). Structure-based design of potent and selective ligands at the four adenosine receptors. *Molecules*, *22*(11). https://doi.org/10.3390/molecules22111945.

Jespers, W., Verdon, G., Azuaje, J., Majellaro, M., Keranen, H., Garcia-Mera, X., et al. (2020). X-ray crystallography and free energy calculations reveal the binding mechanism of A2A adenosine receptor antagonists. *Angewandte Chemie (International Ed. in English)*, *59*(38), 16536–16543. https://doi.org/10.1002/anie.202003788.

Johnston, J. M., Wang, H., Provasi, D., & Filizola, M. (2012). Assessing the relative stability of dimer interfaces in g protein-coupled receptors. *PLoS Computational Biology*, *8*(8). https://doi.org/10.1371/journal.pcbi.1002649, e1002649.

Jung, Y. H., Yu, J., Wen, Z., Salmaso, V., Karcz, T. P., Phung, N. B., et al. (2020). Exploration of alternative scaffolds for P2Y14 receptor antagonists containing a biaryl core. *Journal of Medicinal Chemistry*, *63*(17), 9563–9589. https://doi.org/10.1021/acs.jmedchem.0c00745.

Junker, A., Balasubramanian, R., Ciancetta, A., Uliassi, E., Kiselev, E., Martiriggiano, C., et al. (2016). Structure-based design of 3-(4-aryl-1H-1,2,3-triazol-1-yl)-biphenyl derivatives as P2Y14 receptor antagonists. *Journal of Medicinal Chemistry*, *59*(13), 6149–6168. https://doi.org/10.1021/acs.jmedchem.6b00044.

Kim, S. K., & Jacobson, K. A. (2006). Computational prediction of homodimerization of the A3 adenosine receptor. *Journal of Molecular Graphics & Modelling*, *25*(4), 549–561. https://doi.org/10.1016/j.jmgm.2006.03.003.

Kooistra, A. J., Mordalski, S., Pandy-Szekeres, G., Esguerra, M., Mamyrbekov, A., Munk, C., et al. (2021). GPCRdb in 2021: Integrating GPCR sequence, structure and function. *Nucleic Acids Research*, *49*(D1), D335–D343. https://doi.org/10.1093/nar/gkaa1080.

Kose, M., Gollos, S., Karcz, T., Fiene, A., Heisig, F., Behrenswerth, A., et al. (2018). Fluorescent-labeled selective adenosine A2B receptor antagonist enables competition binding assay by flow cytometry. *Journal of Medicinal Chemistry*, *61*(10), 4301–4316. https://doi.org/10.1021/acs.jmedchem.7b01627.

Lebon, G., Edwards, P. C., Leslie, A. G. W., & Tate, C. G. (2015). Molecular determinants of CGS21680 binding to the human adenosine A2A receptor. *Molecular Pharmacology*, *87*, 907–915. https://doi.org/10.1124/mol.114.097360.

Lebon, G., Warne, T., Edwards, P. C., Bennett, K., Langmead, C. J., Leslie, A. G., et al. (2011). Agonist-bound adenosine A2A receptor structures reveal common features of GPCR activation. *Nature*, *474*(7352), 521–525. https://doi.org/10.1038/nature10136.

Lee, M.-Y., Geiger, J., Ishchenko, A., Han, G. W., Barty, A., White, T. A., et al. (2020). Harnessing the power of an X-ray laser for serial crystallography of membrane proteins crystallized in lipidic cubic phase. *IUCrJ*, *7*, 976–984. https://doi.org/10.1107/S2052252520012701.

Liu, W., Chun, E., Thompson, A. A., Chubukov, P., Xu, F., Katritch, V., et al. (2012). Structural basis for allosteric regulation of GPCRs by sodium ions. *Science*, *337*, 232–236. https://doi.org/10.1126/science.1219218.

Martin-Garcia, J. M., Conrad, C. E., Nelson, G., Stander, N., Zatsepin, N. A., Zook, J., et al. (2017). Serial millisecond crystallography of membrane and soluble protein microcrystals using synchrotron radiation. *IUCrJ*, *4*, 439–454. https://doi.org/10.1107/S205225251700570X.

Martin-Garcia, J. M., Zhu, L., Mendez, D., Lee, M.-Y., Chun, E., Li, C., et al. (2019). High-viscosity injector-based pink-beam serial crystallography of microcrystals at a synchrotron radiation source. *IUCrJ*, *6*, 412–425. https://doi.org/10.1107/S205225251900263X.

Marullo, S., Doly, S., Saha, K., Enslen, H., Scott, M. G. H., & Coureuil, M. (2020). Mechanical GPCR activation by traction forces exerted on receptor N-glycans. *ACS Pharmacology and Translational Science*, *3*(2), 171–178. https://doi.org/10.1021/acsptsci.9b00106.

Matricon, P., Suresh, R. R., Gao, Z. G., Panel, N., Jacobson, K. A., & Carlsson, J. (2021). Ligand design by targeting a binding site water. *Chemical Science*, *12*, 960–968. https://doi.org/10.1039/d0sc04938g.

May, L. T., Bridge, L. J., Stoddart, L. A., Briddon, S. J., & Hill, S. J. (2011). Allosteric interactions across native adenosine-A3 receptor homodimers: Quantification using single-cell ligand-binding kinetics. *The FASEB Journal*, *25*(10), 3465–3476. https://doi.org/10.1096/fj.11-186296.

Melnikov, I., Polovinkin, V., Kovalev, K., Gushchin, I., Shevtsov, M., Shevchenko, V., et al. (2017). Fast iodide-SAD phasing for high-throughput membrane protein structure determination. *Science Advances*, *3*, e1602952. https://doi.org/10.1126/sciadv.1602952.

Moro, S., Hoffmann, C., & Jacobson, K. A. (1999). Role of the extracellular loops of G protein-coupled receptors in ligand recognition: A molecular modeling study of the human $P2Y_1$ receptor. *Biochemistry*, *38*, 3498–3507.

Nass, K., Cheng, R., Vera, L., Mozzanica, A., Redford, S., & Ozerov, D. (2020). Advances in long-wavelength native phasing at X-ray free-electron lasers. *IUCrJ*, *7*, 965–975. https://doi.org/10.1107/S2052252520011379.

Navarro, G., Cordomi, A., Zelman-Femiak, M., Brugarolas, M., Moreno, E., Aguinaga, D., et al. (2016). Quaternary structure of a G-protein-coupled receptor heterotetramer in complex with Gi and Gs. *BMC Biology*, *14*, 26. https://doi.org/10.1186/s12915-016-0247-4.

Neumann, A., Müller, C. E., & Namasivayam, V. (2020). $P2Y_1$-like nucleotide receptors—Structures, molecular modeling, mutagenesis, and oligomerization. *WIREs Computational Molecular Science*, *10*, e1464. https://doi.org/10.1002/wcms.1464.

Pedata, F., Pugliese, A. M., Coppi, E., Dettori, I., Maraula, G., Cellai, L., et al. (2014). Adenosine A2A receptors modulate acute injury and neuroinflammation in brain ischemia. *Mediators of Inflammation*, *2014*, 805198. https://doi.org/10.1155/2014/805198.

Pin, J. P., Kniazeff, J., Prezeau, L., Liu, J. F., & Rondard, P. (2019). GPCR interaction as a possible way for allosteric control between receptors. *Molecular and Cellular Endocrinology*, *486*, 89–95. https://doi.org/10.1016/j.mce.2019.02.019.

Qu, X., Wang, D., & Wu, B. (2020). Chapter 1—Progress in GPCR structure determination. In B. Jastrzebska, & P. S. H. Park (Eds.), *GPCRs* (pp. 3–22). Academic Press.

Rodriguez, D., Chakraborty, S., Warnick, E., Crane, S., Gao, Z. G., O'Connor, R., et al. (2016). Structure-based screening of uncharted chemical space for atypical adenosine receptor agonists. *ACS Chemical Biology*, *11*(10), 2763–2772. https://doi.org/10.1021/acschembio.6b00357.

Rodriguez-Espigares, I., Torrens-Fontanals, M., Tiemann, J. K. S., Aranda-Garcia, D., Ramirez-Anguita, J. M., Stepniewski, T. M., et al. (2020). GPCRmd uncovers the dynamics of the 3D-GPCRome. *Nature Methods*, *17*(8), 777–787. https://doi.org/10.1038/s41592-020-0884-y.

Rucktooa, P., Cheng, R. K. Y., Segala, E., Geng, T., Errey, J. C., Brown, G. A., et al. (2018). Towards high throughput GPCR crystallography: In meso soaking of adenosine A2A receptor crystals. *Scientific Reports*, *8*, 41.

Safdari, H. A., Pandey, S., Shukla, A. K., & Dutta, S. (2018). Illuminating GPCR signaling by cryo-EM. *Trends in Cell Biology*, *28*(8), 591–594. https://doi.org/10.1016/j.tcb.2018.06.002.

Salmaso, V., & Jacobson, K. A. (2020a). In silico drug design for purinergic GPCRs: Overview on molecular dynamics applied to adenosine and P2Y receptors. *Biomolecules*, *10*(6). https://doi.org/10.3390/biom10060812.

Salmaso, V., & Jacobson, K. A. (2020b). Purinergic signaling: Impact of GPCR structures on rational drug design. *ChemMedChem*, *15*(21), 1958–1973. https://doi.org/10.1002/cmdc.202000465.

Segala, E., Guo, D., Cheng, R. K. Y., Bortolato, A., Deflorian, F., Doré, A. S., et al. (2016). Controlling the dissociation of ligands from the adenosine A2A receptor through modulation of salt bridge strength. *Journal of Medicinal Chemistry*, *59*, 6470–6479. https://doi.org/10.1021/acs.jmedchem.6b00653.

Shimazu, Y., Tono, K., Tanaka, T., Yamanaka, Y., Nakane, T., Mori, C., et al. (2019). High-viscosity sample-injection device for serial femtosecond crystallography at atmospheric pressure. *Journal of Applied Crystallography*, *52*, 1280–1288. https://doi.org/10.1107/S1600576719012846.

Sun, B., Bachhawat, P., Chu, M. L., Wood, M., Ceska, T., Sands, Z. A., et al. (2017). Crystal structure of the adenosine A2A receptor bound to an antagonist reveals a potential allosteric pocket. *Proceedings of the National Academy of Sciences of the United States of America*, *114*(8), 2066–2071. https://doi.org/10.1073/pnas.1621423114.

Thevenin, D., Lazarova, T., Roberts, M. F., & Robinson, C. R. (2005). Oligomerization of the fifth transmembrane domain from the adenosine A2A receptor. *Protein Science*, *14*(8), 2177–2186. https://doi.org/10.1110/ps.051409205.

Tosh, D. K., Deflorian, F., Phan, K., Gao, Z. G., Wan, T. C., Gizewski, E., et al. (2012). Structure-guided design of A(3) adenosine receptor-selective nucleosides: Combination of 2-arylethynyl and bicyclo[3.1.0]hexane substitutions. *Journal of Medicinal Chemistry*, *55*(10), 4847–4860. https://doi.org/10.1021/jm300396n.

Tosh, D. K., Salmaso, V., Rao, H., Bitant, A., Fisher, C. L., Lieberman, D. I., et al. (2020). Truncated (N)-methanocarba nucleosides as partial agonists at mouse and human A3 adenosine receptors: Affinity enhancement by N(6)-(2-phenylethyl) substitution. *Journal of Medicinal Chemistry*, *63*(8), 4334–4348. https://doi.org/10.1021/acs.jmedchem.0c00235.

Tosh, D. K., Salmaso, V., Rao, H., Campbell, R., Bitant, A., Gao, Z. G., et al. (2020). Direct comparison of (N)-methanocarba and ribose-containing 2-arylalkynyladenosine derivatives as A_3 receptor agonists. *ACS Medicinal Chemistry Letters*, *11*, 1935–1941.

Toti, K. S., Jain, S., Ciancetta, A., Balasubramanian, R., Chakraborty, S., Surujdin, R., et al. (2017). Pyrimidine nucleotides containing a (S)-methanocarba ring as P2Y6 receptor agonists. *Medicinal Chemistry Communications*, *8*(10), 1897–1908. https://doi.org/10.1039/C7MD00397H.

Townsend-Nicholson, A., Altwaijry, N., Potterton, A., Morao, I., & Heifetz, A. (2019). Computational prediction of GPCR oligomerization. *Current Opinion in Structural Biology*, *55*, 178–184. https://doi.org/10.1016/j.sbi.2019.04.005.

van Rhee, A. M., & Jacobson, K. A. (1996). Molecular architecture of G protein-coupled receptors. *Drug Development Research*, *37*(1), 1–38. https://doi.org/10.1002/(SICI)1098-2299(199601)37:1<1::AID-DDR1>3.0.CO;2-S.

Vasiliauskaite-Brooks, I., Healey, R. D., & Granier, S. (2019). 7TM proteins are not necessarily GPCRs. *Molecular and Cellular Endocrinology*, *491*. https://doi.org/10.1016/j.mce.2019.02.009, 110397.

Velazhahan, V., Ma, N., Pandy-Szekeres, G., Kooistra, A. J., Lee, Y., Gloriam, D. E., et al. (2021). Structure of the class D GPCR Ste2 dimer coupled to two G proteins. *Nature*, *589*(7840), 148–153. https://doi.org/10.1038/s41586-020-2994-1.

Venkatakrishnan, A. J., Deupi, X., Lebon, G., Tate, C. G., Schertler, G. F., & Babu, M. M. (2013). Molecular signatures of G-protein-coupled receptors. *Nature*, *494*(7436), 185–194. https://doi.org/10.1038/nature11896.

Venkatakrishnan, A. J., Ma, A. K., Fonseca, R., Latorraca, N. R., Kelly, B., Betz, R. M., et al. (2019). Diverse GPCRs exhibit conserved water networks for stabilization and activation.

Proceedings of the National Academy of Sciences of the United States of America, *116*(8), 3288–3293. https://doi.org/10.1073/pnas.1809251116.

Wacker, D., Stevens, R. C., & Roth, B. L. (2017). How ligands illuminate GPCR molecular pharmacology. *Cell*, *170*(3), 414–427. https://doi.org/10.1016/j.cell.2017.07.009.

Warne, T., Edwards, P. C., Dore, A. S., Leslie, A. G. W., & Tate, C. G. (2019). Molecular basis for high-affinity agonist binding in GPCRs. *Science*, *364*(6442), 775–778. https://doi.org/10.1126/science.aau5595.

Weinert, T., Olieric, N., Cheng, R., Brünle, S., James, D., & Ozerov, D. (2017). Serial millisecond crystallography for routine room-temperature structure determination at synchrotrons. *Nature Communications*, *8*, 542. https://doi.org/10.1038/s41467-017-00630-4.

White, K. L., Eddy, M. T., Gao, Z. G., Han, G. W., Lian, T., Deary, A., et al. (2018). Structural connection between activation microswitch and allosteric sodium site in GPCR signaling. *Structure*, *26*(2), 259–269.e255. https://doi.org/10.1016/j.str.2017.12.013.

Wold, E. A., Chen, J., Cunningham, K. A., & Zhou, J. (2019). Allosteric modulation of class A GPCRs: Targets, agents, and emerging concepts. *Journal of Medicinal Chemistry*, *62*(1), 88–127. https://doi.org/10.1021/acs.jmedchem.8b00875.

Xu, F., Wu, H., Katritch, V., Han, G. W., Jacobson, K. A., Gao, Z. G., et al. (2011). Structure of an agonist-bound human A2A adenosine receptor. *Science*, *332*(6027), 322–327. https://doi.org/10.1126/science.1202793.

Yu, J., Ciancetta, A., Dudas, S., Duca, S., Lottermoser, J., & Jacobson, K. A. (2018). Structure-guided modification of heterocyclic antagonists of the P2Y14 receptor. *Journal of Medicinal Chemistry*, *61*(11), 4860–4882. https://doi.org/10.1021/acs.jmedchem.8b00168.

Zhang, D., Gao, Z. G., Zhang, K., Kiselev, E., Crane, S., Wang, J., et al. (2015). Two disparate ligand-binding sites in the human P2Y1 receptor. *Nature*, *520*(7547), 317–321. https://doi.org/10.1038/nature14287.

Zhang, J., Zhang, K., Gao, Z. G., Paoletta, S., Zhang, D., Han, G. W., et al. (2014). Agonist-bound structure of the human P2Y12 receptor. *Nature*, *509*(7498), 119–122. https://doi.org/10.1038/nature13288.

Zhang, K., Zhang, J., Gao, Z. G., Zhang, D., Zhu, L., Han, G. W., et al. (2014). Structure of the human P2Y12 receptor in complex with an antithrombotic drug. *Nature*, *509*(7498), 115–118. https://doi.org/10.1038/nature13083.

Zhou, Q., Yang, D., Wu, M., Guo, Y., Guo, W., Zhong, L., et al. (2019). Common activation mechanism of class A GPCRs. *eLife*, *8*. https://doi.org/10.7554/eLife.50279.

CHAPTER

Nanobodies as sensors of GPCR activation and signaling

7

Amal El Daibani[a] and Tao Che[a,b,*]

[a]*Department of Anesthesiology, Washington University School of Medicine, St. Louis, MO, United States*

[b]*Center for Clinical Pharmacology, University of Health Sciences and Pharmacy in St. Louis and Washington University School of Medicine, St. Louis, MO, United States*

**Corresponding author: e-mail address: taoche@wustl.edu*

Chapter outline

Abstract

Nanobodies have emerged as useful tools to study G protein-coupled receptor (GPCR) structure, dynamic, and subcellular localization. Initially, several nanobodies have been developed as chaperones to facilitate GPCR crystallization. To explore their potential as biosensors to monitor receptor activation and dynamics, we here described protocols to characterize nanobody's interaction with GPCRs and their application as probes for protein identification and

Methods in Cell Biology, Volume 166, ISSN 0091-679X, https://doi.org/10.1016/bs.mcb.2021.06.008

visualization on the cellular level. We also introduced a chimeric approach to enable a kappa-opioid receptor derived nanobody to bind to other GPCRs, including orphan GPCRs whose endogenous ligand or intracellular transducers are unknown. This approach provides a reporter assay to identify tool molecules to study the function of orphan GPCRs.

1 Introduction

The G protein-coupled receptor (GPCR) superfamily comprises seven-transmembrane proteins that can sense the stimuli outside the neurons or cells. To date, there are more than 800 GPCRs identified and divided into four major classes (Class A, B, C and F) based on sequence homology and functional similarity (Katritch, Cherezov, & Stevens, 2013; Pierce, Premont, & Lefkowitz, 2002). Among them, about 450 GPCRs are olfactory receptors that belong to the Class A Rhodopsin-like family and mediate functions such as olfaction and light perception (Palczewski, 2006). The rest of the GPCRs (>350) are important components responsible for physiological activities such as vision, cardiovascular, allergy, metabolism, pain sensation and neuronal signaling (Lagerstrom & Schioth, 2008). They do so by binding different neurotransmitters including peptides, lipids, small molecules and proteins. However, there is still a portion of the GPCRs (~120) with unknown endogenous ligands or physiological function called orphan GPCRs. Due to the lack of tool molecules, it is challenging to study these receptors' activation and function in vitro and in vivo.

GPCR activation is initiated by the binding of ligands in the extracellular binding pocket. The chemical information of the ligand is transmitted through allosteric conformational changes to the intracellular region of the receptor and deciphered by coupling to distinct signaling transducers, such as G proteins and arrestins (Flock et al., 2015). The ligand efficacy and the access to downstream effectors are dependent on the conformational complexity of the receptor (Kenakin, 2002). The conformational plasticity provides the basis that different drugs (e.g., agonist, antagonist, inverse agonist) act differentially through the same GPCR and, in particular, is key to functional selectivity (G protein vs arrestins) of the receptor (Kahsai et al., 2011; Manglik et al., 2015; Manglik & Kobilka, 2014; Weis & Kobilka, 2018). Understanding how a GPCR is activated and what determines its functional selectivity would facilitate the development of drugs that produce beneficial effects with reduced side effect profile.

2 Nanobodies as emerging tools to study GPCR activation and signaling

Current assays developed to examine GPCR activation mostly focus on the interaction between GPCRs and the intracellular transducers (e.g., cAMP inhibition, Tango arrestin recruitment, Bioluminescence or Fluorescence resonance energy transfer

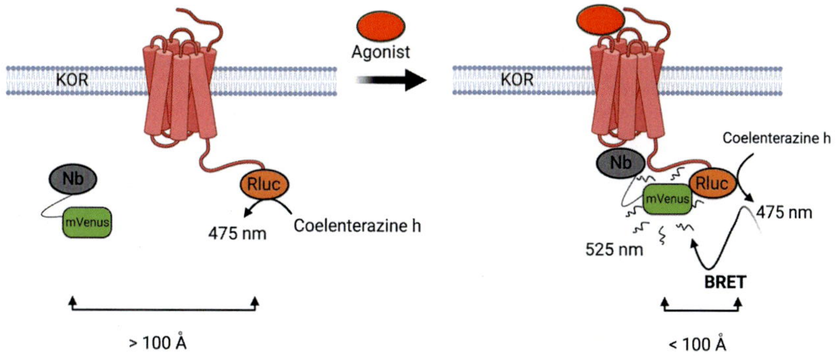

FIG. 1

Scheme of BRET assays measuring interactions between GPCRs and nanobodies. Kappa opioid receptor is tagged with a C-terminal luciferase (Rluc8) that can hydrolyze substrate coelenterazine h and excite at 475 nm. When receptor and nanobody do not form a complex, they are too far away to produce (>100 Å) luminescence. When a mVenus (a GFP variant) tagged nanobody is close to the receptor (<100 Å), the mVenus will be excited and produce luminescence at ~525 nm.

(BRET or FRET) assays). Bioluminescence resonance energy transfer (BRET) is a non-radiative transfer of energy between a genetically fused protein to the bioluminescent donor (e.g., Renilla luciferase (Rluc)) and a fluorescent acceptor protein (e.g., GFP or YFP) that occurs when they are in close proximity (Fig. 1). It is one of the most well-established approaches used to probe the GPCR conformational changes induced by ligand binding. The BRET-based technique has been successfully applied to study the homodimerization and heterodimerization resulting from the ligand binding, to monitor the signaling of GPCR by characterizing the dynamics of protein-protein interactions, and to analyze changes in the activity of GPCR transducerome (Boute, Jockers, & Issad, 2002; Olsen et al., 2020; Salahpour et al., 2012; Stoddart et al., 2015; Szidonya, Cserzo, & Hunyady, 2008). Different types of BRET (BRET1, BRET2, enhanced BRET2, and BRET3) have been developed based on the donor Rluc variant, acceptor fluorescent, and substrate (Bacart, Corbel, Jockers, Bach, & Couturier, 2008; Bertrand et al., 2002; De, Ray, Loening, & Gambhir, 2009; Kocan, Dalrymple, Seeber, Feldman, & Pfleger, 2010). Alongside these mentioned BRET, NanoBRET has also been developed by utilizing an engineered luciferase (NanoLuc), a smaller (19 kDa) and brighter than Rluc, as the donor molecule to conduct BRET proximity assay in real-time and within live cells (Machleidt et al., 2015; Stoddart, Kilpatrick, & Hill, 2018). To further expand our knowledge regarding the molecular mechanism of GPCR activation, a number of nanobodies targeting GPCRs have been developed against a broad range of epitopes and cavities within GPCRs (Manglik, Kobilka, & Steyaert, 2017; Muyldermans, 2013). This method is particularly useful in studying the allosteric modulation of receptor activity and the stabilization of the receptor conformational state. Nanobody is a sub-class of antibody specifically produced from llama or shark. Compared with conventional

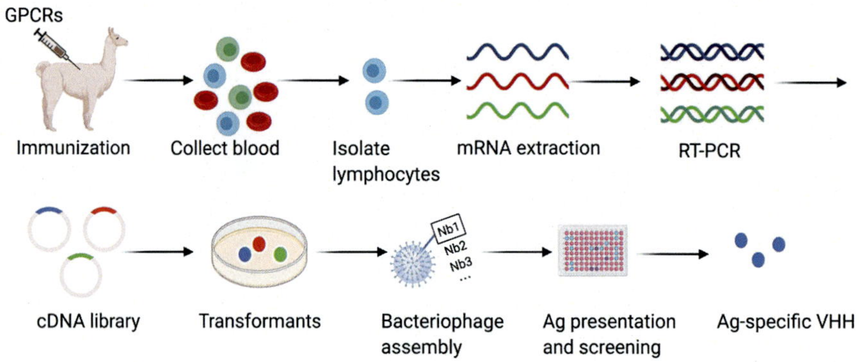

FIG. 2

The workflow of generation of antigen-specific nanobodies. A camelid is immunized with the interested protein (e.g., KORs bound to antagonist or agonist depending on the need of inactive/active state nanobody). Through the construction of cDNA and phage library, $>10^8$ individual nanobody clones could be obtained. Identification of antigen-specific nanobodies can be achieved via bio-panning and validated via various assays such as ELISA. Positive binders can then be enriched and used for biochemical research to identify their binding site and affinity.

antibodies (e.g., IgG, ~150kDa), nanobody has much smaller size (10–15kDa) that only contains the heavy chain of the antibody. To generate antigen-specific nanobody, it is similar as conventional antibody which requires immunization, lymphocyte isolation, cDNA library construction yeast or phage display, panning and screening (Fig. 2).

For example, a broad panel of chemokine receptors (CXCR2, CXCR4, US28 and CXCR7) nanobodies act as antagonists or inverse agonists and have competitively inhibited the chemokine-mediated signaling (Bobkov et al., 2018; Bradley et al., 2015; Heukers et al., 2018; Jahnichen et al., 2010; Maussang et al., 2013; Van Hout et al., 2018). These results have important implications in the development of CXCR4-directed antibodies and US28-targeting nanobodies as novel therapeutics for HIV and tumor, respectively. In addition, nanobodies have been shown to positively modulate the pre-synaptically metabotropic glutamate receptors (mGlu2) by binding to the receptor's active form and promoting the agonist effect in both transfected cells and hippocampal slices derived from rats (Scholler et al., 2017). These findings further indicate the opportunities for therapeutic intervention by using nanobodies to allosterically modulate mGlu2 receptors for a variety of therapeutic effects. Additionally, in the structural studies of the active states of β2 adrenergic receptor (β_2AR), carried out by using the conformational-stabilizing nanobodies approach, have greatly impacted our knowledge in understanding the mechanistic details underlying β_2AR receptor activation by full and partial agonists (Masureel et al., 2018; Rasmussen et al., 2011; Ring et al., 2013; Staus et al., 2016). The stabilization of the muscarinic acetylcholine receptor (M2) receptor by employing nanobody technique facilitates the crystallization of M2 receptor bound to both a selective agonist (iperoxo15) and a positive allosteric modulator (LY2119620), thus improving our understanding regarding the mechanism behind the M2 receptor activation (Kruse et al., 2013). Besides these examples, nanobody technology has further enhanced our knowledge in understanding the structural rearrangements and the pharmacological modulation of other related GPCRs, including mu-opioid receptor (MOR) (Huang et al., 2015) and kappa-opioid receptor (KOR) (Che et al., 2018, 2020).

3 Examples of using nanobodies to probe KOR activation

KOR belongs to the class A GPCRs which produces cellular response upon ligand binding by interacting with Gi/o subfamily proteins, β-arrestins as well as distinct downstream effectors. The binding of the KOR agonist induces the heterotrimeric dissociation, which is initiated by the replacement of GDP with GTP at the G_α subunit. This separation of the G-protein complex into G_α and $G_{\beta\gamma}$ subunits induces the subsequent signaling produced by second messenger including, but not limited to, cyclic AMP, diacylglycerol, and calcium (Bruchas & Chavkin, 2010). KOR agonists have emerged potential in the treatment of pain and pruritus (Che, 2020; Cowan, Kehner, & Inan, 2015; Millan, 1990; Vanderah, 2010), while KOR antagonists and KOR negative allosteric modulators have a promising therapeutic target for anxiety, depression, and addiction (Carlezon & Krystal, 2016). Besides the analgesic effect of KOR agonists, several side effects such as constipation, respiratory depression, hallucinations, and dysphoria are produced from the activation of the Gi/o protein-mediated pathway of KOR (Bruchas & Roth, 2016; Dykstra, Gmerek, Winger, & Woods, 1987; Pfeiffer, Brantl, Herz, & Emrich, 1986; Ranganathan et al., 2012; Tejeda et al., 2013). Consequently, biased agonists that preferentially targeted G-protein over β-arrestin pathways have been the subject of interest (Brust et al., 2016; White et al., 2015). To create new, safer and more effective analgesic and antipruritic therapies, it is important to deeply understand the structural basis of KOR activation and signaling by identifying determinants that are essential for its functional selectivity. As KOR is highly dynamic, a nanobody approach has been developed to intercellularly bind and stabilize the diverse conformational states of the KOR (Che et al., 2018). Nanobodies, which are stable recombinant camelid single-chain domain containing the only heavy-chain antibody, have emerged as powerful platforms for providing insight into the structural and dynamic characterization of various receptors. The nanoscale size, convex shape, and the long three complementarity-determining regions (CDR3) facilitate them to bind in between intracellular loops of GPCRs and stabilize the receptor (Heukers, De Groof, & Smit, 2019; Manglik et al., 2017).

In addition to KOR, the study of nanobodies has highly impacted our knowledge over the past decade in relation to other GPCRs, the largest class of membrane protein that include β_2 adrenergic (β_2AR), M_2 muscarinic (M_2R), and mu-opioid (MOR) receptors (Huang et al., 2015; Kruse et al., 2013; Rasmussen et al., 2011). Briefly, llamas were immunized with purified KOR liposomes bound to the selective KOP agonist, salvinorin A (SalA), to generate nanobodies that can be cloned and expressed (Pardon et al., 2014). The phage display was used for nanobodies screening from the immune libraries. Bioluminescence resonance energy transfer (BRET) technique was then employed by genetically fusing the nanobodies with a fluorescent protein and KOR with Renilla luciferase (Rluc) to verify the binding of nanobodies to the KOR (Che et al., 2018). A nanobody (Nb), Nb39, which has been previously reported to bind to the active site of MOR, unexpectedly showed binding to the active state of KOR, whereas a newly identified Nb6 selectively binds to the inactive state of KOR (Che et al., 2020; Huang et al., 2015) (Fig. 3).

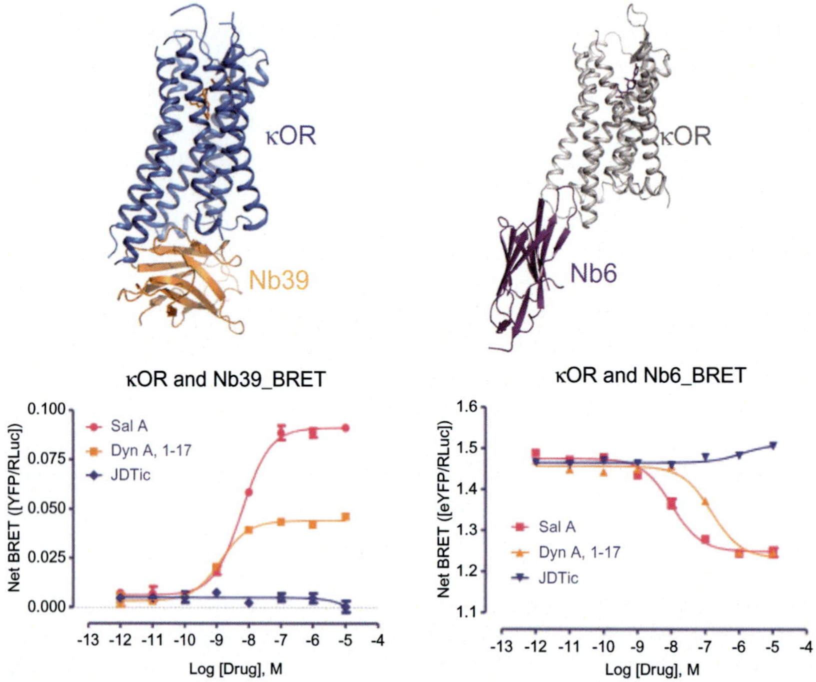

FIG. 3

BRET assays measuring active- or inactive-state stabilizing nanobodies. Nb39 and Nb6 are two nanobodies that can bind and stabilize the active or inactive state of KOR, respectively. Complex structures are adopted from Protein Data Bank ID (6B73 and 6VI4). BRET assays show a different response to agonist treatment between Nb6 and Nb39. Agonists Salvinorin A (Sal A) and Dynorphin A, 1–17 (Dyn A) recruit Nb39 to the receptor by promoting an active state, while the antagonist JDTic does not recruit Nb39. On the other hand, Nb6 recognizes the inactive state of KOR and pre-binds to the receptor when there are no ligands around, leading to high or saturation of the BRET signal. Agonist treatment causes the dissociation of Nb6 from the receptor. JDTic still does not show a significant response as most of the receptors has been bound by the Nb6. Another feature from the KOR-Nb39 curve shows that Dyn A, 1–17 acts as a partial agonist compared to Sal A, supporting that this assay could be used to screen optimal agonists for KOR-Nb complex assembly.

The nanobodies, Nb39 and Nb6, were used as a crystallization chaperone for KOR and consequently, the crystal structures for both active KOR bound to MP1104 and Nb39, along with inactive KOR bound to JDTic (KOR antagonist) and Nb6 were resolved. Additionally, a comparison of the active-state KOR-MP1104-Nb39 complex and inactive-state KOR-JDTic structure highlighted changes in the helices orientation in various positions that include extracellular loop 2 (ECL2), transmembrane helices (TM4, TM5, and TM6), as well as the orthosteric binding pocket (Wu et al., 2012). The overall helical movements of inactive-state KOR-JDTic were comparable with KOR bound to JDTic and Nb6 structure. These helical movements have been reported with other active-state GPCRs including canonical opioid receptors, implying similar mechanisms for GPCRs activation. Notably, the net BRET ratio revealed that Nb39 binding to KOR is in an agonist and efficacy-dependent manner. In the presence of Nb39, high-affinity binding sites for KOR were enhanced while the agonist dissociation rate was reduced as shown in the saturation binding studies, indicating the positive allosteric effect of Nb39. In contrast, the negative allosteric effect of Nb6 was verified by conducting the radioligand binding and ligand association and dissociation assays which revealed that Nb6 negatively modulates the agonist affinity and boosts agonist dissociation. Remarkedly, conformational-specific nanobodies (Nb6 and Nb39) have been utilized as optical biosensors to monitor KOR activation in living cells with high subcellular selectivity and resolution. This illuminates the potential of applying Nb6 and Nb39 to figure out the conformational changes in real-time for other GPCRs, particularly other members of opioid receptors (Che et al., 2020).

In this chapter, we focus on describing the nanobodies methodology and their optimization as a novel instrumental tool to deeply understand the molecular structure and dynamic feature of KOR to assist in designing new therapies with higher selectivity, thereby enhancing their efficacy and avoiding their side effects.

4 Step-by-step protocols

In this section, we mainly focus on three assays: the first is to use BRET assays to measure interactions between GPCRs and inactive/active state nanobodies; the second is to use a chimeric approach to make the KOR-derived Nb6 as a universal biosensor for many other GPCRs which nanobodies or intracellular transducers are unavailable. The third is to use nanobodies as probes in live-cell imaging to monitor cellular receptor distribution. We will use KOR and serotonin 5-HT_{2A} receptor as GPCR representatives.

4.1 Protocols for receptor interaction with intracellular nanobodies

There are several advantages of using BRET assays to measure receptor-nanobody interaction. It is worth pointing out that this protocol is specifically for nanobodies binding to the intracellular regions of GPCRs (Fig. 3). The first advantage is using nanobodies as biosensors to probe receptor inactivation/activation (Staus et al., 2016). The second is to localize intracellular receptors (Che et al., 2020; Irannejad et al., 2013; Stoeber et al., 2018). The third is for the purpose of structural determination, which could help screen optimal ligands that recruit the nanobody and stabilize the receptor-nanobody complex (Che et al., 2018).

4.1.1 Materials and reagents

- pcDNA3.1 (+) Mammalian Expression Vector (ThermoFisher Scientific #V79020)
- human kappa opioid receptor sequence (P41145 (OPRK_HUMAN))
- Renilla Luciferase sequence (Rluc8, Addgene# 87121)
- an enhanced yellow fluorescent protein (mVenus, PMID: 11753368)
- Dulbecco's Modified Eagle Medium (DMEM) (VWR, 45000-306)
- Opti-MEM (Gibco #11058021)
- Penicillin/Streptomycin (Gibco™, 11548876, ThermoFisher)
- Trypsin-EDTA (0.25%), phenol red (Gibco™, 25,200,056, ThermoFisher)
- 96-well white clear bottom cell culture plates (Greiner, #781098)
- FBS (VWR #97068-085)
- Dialyzed FBS (Omega Scientific #FB-03)
- TransIT-2020 (Mirus #MIR5400)
- HEPES (Sigma #H3375)
- HBSS (Invitrogen #14065-056)
- Coelenterazine-h (Promega #S2011)

4.1.2 Protocols

1. Design of KOR-Rluc8 and Nb-mVenus constructs. The plasmid vector for both receptor and nanobody is pcDNA3.1 and expression of proteins is driven by the promotor CMV. The C-terminal of KOR is fused with a Renilla Luciferase sequence (Rluc8). No linker is necessary for KOR as it has a relatively long and flexible C-terminal. For receptors with very short or no C terminal, 3 × or 5 × GS (-Gly-Ser-) could be added between the receptor and Rlu8 sequence. The C terminal of nanobody is tagged with an enhanced yellow fluorescent protein (mVenus). As BRET is a proximity assay, the distance between the excitation and emission domains can significantly affect the signal output. Thus, it is worth screening both N- and C-terminal mVenus tagged and determine which site produces a better signal to noise ratio.
2. Cell preparation and transfection. In a 10-cm cell culture dish (tissue culture treated, Genesee), 3×10^6 HEK 293T cells are plated in a total volume of 10 mL DMEM supplemented with 10% FBS. The cell plate is incubated at 37 °C and 5% CO_2 incubator overnight. The next day, warm up the Opti-MEM at 37 °C. In a sterilized 1.5 mL tube, add 500 μL warmed-up Opti-MEM, then 0.5 μg KOR-Rluc8 and 2.5 μg Nb-mVenus DNAs. Next add 6 μL of TransIT-2020 and mix well. Incubated for 30 min at room temperature. Transfer all volumes to the

10-cm cell culture plate. In the negative control plate, add 0.5 μg KOR-Rlu8 and 2.5 μg pcDNA3.1 and do the rest steps the same as the experimental group. Incubated the transfected plate at 37 °C with 5% CO_2 incubator overnight.

3. Cell plating. Warm up the DMEM supplemented with 1% dialyzed FBS (dFBS) at 37 °C. Use vacuum and aspirating pipet to remove the cell medium, add 2 mL Trypsin-EDTA and incubated 2–3 min at 37 °C or 3–5 min at room temperature to digest the cells. Add 5 mL DMEM/dFBS to stop the Trypsin reaction. Transfer cells to a new 15 mL conical tubes (Genesee) and centrifuge 1500 rpm for 3 min. Use aspirating pipet to remove the supernatant and resuspend the cells in 3 mL DMEM/dFBS medium. Count the cell density. Dilute the cells and plate 25,000–50,000 cells/well in a 96-well white clear bottom cell culture plate. Incubated the plate at 37 °C with 5% CO_2 incubator overnight.
4. Drug stimulation and reading. Prepare serial drug concentrations (e.g., −4, −5, −6…−11, −12 M) in drug buffer (20 mM HEPES pH 7.4, 1 × HBSS). Remember the concentration should be three times higher in the preparation step as it is going to be diluted after adding to the cell plate (see below). Decant the medium in the 96-well cell culture and wash with 100 μL drug buffer. Then add 60 μL of luciferase substrate (coelenterazine h (Promega, 5 mM final concentration in drug buffer)) and incubate for 5 min at dark and room temperature. Then add 30 μL prepared drug solution (final concentration is 1 ×) and incubated for another 5 min. Read the plate in a luminescence reader (e.g., Mithras LB940) at 485 nm (Rluc8) and 530 nm (mVenus) for excitation and emission (1 s/well), respectively. The ratio of mVenus/Rluc is calculated per well. The net BRET ratio is calculated by subtracting the mVenus/Rluc per well from the mVenus/Rluc ratio in the well or negative control. The net BRET ratio is plotted as a function of drug concentration using log(agonist) vs response equation in GraphPad Prism (GraphPad Software) (Fig. 3).

4.2 Protocols of using Nb6 as a biosensor for non-opioid GPCRs

The rationale of using Nb6 as a biosensor for other GPCRs is based on unique interactions between KOR and Nb6 (Che et al., 2020). Based on the crystal structure of KOR bound to Nb6, the CDR3 loop of Nb6 forms interactions with a few residues in the intracellular loop 3 (ICL3) between TM5 and TM6 of KOR. This inspired us that a chimeric receptor including the short sequence at the corresponding place may bind to Nb6. After trying several non-opioid GPCRs (5-HT_{2A}, ET_A, DRD1, 5-HT_{5A}, NTS_1, HRH_2) and two orphan GPCRs (GPR111 GPR149, and GPR151), we confirmed that Nb6 could bind to these GPCRs via a chimeric approach (Che et al., 2020). Here we describe the detailed protocols on how to design chimeric receptors and how to use BRET assays to measure the potential of Nb6 being a biosensor for GPCRs whose nanobodies or transducers are unavailable. For the receptor, we will use the serotonin 5-HT_{2A} receptor as an example.

4.2.1 Materials and reagents

Materials and reagents for this assay are mostly overlapping with the ones in Section 4.1.1 as it is also based on the BRET assay. The key for the design of chimeric receptor is to find optimal site for insertion of the KOR ICL3 region.

4.2.2 Protocol

1. Generation of chimeric receptors. The C-terminal tagged 5-HT_{2A}-Rluc8 is first created. To make chimeric constructs, the human KOR ICL3 sequence (amino acid sequence: ($V^{5.56}$CYTLMILRLKSVRLLSGSREKDRNLRRITRLVLVVV AVFV$^{6.45}$) is inserted into the 5-HT_{2A} ICL3 region and replaces the original sequence between $I^{5.56}$ and $V^{6.45}$ ($I^{5.56}$ and $V^{6.45}$) included, the number is based on the Ballesteros–Weinstein numbering scheme (Ballesteros & Weinstein, 1995)). This inserting sequence has been shown as a universal region workable for most GPCRs we have tested.
2. Cell preparation, transfection and plating. After having the chimeric 5-HT_{2A}^{KOR}-Rluc8 construct, the following protocols for cell preparation and transfection are the same as in Section 4.1. Another experimental group is that cells transfected 5-HT^{2A}-Rluc8 and Nb6-mVenus. Two negative control groups are cells transfected 5-HT_{2A}^{KOR}-Rluc8 or 5-HT_{2A}-Rluc8 only.
3. Measure interactions between chimeric 5-HT_{2A}^{KOR} and Nb6. Drug stimulation (agonist 5-HT) and data recording and analysis are the same as the protocols in Section 4.1 (Fig. 4).
4. Measure kinetics (association and dissociation) of Nb6 at the chimeric 5-HT_{2A} receptor. Cells transfected with 5-HT_{2A}^{KOR}-Rluc/Nb6-YFP or 5-HT_{2A}^{KOR}-Rluc only are plated in parallel on the same plate. Decant the medium and wash the cells with 100 μL drug buffer. Add 60 μL drug buffer including 5 mM (final) coelenterazine h and incubate at dark and room temperature for 5 min. During the drug stimulation, luminescence is recorded in the order: background for 30 s, then add 30 μL agonist 5-HT (Final concentration: 10 μM) and record immediately for 120 s, then add 30 μL antagonist risperidone (final concentration: 10 μM) for 120 s. Cells transfected with 5-HT_{2A}^{KOR}-Rluc only are the background and the net BRET ratio will be the recorded value in the experimental groups subtracted by the background. Data were then plotted in GraphPad Prism (Fig. 4).

4.3 Use nanobodies for live-cell imaging

Because of their small size and reversible interactions with the targets, nanobodies are of special interest for subcellular protein identification and visualization. Studies have also shown that nanobodies could be used as probes for live-cell imaging to look at receptor dynamics and localization in response to agonist or antagonist treatment (Irannejad et al., 2013; Stoeber et al., 2018). Here we describe the protocols of using active (Nb39) or inactive state (Nb6) stabilizing nanobodies to monitor kappa opioid receptor states.

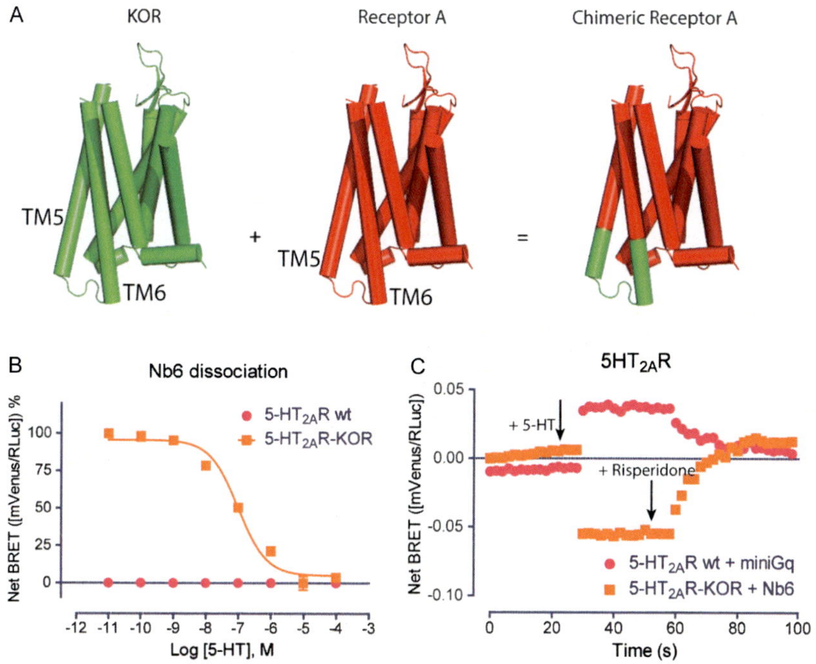

FIG. 4

Nb6 as a biosensor for serotonin 5-HT_{2A} receptor via a chimeric approach. (A) Scheme of chimeric receptor design. The ICL3 loop region of KOR, where the Nb6 binds, is transferred to the target GPCR at the corresponding site. (B) Agonist 5-HT causes the dissociation of Nb6, indicating that the chimeric 5-HT_{2A}^{KOR} receptor could bind to the Nb6, while the wild type 5-HT_{2A} doesn't. (C) Nb6 as a biosensor to monitor 5-HT_{2A} responses to the agonist 5-HT and antagonist risperidone. Nb6 acts similarly as the endogenous Gq protein of 5-HT_{2A}.

This figure has been modified from Che, T., English, J., Krumm, B. E., Kim, K., Pardon, E., Olsen, R. H. J., Wang, S., Zhang, S., Diberto, J. F., Sciaky, N., et al. (2020). Nanobody-enabled monitoring of kappa opioid receptor states. Nature Communications, 11, *1145.*

4.3.1 Materials and reagents

- Plasmids: hKOR-mScarlet, Nb6-mVenus, Nb39-mVenus (Che et al., 2020), GalT-RFP (Irannejad et al., 2013)
- FBS (VWR #97068-085)
- Poly-L-Lysine (Sigma #P2636)
- Sterile Phosphate-Buffered Saline (PBS) for washing step (Gibco™, 14190250)
- HEPES (Sigma #H3375)
- HBSS (Invitrogen #14065-056)
- Bovine Serum Albumin (Sigma #A7030)
- L-ascorbic acid (Sigma #A4403)
- D(+)-Glucose (Sigma #G8270)

4.3.2 Protocols

1. Use sterile tissue culture grade water to prepare 100 μg/mL poly-L-Lysine solution; add 2 mL poly-L-Lysine solution to the center of a MatTek dish (35 mm, MatTek Life Sciences); after 30 min, remove the solution by aspiration and rinse the surface with sterile water.
2. Plate 75,000 HEK 293T cells in pre-coated MatTek dishes in 2 mL DMEM supplemented with 10% FBS; incubate at 37 °C and 5% CO_2 overnight.
3. Transfect the cells with KOR-mScarlet/Nb6-mVenus or KOR-mScarlet/Nb39-mVenus at 4:1 ratio (400 ng KOR and 100 ng Nb in 200 μL warmed Opti-MEM); For Golgi labeling, transfect KOR/Nb39-mVenus/GalT-RFP at 4:1:1 ratio (400 ng KOR, 100 ng Nb39 and 100 ng GalT-RFP in 200 μL warmed Opti-MEM); gently shake the plate a few times and incubate at 37 °C and 5% CO_2 overnight.
4. Remove the medium and wash the cells with PBS twice; add 2 mL modified drug buffer (20 mM HEPES, 1 × HBSS pH 7.4 supplemented with 0.1% BSA, 0.01% L-ascorbic acid and 4.5 g/L Glucose).
5. Record confocal images before and after adding 2 μL agonist or antagonist (final concentration: 10 μM) using an inverted Olympus FV3000RS confocal microscope (The 488 nm diode laser is used for the mVenus excitation and the 561 nm diode laser is used to excite the mScarlet or RFP. The variable slit has a nominal bandwidth of 500–545 nm for the green emission channel and 590–650 nm for the red emission channel) (Fig. 5).

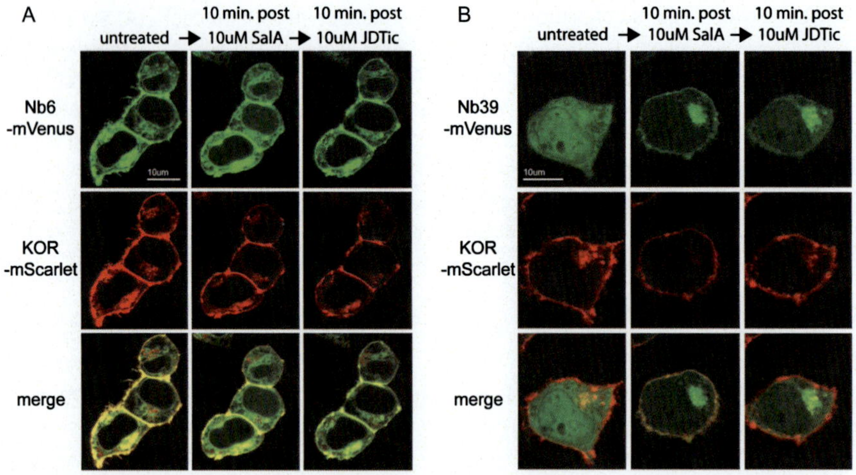

FIG. 5

Nb6 and Nb39 as biosensors to monitor KOR states inside the cells. (A) Confocal live-cell imaging of HEK 293T cells expressing KOR-mScarlet and Nb6-mVenus. 10 μM Sal A or JDTic was added to the cell plate at indicated time points. (B) Confocal live-cell imaging of HEK 293T cells expressing KOR-mScarlet and Nb39-mVenus. 10 μM Sal A or JDTic was added at indicated time points.

This figure has been modified from Che, T., English, J., Krumm, B. E., Kim, K., Pardon, E., Olsen, R. H. J., Wang, S., Zhang, S., Diberto, J. F., Sciaky, N., et al. (2020). Nanobody-enabled monitoring of kappa opioid receptor states. Nature Communications, 11, *1145.*

5 Data analysis

As a non-invasive and small-size protein, nanobody has displayed many advantages in both biomedical and clinical research (de Beer & Giepmans, 2020; Jovcevska & Muyldermans, 2020). It is worth pointing out that much more information could be obtained rather than knowing the receptor-nanobody interactions.

Via cell-based BRET assays, the data will directly tell if the nanobody interacts with targeted receptor, either active or inactive state, by responding to agonist or antagonist treatment. This system could be also used to screen agonists that preferentially recruit the nanobody to the receptor, which is uniquely advantageous for structural purpose. The fluorophore could also be extended to other interested proteins to study large complex formation, e.g., GPCR-G protein-nanobody.

Via a chimeric approach, the KOR-derived Nb6 could be used as a biosensor for GPCRs which nanobodies are not available. Its robust response to agonist or antagonist treatment indicates Nb6 could be potentially applied to study GPCR kinetics in real time. In particular, the chimeric receptor also allows orphan GPCRs to bind to Nb6. This provides a possible platform to screen endogenous or exogenous ligands that can bind to the orphan GPCRs, a process called deorphanization or reverse pharmacology (Pardon et al., 2018). The major difficulty of "deorphanization" is the lack of tool molecules that enable further receptor signaling and function studies.

Considering that the chimeric approach requires the replacement of ICL3 of interested GPCR, the function of chimeric receptors may change. To look at the functional consequence of the chimeric receptor, we recommend testing the G protein activation and arrestin recruitment and comparing it with the wild type receptor. If the chimeric receptor activity has been significantly changed, making less modification of the ICL3 region (e.g., use shorter KOR ICL3 insert) could be attempted. Such example could be found in the design of chimeric 5-HT_{5A} receptor (Che et al., 2020).

Via live-cell imaging, intracellular-expressing nanobodies (intrabodies) could be reversible biosensors to robustly detect the target protein. Here we have shown the co-localization of nanobodies and receptors on the plasm membrane and Golgi; others have demonstrated that nanobodies could also be directed to several cellular compartments, such as endosome or endoplasmic reticulum (Stoeber et al., 2018). It is expected that antigen-specific nanobodies can target endogenous expression levels of proteins and facilitate the functional study of protein in its native environment. Several GPCR-bound nanobodies have been shown as allosteric entities that could positively or negatively modulate the function of GPCRs (Che et al., 2020; Staus et al., 2016). Highly specific nanobodies could be used as in vivo probes to interrogate the receptor function in situ by enhancing or inhibiting the receptor through positive or negative allosteric modulation.

6 Conclusion

In this chapter, we provide three examples of using nanobodies as probes to monitor GPCR interaction, state conversion, and intracellular localization. We also provide strategies of applying available nanobodies to other GPCRs which nanobodies are unavailable through chimeric approaches. The unique properties of a nanobody (small size, enhanced stability) have significantly expanded their application from fundamental research to clinical intervention. We believe that the protocol illustrated here will guide the molecular characterization of nanobodies and explore their potential for different purposes.

References

Bacart, J., Corbel, C., Jockers, R., Bach, S., & Couturier, C. (2008). The BRET technology and its application to screening assays. *Biotechnology Journal*, *3*, 311–324.

Ballesteros, J. A., & Weinstein, H. (1995). Integrated methods for the construction of three-dimensional models and computational probing of structure-function relations in G protein-coupled receptors. *Methods in Neurosciences*, *25*, 366.

Bertrand, L., Parent, S., Caron, M., Legault, M., Joly, E., Angers, S., et al. (2002). The BRET2/arrestin assay in stable recombinant cells: A platform to screen for compounds that interact with G protein-coupled receptors (GPCRS). *Journal of Receptor and Signal Transduction Research*, *22*, 533–541.

Bobkov, V., Zarca, A. M., Van Hout, A., Arimont, M., Doijen, J., Bialkowska, M., et al. (2018). Nanobody-Fc constructs targeting chemokine receptor CXCR4 potently inhibit signaling and CXCR4-mediated HIV-entry and induce antibody effector functions. *Biochemical Pharmacology*, *158*, 413–424.

Boute, N., Jockers, R., & Issad, T. (2002). The use of resonance energy transfer in high-throughput screening: BRET versus FRET. *Trends in Pharmacological Sciences*, *23*, 351–354.

Bradley, M. E., Dombrecht, B., Manini, J., Willis, J., Vlerick, D., De Taeye, S., et al. (2015). Potent and efficacious inhibition of CXCR2 signaling by biparatopic nanobodies combining two distinct modes of action. *Molecular Pharmacology*, *87*, 251–262.

Bruchas, M. R., & Chavkin, C. (2010). Kinase cascades and ligand-directed signaling at the kappa opioid receptor. *Psychopharmacology*, *210*, 137–147.

Bruchas, M. R., & Roth, B. L. (2016). New technologies for elucidating opioid receptor function. *Trends in Pharmacological Sciences*, *37*, 279–289.

Brust, T. F., Morgenweck, J., Kim, S. A., Rose, J. H., Locke, J. L., Schmid, C. L., et al. (2016). Biased agonists of the kappa opioid receptor suppress pain and itch without causing sedation or dysphoria. *Science Signaling*, *9*, ra117.

Carlezon, W. A., Jr., & Krystal, A. D. (2016). Kappa-opioid antagonists for psychiatric disorders: From bench to clinical trials. *Depression and Anxiety*, *33*, 895–906.

Che, T. (2020). Advances in the treatment of chronic pain by targeting GPCRs. *Biochemistry*, *60*, 1401–1412.

Che, T., English, J., Krumm, B. E., Kim, K., Pardon, E., Olsen, R. H. J., et al. (2020). Nanobody-enabled monitoring of kappa opioid receptor states. *Nature Communications*, *11*, 1145.

Che, T., Majumdar, S., Zaidi, S. A., Ondachi, P., McCorvy, J. D., Wang, S., et al. (2018). Structure of the nanobody-stabilized active state of the kappa opioid receptor. *Cell*, *172*, 55–67 e15.

Cowan, A., Kehner, G. B., & Inan, S. (2015). Targeting itch with ligands selective for kappa opioid receptors. *Handbook of Experimental Pharmacology*, *226*, 291–314.

de Beer, M. A., & Giepmans, B. N. G. (2020). Nanobody-based probes for subcellular protein identification and visualization. *Frontiers in Cellular Neuroscience*, *14*, 573278.

De, A., Ray, P., Loening, A. M., & Gambhir, S. S. (2009). BRET3: A red-shifted bioluminescence resonance energy transfer (BRET)-based integrated platform for imaging protein-protein interactions from single live cells and living animals. *The FASEB Journal*, *23*, 2702–2709.

Dykstra, L. A., Gmerek, D. E., Winger, G., & Woods, J. H. (1987). Kappa opioids in rhesus monkeys. I. Diuresis, sedation, analgesia and discriminative stimulus effects. *The Journal of Pharmacology and Experimental Therapeutics*, *242*, 413–420.

Flock, T., Ravarani, C. N. J., Sun, D., Venkatakrishnan, A. J., Kayikci, M., Tate, C. G., et al. (2015). Universal allosteric mechanism for Gα activation by GPCRs. *Nature*, *524*, 173–179.

Heukers, R., De Groof, T. W. M., & Smit, M. J. (2019). Nanobodies detecting and modulating GPCRs outside in and inside out. *Current Opinion in Cell Biology*, *57*, 115–122.

Heukers, R., Fan, T. S., de Wit, R. H., van Senten, J. R., De Groof, T. W. M., Bebelman, M. P., et al. (2018). The constitutive activity of the virally encoded chemokine receptor US28 accelerates glioblastoma growth. *Oncogene*, *37*, 4110–4121.

Huang, W. J., Manglik, A., Venkatakrishnan, A. J., Laeremans, T., Feinberg, E. N., Sanborn, A. L., et al. (2015). Structural insights into mu-opioid receptor activation. *Nature*, *524*, 315–321.

Irannejad, R., Tomshine, J. C., Tomshine, J. R., Chevalier, M., Mahoney, J. P., Steyaert, J., et al. (2013). Conformational biosensors reveal GPCR signalling from endosomes. *Nature*, *495*, 534–538.

Jahnichen, S., Blanchetot, C., Maussang, D., Gonzalez-Pajuelo, M., Chow, K. Y., Bosch, L., et al. (2010). CXCR4 nanobodies (VHH-based single variable domains) potently inhibit chemotaxis and HIV-1 replication and mobilize stem cells. *Proceedings of the National Academy of Sciences of the United States of America*, *107*, 20565–20570.

Jovcevska, I., & Muyldermans, S. (2020). The therapeutic potential of nanobodies. *BioDrugs*, *34*, 11–26.

Kahsai, A. W., Xiao, K., Rajagopal, S., Ahn, S., Shukla, A. K., Sun, J., et al. (2011). Multiple ligand-specific conformations of the β2-adrenergic receptor. *Nature Chemical Biology*, *7*, 692–700.

Katritch, V., Cherezov, V., & Stevens, R. C. (2013). Structure-function of the G protein-coupled receptor superfamily. *Annual Review of Pharmacology*, *53*, 531–556.

Kenakin, T. (2002). Efficacy at G-protein-coupled receptors. *Nature Reviews. Drug Discovery*, *1*, 103–110.

Kocan, M., Dalrymple, M. B., Seeber, R. M., Feldman, B. J., & Pfleger, K. D. (2010). Enhanced BRET technology for the monitoring of agonist-induced and agonist-independent interactions between GPCRs and β-arrestins. *Frontiers in Endocrinology*, *1*, 12.

Kruse, A. C., Ring, A. M., Manglik, A., Hu, J., Hu, K., Eitel, K., et al. (2013). Activation and allosteric modulation of a muscarinic acetylcholine receptor. *Nature*, *504*, 101–106.

Lagerstrom, M. C., & Schioth, H. B. (2008). Structural diversity of G protein-coupled receptors and significance for drug discovery. *Nature Reviews. Drug Discovery*, *7*, 339–357.

Machleidt, T., Woodroofe, C. C., Schwinn, M. K., Mendez, J., Robers, M. B., Zimmerman, K., et al. (2015). NanoBRET—A novel BRET platform for the analysis of protein-protein interactions. *ACS Chemical Biology*, *10*, 1797–1804.

Manglik, A., Kim, T. H., Masureel, M., Altenbach, C., Yang, Z., Hilger, D., et al. (2015). Structural insights into the dynamic process of β2-adrenergic receptor signaling. *Cell*, *161*, 1101–1111.

Manglik, A., & Kobilka, B. (2014). The role of protein dynamics in GPCR function: Insights from the β2AR and rhodopsin. *Current Opinion in Cell Biology*, *27*, 136–143.

Manglik, A., Kobilka, B. K., & Steyaert, J. (2017). Nanobodies to study G protein-coupled receptor structure and function. *Annual Review of Pharmacology and Toxicology*, *57*, 19–37.

Masureel, M., Zou, Y., Picard, L. P., van der Westhuizen, E., Mahoney, J. P., Rodrigues, J., et al. (2018). Structural insights into binding specificity, efficacy and bias of a β2AR partial agonist. *Nature Chemical Biology*, *14*, 1059–1066.

Maussang, D., Mujic-Delic, A., Descamps, F. J., Stortelers, C., Vanlandschoot, P., Stigter-van Walsum, M., et al. (2013). Llama-derived single variable domains (nanobodies) directed against chemokine receptor CXCR7 reduce head and neck cancer cell growth in vivo. *The Journal of Biological Chemistry*, *288*, 29562–29572.

Millan, M. J. (1990). Kappa-opioid receptors and analgesia. *Trends in Pharmacological Sciences*, *11*, 70–76.

Muyldermans, S. (2013). Nanobodies: Natural single-domain antibodies. *Annual Review of Biochemistry*, *82*, 775–797.

Olsen, R. H. J., DiBerto, J. F., English, J. G., Glaudin, A. M., Krumm, B. E., Slocum, S. T., et al. (2020). TRUPATH, an open-source biosensor platform for interrogating the GPCR transducerome. *Nature Chemical Biology*, *16*, 841–849.

Palczewski, K. (2006). G protein-coupled receptor rhodopsin. *Annual Review of Biochemistry*, *75*, 743–767.

Pardon, E., Betti, C., Laeremans, T., Chevillard, F., Guillemyn, K., Kolb, P., et al. (2018). Nanobody-enabled reverse pharmacology on G-protein-coupled receptors. *Angewandte Chemie (International Ed. in English)*, *57*, 5292–5295.

Pardon, E., Laeremans, T., Triest, S., Rasmussen, S. G., Wohlkonig, A., Ruf, A., et al. (2014). A general protocol for the generation of nanobodies for structural biology. *Nature Protocols*, *9*, 674–693.

Pfeiffer, A., Brantl, V., Herz, A., & Emrich, H. M. (1986). Psychotomimesis mediated by kappa opiate receptors. *Science*, *233*, 774–776.

Pierce, K. L., Premont, R. T., & Lefkowitz, R. J. (2002). Seven-transmembrane receptors. *Nature Reviews. Molecular Cell Biology*, *3*, 639–650.

Ranganathan, M., Schnakenberg, A., Skosnik, P. D., Cohen, B. M., Pittman, B., Sewell, R. A., et al. (2012). Dose-related behavioral, subjective, endocrine, and psychophysiological effects of the kappa opioid agonist Salvinorin A in humans. *Biological Psychiatry*, *72*, 871–879.

Rasmussen, S. G., Choi, H. J., Fung, J. J., Pardon, E., Casarosa, P., Chae, P. S., et al. (2011). Structure of a nanobody-stabilized active state of the β2 adrenoceptor. *Nature*, *469*, 175–180.

Ring, A. M., Manglik, A., Kruse, A. C., Enos, M. D., Weis, W. I., Garcia, K. C., et al. (2013). Adrenaline-activated structure of β2-adrenoceptor stabilized by an engineered nanobody. *Nature*, *502*, 575–579.

Salahpour, A., Espinoza, S., Masri, B., Lam, V., Barak, L. S., & Gainetdinov, R. R. (2012). BRET biosensors to study GPCR biology, pharmacology, and signal transduction. *Frontiers in Endocrinology*, *3*, 105.

Scholler, P., Nevoltris, D., de Bundel, D., Bossi, S., Moreno-Delgado, D., Rovira, X., et al. (2017). Allosteric nanobodies uncover a role of hippocampal mGlu2 receptor homodimers in contextual fear consolidation. *Nature Communications*, *8*, 1967.

Staus, D. P., Strachan, R. T., Manglik, A., Pani, B., Kahsai, A. W., Kim, T. H., et al. (2016). Allosteric nanobodies reveal the dynamic range and diverse mechanisms of G-protein-coupled receptor activation. *Nature*, *535*, 448–452.

Stoddart, L. A., Johnstone, E. K. M., Wheal, A. J., Goulding, J., Robers, M. B., Machleidt, T., et al. (2015). Application of BRET to monitor ligand binding to GPCRs. *Nature Methods*, *12*, 661–663.

Stoddart, L. A., Kilpatrick, L. E., & Hill, S. J. (2018). NanoBRET approaches to study ligand binding to GPCRs and RTKs. *Trends in Pharmacological Sciences*, *39*, 136–147.

Stoeber, M., Jullie, D., Lobingier, B. T., Laeremans, T., Steyaert, J., Schiller, P. W., et al. (2018). A genetically encoded biosensor reveals location bias of opioid drug action. *Neuron*, *98*, 963–976 e965.

Szidonya, L., Cserzo, M., & Hunyady, L. (2008). Dimerization and oligomerization of G-protein-coupled receptors: Debated structures with established and emerging functions. *The Journal of Endocrinology*, *196*, 435–453.

Tejeda, H. A., Counotte, D. S., Oh, E., Ramamoorthy, S., Schultz-Kuszak, K. N., Backman, C. M., et al. (2013). Prefrontal cortical kappa-opioid receptor modulation of local neurotransmission and conditioned place aversion. *Neuropsychopharmacology*, *38*, 1770–1779.

Van Hout, A., Klarenbeek, A., Bobkov, V., Doijen, J., Arimont, M., Zhao, C., et al. (2018). CXCR4-targeting nanobodies differentially inhibit CXCR4 function and HIV entry. *Biochemical Pharmacology*, *158*, 402–412.

Vanderah, T. W. (2010). Delta and kappa opioid receptors as suitable drug targets for pain. *The Clinical Journal of Pain*, *26*(Suppl. 10), S10–S15.

Weis, W. I., & Kobilka, B. K. (2018). The molecular basis of G protein-coupled receptor activation. *Annual Review of Biochemistry*, *87*, 897–919.

White, K. L., Robinson, J. E., Zhu, H., DiBerto, J. F., Polepally, P. R., Zjawiony, J. K., et al. (2015). The G protein-biased kappa-opioid receptor agonist RB-64 is analgesic with a unique spectrum of activities in vivo. *The Journal of Pharmacology and Experimental Therapeutics*, *352*, 98–109.

Wu, H., Wacker, D., Mileni, M., Katritch, V., Han, G. W., Vardy, E., et al. (2012). Structure of the human kappa-opioid receptor in complex with JDTic. *Nature*, *485*, 327–332.

CHAPTER

8

Confocal and TIRF microscopy based approaches to visualize arrestin trafficking in living cells

Frédéric Gaëtan Jean-Alphonse[a,b] and Silvia Sposini[c,d,*]

[a]*CNRS, IFCE, INRAE, Université de Tours, PRC, Nouzilly, France*

[b]*Université Paris-Saclay, Inria, Inria Saclay-Île-de-France, Palaiseau, France*

[c]*Department of Metabolism, Digestion and Reproduction, Institute of Reproductive and Developmental Biology, Imperial College London, London, United Kingdom*

[d]*University of Bordeaux, CNRS, Interdisciplinary Institute for Neuroscience, Bordeaux, France*

**Corresponding author: e-mail address: silvia.sposini@u-bordeaux.fr*

Chapter outline

Methods in Cell Biology, Volume 166, ISSN 0091-679X, https://doi.org/10.1016/bs.mcb.2021.06.009

Abstract

Arrestins are key proteins that serve as versatile scaffolds to control and mediate G protein coupled receptors (GPCR) activity. Arrestin control of GPCR functions involves their recruitment from the cytosol to plasma membrane-localized GPCRs and to endosomal compartments, where they mediate internalization, sorting and signaling of GPCRs. Several methods can be used to monitor trafficking of arrestins; however, live fluorescence imaging remains the method of choice to both assess arrestin recruitment to ligand-activated receptors and to monitor its dynamic subcellular localization. Here, we present two approaches based on Total Internal Fluorescence (TIRF) microscopy and confocal microscopy to visualize arrestin trafficking in live cells in real time and to assess their co-localization with the GPCR of interest and their localization at specific subcellular locations.

1 Introduction

Arrestins are key proteins that serve as versatile scaffolds to control and mediate G protein coupled receptors (GPCR) activity. Due to their flexible structure, they can assume diverse structural conformations that permit interaction with a variety of cellular components including members of the endocytic machinery, signaling molecules or lipid components of the membrane (Xiao et al., 2007). Thus, they are involved in diverse cellular functions, however, they are best characterized for their role in the desensitization process of the GPCR-induced G protein activity and in serving as a scaffolding platforms for many proteins, including signaling molecules. In most cases, arrestin control of GPCR functions involves their recruitment from the cytosol to the plasma membrane-localized GPCRs to interfere with G protein-GPCR coupling and promote receptor internalization.

In this chapter we will focus on arrestin trafficking to the plasma membrane (PM) and to endosomes in response to receptor activation and internalization. We describe two approaches to visualize these steps through imaging of live cells through fluorescence microscopy.

2 Arrestins roles in GPCR trafficking and signaling

G protein coupled receptors (GPCRs) represent the vastest superfamily of membrane receptor proteins. Due to their widespread expression in virtually every tissue of the human body, they mediate a wide range of physiological responses, including endocrine physiology regulation, metabolism, neuromodulation, pain perception, reproduction. The canonical mechanism for GPCRs to translate messages from the extracellular environment is via signal transduction through coupling to distinct heterotrimeric G proteins. However, given that an individual cell can express more than 100 GPCRs (Hakak, Shrestha, Goegel, Behan, & Chalmers, 2003; Vassilatis et al., 2003), an outstanding question still to be answered is how such a limited number of G protein pathways can mediate diverse responses in different tissues. Two examples of how GPCR signaling is diversified are membrane trafficking and interaction with adaptor proteins other than G proteins that act as molecular modulators of GPCR function. Arrestins have been shown to display pivotal roles in both mechanisms. Four isoforms of arrestin have been identified: arrestin-1 and -4 are restricted to the visual system and accordingly named "visual arrestins" (Wilden, Wüst, Weyand, & Kühn, 1986), whereas the other two isoforms, arrestin-2 and -3, also called β-arrestin1 and 2, are ubiquitously distributed (Attramadal et al., 1992; Lohse, Benovic, Codina, Caron, & Lefkowitz, 1990). Arrestins are multifunctional and versatile proteins, serving as scaffold for a multitude of proteins including signaling and trafficking molecules and thus controlling various cellular functions (Crépieux et al., 2017; Xiao et al., 2007). Originally, their name was given by their ability to turn off or arrest G protein-mediated signaling by impeding coupling between GPCRs and G proteins in a process called desensitization. In contrast to the membrane-bound G proteins, arrestins are mainly cytosolic, and in most cases, their interaction with ligand-activated receptors is dependent on phosphorylation of the receptor's cytosolic interface mediated by GPCR-serine/threonine kinases (GRK) activity. Indeed, ligand binding and G protein coupling is rapidly followed by GPCR phosphorylation, which results in the increased affinity for arrestin, leading to arrestin trafficking from the cytoplasm to PM localized GPCRs. Distinct phosphorylation profiles of the receptor, or the so called "phosphorylation barcodes" or "phospho-signature," have important consequences not only on the relative affinity between arrestins and receptors, but also on the arrestin conformation and subsequent functions (Nobles et al., 2011). GPCRs C-terminus pattern of phosphorylation is typically associated with either transient or long-lasting interactions with arrestin, impacting on arrestin subcellular localization and signaling (Nobles et al., 2011; Pavlos & Friedman, 2017; Tobin, Butcher, & Kong, 2008). Phosphorylation signature is associated with diverse modes of β-arrestin binding to the receptor, either through the core (ICLs), the C-tail or both (Shukla et al., 2014).

Interestingly, considering that only two isoforms of arrestins are exploited by GPCRs in comparison to the tens of G protein types, one must really appreciate the capacity of β-arrestins to sense, through their N-terminal domain, the different phosphorylation forms of the receptor (Zhou et al., 2017). This determines conformational changes in the β-arrestins themselves resulting in "conformational signatures" recognized by different downstream effectors as another way to diversify GPCR activity (Lee et al., 2016; Nuber et al., 2016).

β-arrestin canonical roles in desensitization are not limited to uncoupling receptor and G protein and G protein-dependent signal termination. β-arrestins play a key role in receptor clustering into clathrin-coated pits (CCPs), the first step of GPCR internalization, as well as mediating signaling from the PM (Eichel et al., 2018). Following accumulation of GPCRs complexes in clathrin structures, GPCRs are targeted to defined endocytic cellular locations known as endosomes or sorting endosomes, where they undergo distinct post-endocytic fates. GPCRs can either be recycled back to the PM, traffic to lysosomes for degradation and permanent signal termination or, in a few cases, trafficked to the Trans-Golgi Network (Godbole, Lyga, Lohse, & Calebiro, 2017; Hanyaloglu & von Zastrow, 2008). As part of theses GPCR endocytic trafficking, the nature of receptor interaction with arrestins, short or prolonged, is often associated with distinct residence time in the endosomal system, with a more prolonged endosomal lifetime for GPCRs tightly bound to arrestins. This led to the classification of GPCRs into two main categories or classes; the class A GPCRs, such as the β2 adrenergic receptor β2AR and the μ opioid receptor MOR, are GPCRs that bind β-arrestin2 with higher affinity than β-arrestin1 but dissociate from arrestins at the PM and do not traffic with them into endocytic vesicles. In contrast, class B receptors (angiotensin II type 1A receptor AT1R, neurotensin receptor 1, vasopressin V2 receptor V2R) bind both β-arrestin isoforms with similar high affinities and remain in complex with β-arrestins along their intracellular trafficking into endocytic vesicles (Oakley, Laporte, Holt, Caron, & Barak, 2000). Indeed, receptor C-tail phospho-signature is largely responsible these trafficking events. The presence of β-arrestins at endosomal compartments triggered the discovery that these adaptor proteins play important modulatory roles not only in GPCR post-endocytic trafficking, but also in signaling from these endocytic compartments (Lobingier & von Zastrow, 2019; Pavlos & Friedman, 2017; Peterson & Luttrell, 2017; Vilardaga, Jean-Alphonse, & Gardella, 2014), expanding our knowledge on the spatial and temporal control of GPCR activity.

Yet, we need to precise that not all GPCRs behave in the same way in regard to their interaction and trafficking with arrestins. Several approaches allow to monitor trafficking of arrestins including several microscopy techniques, Bioluminescence Resonance Energy Transfer (BRET) and Fluorescence Resonance Energy Transfer (BRET) (Cao, Namkung, & Laporte, 2019), however, among them, live fluorescence imaging remains the method of choice to both assess arrestin recruitment to ligand-activated receptors and to monitor its dynamic subcellular localization.

3 β-arrestin trafficking to the PM

Upon binding to active and phosphorylated receptors, arrestins undergo structural modifications that allow them to reach an active state. In this conformation, arrestins are able to interact with other proteins present at the PM, such as AP-2 and clathrin, and with the lipid bilayer itself.

Indeed, β-arrestins interact with a series of phosphoinositides. Specifically, interaction with IP6, PI(3,4,5)P_3, PI(4,5)P_2, IP_4 and IP_3 have been reported, with arrestin-IP6 interaction displaying the highest affinity (Gaidarov, Krupnick, Falck, Benovic, & Keen, 1999). IP6 is a soluble phosphoinositide which regulates arrestin functions by modulating its oligomerization. It has been proposed that via promoting β-arrestin1 and 2 homo- and hetero-oligomerization, IP6 sequestrates arrestin molecules in the cytoplasm, thus playing a negative role in arrestin trafficking to the PM (Milano, Kim, Stefano, Benovic, & Brenner, 2006). While mutation of IP6 interaction sites did not alter arrestin ability to interact with the endocytic machinery, mutation of three specific residues in arrestin's C-domain responsible for phosphoinositides binding abolished β2AR recruitment to CCPs and endocytosis, yet interaction with clathrin was preserved (Eichel et al., 2018; Gaidarov et al., 1999). Furthermore, interaction with PI(4,5)P_2 was shown to be essential to maintain arrestin clustered at CCPs after dissociation from its activating class A GPCR, such as β2AR or β1AR, but not when associated to a class B GPCR, such as V2R, as this class of receptors exhibit prolonged interaction with arrestin even after their internalization (Eichel et al., 2018).

Direct evidence for arrestin engagement with the PM come from even more recent studies. Cryo-electron microscopy revealed the structure of β-arrestin1 in complex with M2 muscarinic receptor reconstituted in lipid nanodiscs and showed that the β-arrestin1 C-domain edge interacts with the phospholipid bilayer of the nanodisc (Staus et al., 2020). This interaction stabilizes the active conformation of β-arrestin1 and, in addition to receptor phosphorylation, it is required to achieve high affinity coupling with the receptor transmembrane bundle (Staus et al., 2020). These findings expand and corroborate the data obtained by crystallography (Kang et al., 2015) and molecular dynamics simulations (Lally, Bauer, Selent, & Sommer, 2017) conducted on arrestin1 and rhodopsin complex, which proposed for the first time that the C-domain of arrestin1 functions as a membrane anchor and that this anchoring function is essential for the specificity and affinity of arrestin binding to the receptor (Lally et al., 2017).

It is interesting to note that, while other GPCR-interacting proteins engage with the PM through lipid anchors, arrestins do not show such anchors, yet interact with the lipid bilayer in their own peculiar fashion. Post-translation modifications such as phosphorylation and ubiquitination play important roles in modulating the trafficking of arrestins and their functions. Several phosphorylated residues have been identified in the β-arrestin sequences, including Ser^{412} for β-arrestin1 (Barthet et al., 2007; Lin et al., 1997; Lin, Miller, Luttrell, & Lefkowitz, 1999) and Thr^{276}, Ser^{361} and Thr^{383} for β-arrestin2 (Lin et al., 2002; Paradis et al., 2015). These phosphorylation events affect GPCR internalization and/or sequestration and, consequently, steady-state level of GPCR cell-surface expression: phosphorylation of β-arrestin1 at Ser^{412} by Erk1/2 as well as the phosphorylation of β-arrestin2 at both Ser^{361} and Thr^{383} reduce their ability to induce internalization of β2AR (Lin et al., 1997, 1999, 2002), whereas phosphorylation of β-arrestin2 at Ser^{14} and Thr^{276} promotes intracellular sequestration of CXCR4 receptor (Paradis et al., 2015). These results indicate that β-arrestin2 phosphorylation exerts contrasting effects on GPCR trafficking that may depend on the nature of the phosphorylated residue(s) and of the GPCR.

Soon after their activation induced by the interaction with a ligand-bound GPCR, arrestins become ubiquitinated. Attachment of ubiquitin to lysine residues of arrestins is performed by E3 ubiquitin ligases like Nedd4 or Mdm2 (Shenoy et al., 2008). Ubiquitination of arrestins can be transient, with ubiquitin moieties removed by deubiquitinases before the receptor is endocytosed, or persistent, in the case of receptor and arrestin co-internalization. It was indeed demonstrated that trafficking of arrestin to endosomes together with the class B receptor V2R requires arrestin to be ubiquitinated (Shenoy & Lefkowitz, 2003). As a matter of fact, an arrestin mutant lacking ubiquitination sites forms an unstable complex even with class B GPCRs, while an arrestin chimera permanently ligated to ubiquitin preserves its interaction with class A GPCRs even after internalization (Shenoy & Lefkowitz, 2003). Interestingly, ubiquitinases do not only perform post-translational modifications of arrestins, but they use arrestins as scaffold to ubiquitinate other substrates that are in complex with arrestins, namely receptors such as the β2AR and the class C metabotropic glutamate receptor 7 (Lee et al., 2019; Shenoy, McDonald, Kohout, & Lefkowitz, 2001).

4 β-arrestin actions from the PM

β-arrestins mediate GPCR endocytosis through the binding of both receptors and the CCP proteins clathrin and its adaptor AP2. Once GPCRs are concentrated in CCPs, the scission of clathrin-coated vesicles from the PM is achieved by the large GTPase dynamin, resulting in GPCR internalization to endosomes. Although β-arrestins appear to be fundamental in GPCR endocytosis, β-arrestin-independent GPCR internalization has been reported for receptors such as the Protease-Activated receptor

4 PAR4 (Smith et al., 2016; van Koppen & Jakobs, 2004). Thus, GPCRs greatly differ in how they utilize this GRK/arrestin/CCP model for their desensitization and internalization.

Although GRK and β-arrestin induce rapid desensitization, several studies indicate that these common internalization and desensitization mechanisms have expanding roles in the spatio-temporal control of GPCR signaling. For example, there is known heterogeneity in CCP function, where different GPCRs can be organized within distinct subsets of CCPs to dictate subsequent post-endocytic fate between recycling or lysosomal pathways (Mundell, Luo, Benovic, Conley, & Poole, 2006). Such divergent sorting of GPCRs at the PM was mechanistically dependent on GRK/arrestin (for CCPs directing receptors to the recycling pathway) and receptor phosphorylation by second messenger kinases (for lysosomal sorting).

As mentioned above, it was long admitted that the main functions of arrestins were to interfere with G protein coupling, to target receptors to CCPs and facilitating their internalization and promoting further desensitization. However, because of the nature of arrestins to serve as a scaffold for signaling molecules, it becomes more obvious that GPCRs can induce an arrestin-dependent signaling (Gurevich & Gurevich, 2018; Peterson & Luttrell, 2017). Indeed, recruitment of arrestins to receptors leads to a number of structural changes that reveal binding sites, allowing effectors such as JNK, ERK1/2, MEK, STATs to assemble into a signalosome (Chen, Iverson, & Gurevich, 2018).

For receptors such as the MOR, receptor-arrestin residency time at CCPs plays a role in the magnitude of arrestin-mediated signaling and is dependent on the nature of the MOR ligand (Weinberg, Zajac, Phan, Shiwarski, & Puthenveedu, 2017). Interestingly, another recent study has demonstrated that arrestin-mediated signal transduction can take place even in absence of arrestin-receptor clusters. Indeed, although activated β1AR poorly clusters compared to GPCRs such as the β2AR, it robustly recruits β-arrestin2 to CCPs without detectable receptor localized to CCPs and this CCP-bound β-arrestin was shown to activate a prolonged MAPK signal profile (Eichel, Jullié, & von Zastrow, 2016). In another study, the same authors propose that in this "activation at a distance" model, arrestin is captured at the PM through interactions with phospholipids after a brief interaction with the activated receptor's core, independently of the C-tail phosphorylation status (Eichel et al., 2018). Atomic-level simulations of arrestin have indeed proved that arrestin binding to either receptor's core or C-tail can independently lead to arrestin activation (Latorraca et al., 2018).

The idea that both core and C-tail interactions are necessary for receptor endocytosis and signaling has been further challenged by another study where V2R was shown to efficiently internalize and activate ERK response thanks to the engagement between β-arrestin1 and the sole receptor tail (Kumari et al., 2017).

It must also be mentioned that, although arrestins play the role of positive mediators of MAPK phosphorylation for the majority of GPCRs, they can also serve as negative regulators of signaling. In the case of AT1R, knock-down of β-arrestin1, but not β-arrestin2, caused an increase in the phosphorylation of ERK1/2 (Ahn, Wei, Garrison, & Lefkowitz, 2004). The molecular mechanism behind the either promoting or inhibiting role of arrestins in ERK1/2 activation has been recently elucidated by Baidya et al., who propose that spatial positioning of phosphorylation sites in the receptors recognized by arrestins determine the direction of arrestin contribution to this signaling pathway (Baidya, Kumari, Dwivedi-Agnihotri, Pandey, Chaturvedi, et al., 2020a; Baidya, Kumari, Dwivedi-Agnihotri, Pandey, Sokrat, et al., 2020b). Moreover, the spatial positioning of single phosphorylation sites has been recognized to have a broader role on arrestin physiology, determining the physical recruitment of arrestins to the receptor as well as their role on receptor trafficking (Dwivedi-Agnihotri et al., 2020).

5 β-arrestin actions from endocytic compartments

Following internalization, GPCRs are targeted to defined cellular locations, thus mediating distinct signaling fates (Hanyaloglu & von Zastrow, 2008). Although activation at the PM has always been considered as the main source of GPCR signaling, not all internalized receptors are quiescent. In fact, as mentioned before, it was long known that arrestin-mediated signaling can occur from endosomes, however, the discovery that some GPCRs can prolong G protein-mediated signaling from endosomes has challenged our current view on the role of receptor internalization (Pavlos & Friedman, 2017). To date, a variety of intracellular compartments have been described as potential signaling hubs for GPCRs, including endosomes and Golgi apparatus (Irannejad & von Zastrow, 2014). We will focus on endosomal signaling as these endomembranes exhibit co-presence of receptor and arrestins.

Endosomal G protein signaling was first described for class B GPCRs that stably interact and co-internalize with arrestins in endosomes, such as the parathyroid hormone receptor (PTH1R), the thyroid stimulated hormone receptor (TSHR) and vasopressin receptor type 2 (V2R) (Calebiro et al., 2009; Feinstein et al., 2013; Ferrandon et al., 2009; Wehbi et al., 2013). Since then, endosomal signaling by class A GPCRs, characterized by a much weaker interaction with arrestins and a more rapid dissociation, was also demonstrated to occur following internalization. However, the sustained signaling from endocytic compartments by these receptors seems to be mediated by a "conventional-like" mechanism that occurs at the PM and in a β-arrestin independent manner (Irannejad et al., 2013). For receptors forming a stable complex with β-arrestin, β-arrestin co-traffics with receptors from the cell surface to endosomes and, in opposition to its desensitization role at the PM, β-arrestin supports G protein-mediated signaling rather than interfering with it (Jean-Alphonse et al., 2017; Wehbi et al., 2013). Recent advances in structural studies allowed to better understand the apparent contradictory arrestin functions at the PM and in the endosome. Despite the long lasting interaction with arrestin, Cryo-EM analysis of the β2AR-V2R C-tail chimera showed that arrestin can adopt different conformation within the receptor, including a "tail conformation" in which arrestin can form a stable complex by only engaging through the receptor C-tail (Shukla et al., 2014); this conformation opens up a space for a G protein to interact with the receptor's core leading to the formation of a GPCR/G protein/β-arrestin "super-complex" or "megaplex" at the level of individual endosomes (Jean-Alphonse et al., 2017; Nguyen et al., 2019; Thomsen et al., 2016).

Apart from playing pivotal roles in sustaining class B receptors signaling, endosomal arrestin has been shown to regulate the post-endocytic sorting of several GPCRs. For example, β-arrestin1 mediates chemokine receptor CXCR4 degradation by promoting its trafficking from early endosomes to lysosomes, via an interaction between arrestin and the E3 ubiquitin ligase atrophin-interacting protein 4 (AIP4) which occurs at the endosomes (Bhandari, Trejo, Benovic, & Marchese, 2007). For receptors that are sorted to the recycling pathway such as the bradykinin receptor 2 B2R, it has been proposed that, upon co-internalization into endosomes, β-arrestin2 is required to dissociate from the receptor for its recycling back to the PM. Stabilization of the arrestin-B2R complex results in receptor trapping in endosomes (Simaan, Bedardgoulet, Fessart, Gratton, & Laporte, 2005). A study conducted on natural variants of the class B AT1R showed that it is the ability of β-arrestin2 to form long-lived complexes with receptors in endosomes, rather than at the PM, to dictate both the MAPK signaling output as well as the post-endocytic trafficking of the receptor. Indeed, receptor variants that could not form stable complexes with arrestin at endosomes, were recycled to the PM mimicking class A receptors behavior (Cao et al., 2020).

6 Significance of arrestin trafficking

β-arrestins have been intensively studied for more than two decades. Their control on both GPCR signaling and trafficking fate, and their physiological consequences, are well documented for many but not all GPCRs. It is now obvious that arrestin functions rely on the nature and the duration of their interaction with receptors as well as the particular conformation adopted at a time. They are also critically dependent on the nature of the assembly with signaling or trafficking molecules. Additionally, arrestin subcellular distribution and trafficking are also key features to take into consideration as they strongly correlate with control and fine-tuning of GPCR signaling activities in both spatial and temporal manners. These various aspects of arrestin are particularly critical for the development and characterization of new ligands targeting GPCRs. Indeed, because some ligands can display "functional selectivity" properties that favor/stabilized specific receptor conformations, they can also lead to important functional consequences on both arrestin signaling, and on the spatial and temporal distribution of both arrestin and receptors. Therefore, experimental approaches suitable to assess arrestin trafficking at the subcellular level allow to better understand how arrestin dynamics correlate with GPCR activity and by extension with cell behavior.

7 Overview of the protocols

Here, we describe two main methods to visualize β-arrestin trafficking in living cells, following GPCR activation. Both methods are based on imaging by fluorescence microscopy. The first method uses Total Internal Reflection Fluorescence (TIRF) microscopy; TIRF employs the phenomenon of the evanescent wave to excite only the fluorophores that reside in a range of 100 nm along the z axes from the incident light, meaning from the PM of the cells that is touching the glass coverslip. Apart from high axial resolution, TIRF achieves outstanding signal-to-noise ratio, making it the technique of choice for the visualization of PM associated processes, such as arrestin recruitment from the cytosol to membrane-bound GPCRs. The second method uses confocal microscopy; by focusing the laser light onto a specific point within the sample, cells are optically sliced at different focal planes. This allows to obtain sharp images of fluorescent objects at different depths within the cell, such as endosomes. In order to perform live-cell fluorescence imaging, fluorescent tags on the proteins of interest are required. This means that, unless genome-edited cell lines are used (achieved by CRISPR/Cas9 mediated knock-in), overexpression of the protein of interest needs to be performed. According to the specifics of the available microscopes (number and wavelengths of exciting lasers, optical filters, speed of the cameras, etc.) one can perform imaging of multiple proteins tagged with different fluorophores; one example is shown in Fig. 1, where 3-color imaging of receptor,

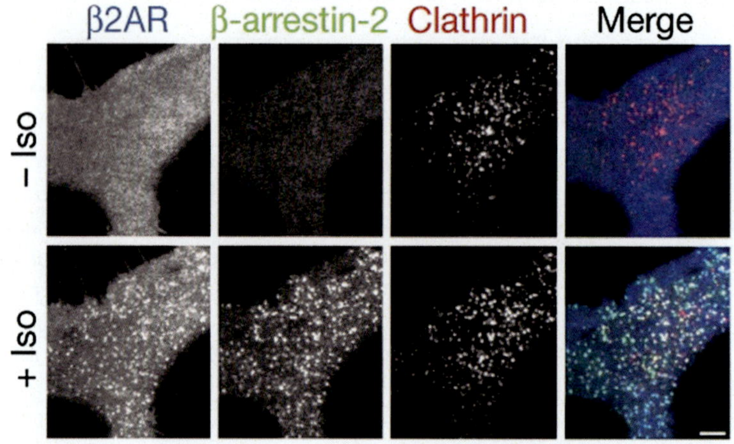

FIG. 1

Live-cell TIRF microscopy shows trafficking of β-arrestin-2 to the PM and cluster formation upon agonist stimulation (isoproterenol) of β2AR. HEK293 cells transiently transfected with Flag-β2AR (blue), β-arrestin-2-GFP (green) and clathrin-light-chain–DsRed (red) before and after treatment with isoproterenol (ISO, 1 μM). Scale bar = 1 μm.

Adapted from Eichel, K., Jullié, D., Barsi-Rhyne, B., Latorraca, N. R., Masureel, M., Sibarita, J.-B., Dror, R. O., & von Zastrow, M. (2018). Catalytic activation of β-arrestin by GPCRs. Nature, *557(7705), 381–386. https://doi.org/10.1038/s41586-018-0079-1.*

arrestin and clathrin is achieved in the same sample at the same time. Instead of clathrin, other adaptor proteins can be visualized, as wells as intrabodies (Baidya, Kumari, Dwivedi-Agnihotri, Pandey, Chaturvedi, et al., 2020a; Baidya, Kumari, Dwivedi-Agnihotri, Pandey, Sokrat, et al., 2020b), organelle markers and phospholipids.

8 Step-by-step methods

8.1 HEK293 cell culture

8.1.1 Materials and reagents

- Sterile Phosphate Buffered Saline (PBS)1 ×
- Trypsin/EDTA (Thermo Fisher, cat. no. 25300054)
- DMEM (Thermo Fisher, cat. no. 31966047)
- Penicillin/Streptomycin (Thermo Fisher, cat. no. 15070063)
- Fetal Bovine Serum (FBS) (Thermo fisher, cat. no. A4766801)
- Lipofectamine 2000 (Thermo Fisher, cat. no. 11668030)
- Opti-MEM (Thermo Fisher, cat. no. 31985062)
- 18-mm glass coverslips, 35 mm glass bottom dishes or glass bottom chambered cell culture slides No. 1.0 glass thickness (MatTek Corporation)
- T75 flasks, sterile
- 6- and 12-well plates, sterile
- cDNAs of interest; e.g., Flag-tagged β2AR, fluorescently tagged (YFP, GFP, etc.) β-arrestin 1 or 2
- Poly-D-Lysine (PDL) (Thermo Fisher, cat. no. A3890401)

8.1.2 Procedure: Passaging cells

Timing: 10–30 min

a. Take T75 flask containing confluent HEK293 cells from the incubator and visually inspect health and density of the cells under a transmitted light microscope.
b. Bring flask under the hood and aspirate medium.
c. Gently add 5 mL of pre-warmed PBS 1 × to the flask, on the side opposite to the one cells are plated, rock the flask and aspirate the PBS.
d. Add 1 mL of pre-warmed trypsin and incubate flask at 37 °C for 2–3 min.
e. Visually inspect the flask to check that cells have detached.
f. Add 10 mL of pre-warmed DMEM supplemented with 10% (v/v) FSB, 1% (v/v) penicillin/streptomycin and resuspend.
g. Transfer the cell suspension in a 15 mL tube and centrifuge for 5 min at 1000 rpm.
h. Aspirate the supernatant and resuspend the pellet in 10 mL of DMEM supplemented with 10% (v/v) FSB, 1% (v/v) penicillin/streptomycin.
i. Count cells using a counting chamber.
j. Transfer 360,000 cells to each well of a 6-well plate containing 2 mL of medium in each well. The number of wells to use is decided according to the number of different transfections keeping in mind that 1 well corresponds to 1 transfection.

Note

All steps should be conducted under sterile conditions.

8.1.3 Procedure: Transfecting cells

Timing: 20–30 min

a. Take 6-well plate containing HEK293 cells plated the day before from the incubator and visually inspect health and density of the cells under a transmitted light microscope. Cells should be 60% confluent. Put the plate back in the incubator.
b. Prepare a transfection mix containing Opti-MEM, cDNAs of interest and Lipofectamine 2000 and according to the manufacturer's instructions.
c. Add transfection mix directly onto the cells.

Notes

i All steps should be conducted under sterile conditions.
ii The GPCR of interest should be N-terminally tagged.

8.1.4 Procedure: Plating cells on PDL coated glass coverslips

Timing: 40–60 min

a. Take a 12-well plate containing 18 mm coverslips, 35 mm glass bottom dishes or glass bottom chambered slides and add enough volume of PDL to cover the surface (e.g., 0.3 mL for a well of a 12 well plate).
b. Incubate at room temperature for half an hour.
c. Aspirate PDL and wash with sterile PBS or water three times.
d. Let the plate dry under the hood.
e. In the meanwhile, take the 6-well plate containing HEK293 cells transfected the day before from the incubator and visually inspect health and density of the cells under a transmitted light microscope.
f. Bring plate under the hood and aspirate medium.
g. Gently add 0.5 mL of pre-warmed PBS 1 × to each well and aspirate.
h. Add 0.2 mL of pre-warmed trypsin and incubate plate at 37 °C for 1–2 min.
i. Visually inspect the plate to check that cells have detached.
j. Add 2 mL of pre-warmed DMEM supplemented with 10% (v/v) FSB, 1% (v/v) penicillin/streptomycin and resuspend.
k. Transfer the cell suspension into PDL coated wells, in order to obtain 40% confluent cells on the day of the experiment.

Alternatives

i. As an alternative to PDL coating, PLL or another coating reagent compatible with cell live imaging can be used. Coating helps cells to adhere to glass surfaces, thus it prevents cell detachment during washing steps. No coating is also an option, but extra care needs to be taken when handling the HEK293 cells to avoid cell loss.
ii. Instead of glass coverslips, 35 mm glass bottom dishes or glass bottom chambered cell culture slides could be used.

Notes

i. All steps should be conducted under sterile conditions.
ii. Cell should not be over confluent on the day of the experiment to be able to image individual cells.

8.1.5 Optimization and troubleshooting

We recommend to optimize the amount of DNA for transfection trying different conditions. As a starting point, we suggest 2 μg of DNA of receptor and 0.5 μg of β-arrestin for each well of a six well plate; however, we are aware of successful results using up to 10–100 times less DNA.

8.2 Monitoring arrestin trafficking via TIRF microscopy

8.2.1 Materials and reagents

- Fluorescent beads (Tetraspeck microspheres, 0.2 μm; Thermo Fisher, cat. no. T7280).
- Immersion oil (Cargille, cat. no. 16245).
- Imaging medium; e.g., Phenol red-free DMEM (Thermo Fisher, cat. no. 21063029), FluoroBrite DMEM (Thermo Fisher, cat. no. A1896701) or HEPES Buffer Solution (NaCl 135 mM, KCl 5 mM, $CaCl_2$ 1.8 mM, $MgCl_2$ 0.4 mM, Glucose 1 mM, HEPES 20 mM).
- Ligand used to stimulate internalization of receptor of interest; e.g., Isoproterenol (Merck, cat. no. I6504) in the case of β2AR.
- Mouse M1-anti Flag antibody conjugated to fluorophore; e.g., AlexaFluor488 conjugated FLAG-tag monoclonal antibody (Thermo Fisher, cat. no. MA1-142-A488).

8.2.2 Procedure

a. One hour before imaging, switch on TIRF set-up to allow it to reach working temperature and stabilize. Set the heating system to 37 °C.
b. Test TIRF set-up for alignment and uniformity of illumination by using fluorescent beads. Acquire an image of the beads in the two channels used for live-cell imaging.
c. Take the 12-well plate containing HEK293 cells plated the day before from the incubator and visually inspect health and density of the cells under a transmitted light microscope.
d. If cells look healthy (e.g., not round, not over confluent), remove medium and gently rinse with serum free, red phenol free, pre-warmed imaging medium.
e. Incubate cells for 10 min at 37 °C with M1-anti FLAG antibody conjugated to fluorophore diluted 1:500 in imaging medium.
f. Rinse cells three times, mount the coverslip onto a live-imaging chamber, add a volume of imaging medium to prevent cells from drying.
g. Clean both objective and bottom of coverslip with lens paper and a drop of ethanol.

h. Add a drop of oil on the microscope objective and quickly transfer the live-imaging chamber on it.

i. Look for the cell of interest, which ideally expresses normal levels of receptor and low level of arrestin.

Receptor should have a uniform distribution throughout the surface of the cell that is touching the glass coverslip. The receptor channel should guide the experimenter in adjusting both TIRF angle and focus if necessary. Arrestin should appear very faint, as prior to receptor stimulation it is localized to the cytosol thus not clearly visible in the TIRF field.

j. Start recording. Record 3–5 min of basal, then add ligand to the chamber to reach the desired final concentration (e.g., for Isoproterenol a final concentration of 10 μM was used) and continue imaging for further 10 min. We recommend imaging at a frame rate of 0.5 Hz (one frame every 2 s) to preserve cells from photo-damage, considering the length of the recording.

Notes

i It is important to choose cells with low expression of arrestin to be able to visualize its recruitment at the PM. If levels at the PM are already high before stimulation with receptor ligand, arrestin recruitment will hardly be observed.

ii In order to achieve optimal TIRF imaging, air bubbles should not be present between the oil and the glass coverslip.

iii Only one cell per coverslip can be imaged, to be able to record the same cell before and after stimulation.

8.2.3 Expected outcomes

Arrestin recruitment should appear evident within the first couple of minutes after agonist addition. It appears as an increase in fluorescence in the TIRF field as a consequence of arrestin translocation from the cytosol to the PM. Clusters should also appear; these could be co-localized to receptor clusters (e.g., β2AR) or not, in the case of receptors that poorly cluster in response to agonist stimulation (e.g., β1AR) (Eichel et al., 2018).

8.2.4 Analysis

We recommend talking to experts in the field, for example, members of microscopy facilities, to help you find the best and most rigorous way to analyze your images. We also recommend to review the literature for general and more specific advice on fluorescence microscopy (Lee & Kitaoka, 2018).

- *Fluorescence intensity*

To measure arrestin recruitment to the PM, we suggest measuring the fluorescence of the arrestin channel before and after stimulation with the receptor's ligand within each cell, through the use of a mask. This can be easily done using the open source software ImageJ (or Fiji) but can also be achieved using alternative image analysis software. Using ImageJ, a mask must be drawn around each cell of interest using the freehand selection tool. Then, go to "Analyze" menu and select "Measure". After having calculated the fluorescence within the cell, repeat the same steps for an area outside the cell; this will your background fluorescence, which needs to be subtracted from the fluorescence values calculated for your cells.

- *Number and lifetime of arrestin clusters*

For this type of analysis, we suggest referring to Eichel et al. (2018) and Weinberg et al. (2017). Quantitative assessment of arrestin clusters and lifetime can be achieved by manual analysis using ImageJ. Qualitative assessments can be achieved through Maximum Intensity Projections ("Image–Stacks–Z project" path in ImageJ) for the number of clusters and through Kymographs ("Image–Stacks–Reslice" path in ImageJ or by using KYMOMAKER (Chiba, Shimada, Kinjo, Suzuki, & Uchida, 2014)) for clusters lifetimes.

- *Co-localization between arrestin, receptor clusters and other markers*

Co-localization can be assessed by manually selecting endosomes in the receptor channel to obtain a list of regions of interest (ROIs) using the selection tools in ImageJ, and the checking if in those specific regions fluorescence from the arrestin channel is also present. Furthermore, this can be corroborated by line scanning analysis complemented by some coefficients that measure the degree of co-localization, such as Pearson's correlation coefficient (PCC) or Mander's overlapping coefficient (MOC) (Bolte & Cordelières, 2006; Dunn, Kamocka, & McDonald, 2011). As endosomes are identified as objects, we recommend using object-based approaches. Different ImageJ plugins are available to quantitatively assess co-localization, such as JACoP (Bolte & Cordelières, 2006).

8.2.5 Advantages

TIRF is probably the best technique to visualize events that take place at the PM of cells. This is due to the high signal-to-noise ratio that this technique achieves, thanks to the fact that only a few nanometers inside the cell are excited by the incident laser, thus avoiding the visualization of all fluorescent molecules that reside and move inside the cell.

8.2.6 Limitations

Because of the particular set-up of this technique, it is difficult to visualize intracellular cellular compartments other than the PM. In some exception, organelles residing underneath the PM can be observed. However, slight modification of the imaging set up, such as variation of the angle with which the laser hits the sample, one can achieve imaging of slighter inner depths into the cell, the so called highly inclined and laminated optical sheet (HILO) imaging.

8.2.7 Optimization and troubleshooting

If arrestin recruitment is not observed, this might be due to expression levels of arrestin being either too high or too low. We recommend choosing cells where a faint fluorescence is visible do to the presence of arrestin in the cytoplasm and thus only mildly visible in the TIRF field. If the fluorescence of the arrestin channel is already high before ligand addition, it's most likely due to the fact that the cell is expressing such high levels of arrestin that it is already localized at the PM, thus ligand stimulation may not provoke any detectable change in its subcellular localization. On the other hand, choosing cells with no detectable levels of fluorescence in the arrestin channel, could lead to no detection of arrestin even after ligand stimulation; this is due to the fact that either the cell is not expressing arrestin at all, or it is expressing it at such low levels that are below the detection limit. We recommend setting up a pilot experiment and image cells with different amounts of arrestin, to understand which are the levels that lead to the best detection of arresting recruitment and clustering.

Other possibilities include that some particular receptors may preferentially bind to a specific isoform of arrestin or have very transient or no interaction at all with arrestins. We suggest reviewing the literature and/or trying arrestin recruitment with either isoform or with a receptor model such as β2AR or V2R.

8.2.8 Safety considerations

A laser safety course should be taken before conducting experiments using light microscopy.

8.3 Monitoring arrestin trafficking via confocal microscopy

8.3.1 Materials and reagents

- Imaging medium; e.g., Phenol red-free DMEM (Thermo Fisher cat. no. 21063029), FluoroBrite DMEM (Thermo Fisher cat. no. A1896701) or HEPES Buffer Solution (NaCl 135 mM, KCl 5 mM, $CaCl_2$ 1.8 mM, $MgCl_2$ 0.4 mM, Glucose 1 mM, HEPES 20 mM).
- Ligand used to stimulate internalization of receptor of interest; e.g., Arginine Vasopressin AVP (Merck cat. no. V9879) in the case of V2R.
- Mouse M1-anti Flag antibody conjugated to fluorophore; e.g., AlexaFluor488 conjugated FLAG-tag monoclonal antibody (Thermo Fisher cat. no. MA1-142-A488).

8.3.2 Procedure

a. One hour before imaging, switch on confocal set-up to allow it to reach working temperature and stabilize. Set the heating system to 37 °C.

b. Take the 12-well plate containing HEK293 cells plated the day before from the incubator and visually inspect health and density of the cells under a transmitted light microscope.

c. If cells look healthy (e.g., not round, not over confluent), remove medium and gently rinse with serum free, red phenol free, pre-warmed imaging medium.
d. Incubate cells for 10 min at 37 °C with M1-anti FLAG antibody conjugated to fluorophore diluted 1:500 in imaging medium.
e. Rinse cells three times, mount the coverslip onto a live-imaging chamber, add a volume of imaging medium to prevent cells from drying.
f. Clean both objective and bottom of coverslip with lens paper and a drop of ethanol.
g. Add a drop of oil on the microscope objective and quickly transfer the live-imaging chamber on it.
h. Look for the cell of interest, which ideally expresses normal levels of receptor and low level of arrestin.

Receptor should be visible on the border of the cell, uniformly distributed. Arrestin should be characterized by a faint, uniform, cytosolic distribution.
i. Start recording. Record 3–5 min of basal, then add ligand to the chamber to reach the desired final concentration and continue imaging for further 30 min. We recommend imaging at a frame rate of 0.5 Hz (one frame every 2 s) to preserve cells from photo-damage, considering the length of the recording.

8.3.3 Expected outcomes

Arrestin recruitment to the PM should appear within the first couple of minutes after agonist addition but it is not always observed. Following several minutes, receptor should start to be internalized into endosomes, and arrestin recruitment may appear or not in the same structures. However, different GPCRs respond to ligand stimulation resulting in different times of receptor endocytosis and endosomal trafficking, which is also reflected in variable times of arrestin trafficking from the cytosol to the PM and to endosomal compartments.

8.3.4 Analysis

For arrestin recruitment to the PM, we suggest running line scan analysis using, for example, PeakFinder tool of ImageJ.

For arrestin trafficking to endosomes, line scan analysis can also be used, complemented by some coefficients that measure the degree of co-localization. As endosomes are identified as objects, we recommend using object-based approaches. Different ImageJ plugins are available to quantitatively assess co-localization, such as JACoP (Bolte & Cordelières, 2006).

8.3.5 Advantages

With this approach formation of endosomes and recruitment of arrestin at both PM and endosomal compartment can be easily visualized.

8.3.6 Limitations

Arrestin trafficking to endosomes might be hard to visualize for class A GPCRs (Oakley et al., 2000).

8.3.7 Optimization and troubleshooting

This approach requires overexpression of both tagged arrestin and GPCR. As an alternative to limit the effect of arrestin overexpression is to use the recently developed genetically encoded intrabody sensors for βarrrestin1 (Baidya, Kumari, Dwivedi-Agnihotri, Pandey, Chaturvedi, et al., 2020a; Baidya, Kumari, Dwivedi-Agnihotri, Pandey, Sokrat, et al., 2020b). These sensors have been proven to effectively and reliably report localization and trafficking of β-arrestin1 bound to chimeric GPCRs characterized by the C-tail of V2R, including D2V2R and B2AV2R, or to a set of native GPCRs such as C5a receptor 1, the neurotensin receptor 1, the muscarinic acetylcholine receptor subtype 2 M2R, and the atypical chemokine receptor subtype 2 (Baidya, Kumari, Dwivedi-Agnihotri, Pandey, Chaturvedi, et al., 2020a; Baidya, Kumari, Dwivedi-Agnihotri, Pandey, Sokrat, et al., 2020b). In the latter circumstance, only the intrabody sensor needs to be transfected into the cell type of choice, a procedure which has been demonstrated to leave unchanged the trafficking and signaling profile of the receptor of interest, limiting the need of overexpressing both the receptor and arrestin.

8.3.8 Safety considerations

A laser safety course should be taken before conducting experiments using light microscopy.

9 Additional methods

The methods presented in this chapter represent just two ways to obtain information about arrestin trafficking upon GPCR activation and internalization into endosomes in living cells in real time. The main information that can be extrapolated from these experimental approaches are: (1) the time of arrestin recruitment after ligand addition to both membrane and endosomes, (2) the number of clusters formed at the PM and their lifetime, (3) the co-presence of receptor, arrestin and, if desired, a third component (e.g., clathrin, endosomal markers, phospholipids) at those subcellular locations. These methods can be adapted to the researcher needs to answer different questions and obtain various types of information. For example, by adding to the protocol for live imaging by confocal microscopy a fixation step at defined time point pre- and post-stimulation, samples can be further probed using antibodies against key components of cellular compartments and machineries without the need of overexpression. If antibodies against the Early Endosomal Antigen 1 (EEA1) or APPL1 are used, one can discriminate whether arrestin is recruited to early endosomes or very early endosomes, respectively (Baidya, Kumari, Dwivedi-Agnihotri, Pandey, Chaturvedi, et al., 2020a; Baidya, Kumari, Dwivedi-Agnihotri, Pandey, Sokrat, et al., 2020b; Sposini et al., 2017). Similarly, one can use antibodies against proteins typically involved in receptor clustering and endocytosis such clathrin, dynamin, actin, to determine their presence at arrestin clusters. This variant of the protocol is compatible with both TIRF and confocal imaging, although attention needs to be payed to the sample preparation for each type of microscopy technique, as these have different requirements (Jonkman, Brown, Wright, Anderson, & North, 2020; Mattheyses, Simon, & Rappoport, 2010).

Other methods to be mentioned for their potential to provide detailed information are super resolution imaging techniques. These include photo activated localization microscopy (PALM), stochastic optical reconstruction microscopy (STORM) and structured illumination microscopy (SIM). They have just started to be applied to the study of arrestins, due to the complexity of the imaging set-ups, sample preparation or image analysis. Yet, they can supply detailed knowledge on the biology of arrestins, such as their nanoscale organization and dynamics at the single molecule level. However, only PALM is fully compatible with live-cell imaging, while STORM requires cell fixation due to the toxic buffers required for photo blinking, and SIM is currently too time consuming to image in real time highly dynamic processes such as arrestin trafficking.

In a study combining PALM/STORM/SIM with transmission electron microscopy to study the localization of clathrin-associated proteins during clathrin-coated structures (CCS) formation, β-arrestin2 was found to localize both and the edge and the center of clathrin lattices, with these two subpopulations changing their concentration according to the evolution of the CCS from flat to domed (Sochacki, Dickey, Strub, & Taraska, 2017). In another study focused on the chemokine receptor 5 CCR5, STORM combined with coordinate and image-based cluster analysis was applied to quantitatively assess the clustering of β-arrestin1 upon stimulation of CCR5 with a panel of ligands (Truan et al., 2013). More recently, single particle tracking PALM (sptPALM) was employed to quantitatively assess the dynamics of β-arrestin2 recruitment and binding at CCSs following either β1AR or β2AR activation. This study highlighted the presence of two populations of β-arrestin2, one mobile and one immobile, with the immobile fraction representing β-arrestin2 bound to CCSs irrespective of the activating β-adrenergic receptor (Eichel et al., 2018).

10 Summary

Arrestins are key modulators of GPCRs signaling and trafficking; their action is mediated through various mechanisms, with intracellular trafficking as an essential feature. In this chapter we provide a step-by-step protocol for the visualization of arrestin trafficking in real time in living HEK293 cells using two optical approaches, TIRF and confocal microscopy. These approaches allow to visualize arrestin dynamics at the PM, or in the cytosol and at the PM, respectively; thus, the choice of which approach to use depends on both the research aims and hardware availability, keeping in mind that the results obtained are complementary and can greatly benefit from quantitative analyses. The protocols presented here apply to HEK293 cells and imply overexpression of GPCRs and fluorescently tagged arrestin; however, they may be suitable for other cell types, as well as be adapted to the researcher's needs. Considering their general accessibility and the wealth of information that they can provide, we consider these methods as powerful tools to explore arrestin trafficking within the cell.

Acknowledgments

We would like to thank Dr. Aylin Hanyaloglu at Imperial College London for her training on live- and fixed-cell confocal imaging of GPCR endosomal trafficking and for providing material and expertise to carry out the experiment shown in Fig. 2 in her laboratory. We also thank Dr. Arun Shukla for providing Flag-V2R and β-arrestin-1-YFP plasmids used in Fig. 2. This work was supported by a Sir Henry Wellcome Postdoctoral Fellowship (Wellcome Trust) awarded to S.S. (218650/Z/19/Z).

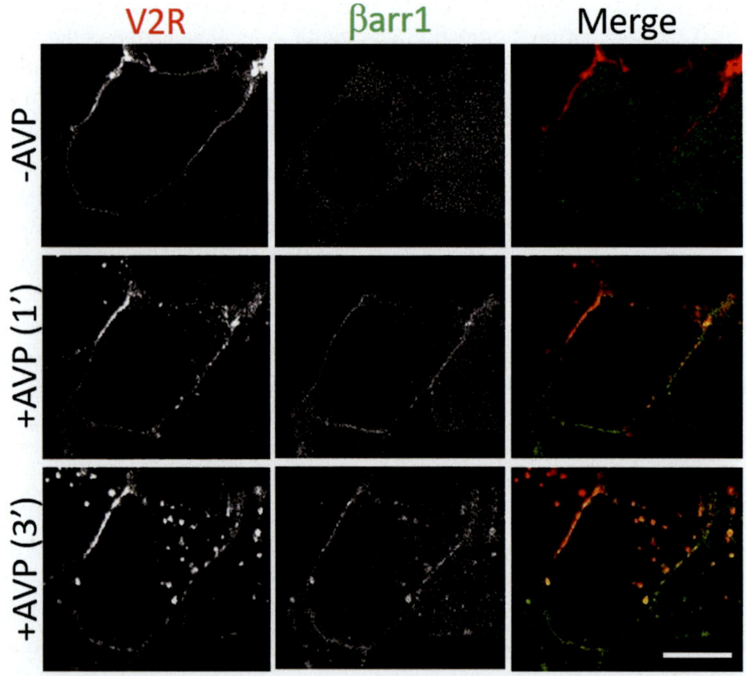

FIG. 2

Live-cell confocal microscopy shows trafficking of β-arrestin-1 to the PM and endosomes following agonist stimulation of V2R. HEK293 cells transiently transfected with Flag-V2R and β-arrestin-1-YFP were stained with M1 anti Flag antibody conjugated to AlexaFluor555 and imaged live using a SP5 Leica confocal microscope at 1 Hz. Arrestin recruitment is observed after 1′ of agonist angiotensin vasopressin (AVP, 100 nM). Scale bar = 5 μm.

References

Ahn, S., Wei, H., Garrison, T. R., & Lefkowitz, R. J. (2004). Reciprocal regulation of angiotensin receptor-activated extracellular signal-regulated kinases by β-arrestins 1 and 2. *Journal of Biological Chemistry*, *279*(9), 7807–7811. https://doi.org/10.1074/jbc.C300443200.

Attramadal, H., Arriza, J. L., Aoki, C., Dawson, T. M., Codina, J., Kwatra, M. M., et al. (1992). Beta-arrestin2, a novel member of the arrestin/beta-arrestin gene family. *Journal of Biological Chemistry*, *267*(25), 17882–17890. https://doi.org/10.1016/S0021-9258(19)37125-X.

Baidya, M., Kumari, P., Dwivedi-Agnihotri, H., Pandey, S., Chaturvedi, M., Stepniewski, T. M., et al. (2020a). Key phosphorylation sites in GPCRs orchestrate the contribution of β-arrestin 1 in ERK 1/2 activation. *EMBO Reports*, *21*(9). https://doi.org/10.15252/embr.201949886.

Baidya, M., Kumari, P., Dwivedi-Agnihotri, H., Pandey, S., Sokrat, B., Sposini, S., et al. (2020b). Genetically encoded intrabody sensors report the interaction and trafficking of β-arrestin 1 upon activation of G-protein–coupled receptors. *Journal of Biological Chemistry*, *295*(30), 10153–10167. https://doi.org/10.1074/jbc.RA120.013470.

Barthet, G., Framery, B., Gaven, F., Pellissier, L., Reiter, E., Claeysen, S., et al. (2007). 5-Hydroxytryptamine4 receptor activation of the extracellular signal-regulated kinase pathway depends on Src activation but not on G protein or [N]L-arrestin signaling. *Molecular Biology of the Cell*, *18*, 13.

Bhandari, D., Trejo, J., Benovic, J. L., & Marchese, A. (2007). Arrestin-2 interacts with the ubiquitin-protein isopeptide ligase atrophin-interacting protein 4 and mediates endosomal sorting of the chemokine receptor CXCR4. *Journal of Biological Chemistry*, *282*(51), 36971–36979. https://doi.org/10.1074/jbc.M705085200.

Bolte, S., & Cordelières, F. P. (2006). A guided tour into subcellular colocalization analysis in light microscopy. *Journal of Microscopy*, *224*(3), 213–232. https://doi.org/10.1111/j.1365-2818.2006.01706.x.

Calebiro, D., Nikolaev, V. O., Gagliani, M. C., de Filippis, T., Dees, C., Tacchetti, C., et al. (2009). Persistent cAMP-signals triggered by internalized G-protein–coupled receptors. *PLoS Biology*, *7*(8), e1000172. https://doi.org/10.1371/journal.pbio.1000172.

Cao, Y., Kumar, S., Namkung, Y., Gagnon, L., Cho, A., & Laporte, S. A. (2020). Angiotensin II type 1 receptor variants alter endosomal receptor–β-arrestin complex stability and MAPK activation. *Journal of Biological Chemistry*, *295*(38), 13169–13180. https://doi.org/10.1074/jbc.RA120.014330.

Cao, Y., Namkung, Y., & Laporte, S. A. (2019). Methods to monitor the trafficking of β-arrestin/G protein-coupled receptor complexes using enhanced bystander BRET. In M. G. H. Scott, & S. A. Laporte (Eds.), *Beta-Arrestins: Methods and Protocols* (pp. 59–68). New York: Springer. https://doi.org/10.1007/978-1-4939-9158-7_3.

Chen, Q., Iverson, T. M., & Gurevich, V. V. (2018). Structural basis of arrestin-dependent signal transduction. *Trends in Biochemical Sciences*, *22*, 29636212. https://doi.org/10.1016/j.tibs.2018.03.005.

Chiba, K., Shimada, Y., Kinjo, M., Suzuki, T., & Uchida, S. (2014). Simple and direct assembly of kymographs from movies using KYMOMAKER. *Traffic*, *15*(1), 1–11. https://doi.org/10.1111/tra.12127.

Crépieux, P., Poupon, A., Langonné-Gallay, N., Reiter, E., Delgado, J., Schaefer, M. H., et al. (2017). A comprehensive view of the β-arrestinome. *Frontiers in Endocrinology*, *8*, 32. 28321204. https://doi.org/10.3389/fendo.2017.00032.

Dunn, K. W., Kamocka, M. M., & McDonald, J. H. (2011). A practical guide to evaluating colocalization in biological microscopy. *American Journal of Physiology-Cell Physiology*, *300*(4), C723–C742. https://doi.org/10.1152/ajpcell.00462.2010.

Dwivedi-Agnihotri, H., Chaturvedi, M., Baidya, M., Stepniewski, T. M., Pandey, S., Maharana, J., et al. (2020). Distinct phosphorylation sites in a prototypical GPCR differently orchestrate β-arrestin interaction, trafficking, and signaling. *Science Advances*, *6*(37), eabb8368, 32917711. https://doi.org/10.1126/sciadv.abb8368.

Eichel, K., Jullié, D., Barsi-Rhyne, B., Latorraca, N. R., Masureel, M., Sibarita, J.-B., et al. (2018). Catalytic activation of β-arrestin by GPCRs. *Nature*, *557*(7705), 381–386. https://doi.org/10.1038/s41586-018-0079-1.

Eichel, K., Jullié, D., & von Zastrow, M. (2016). β-Arrestin drives MAP kinase signalling from clathrin-coated structures after GPCR dissociation. *Nature Cell Biology*, *18*(3), 303–310. https://doi.org/10.1038/ncb3307.

Feinstein, T. N., Yui, N., Webber, M. J., Wehbi, V. L., Stevenson, H. P., King, J. D., et al. (2013). Noncanonical control of vasopressin receptor type 2 signaling by retromer and arrestin. *Journal of Biological Chemistry*, *288*(39), 27849–27860. https://doi.org/10.1074/jbc.M112.445098.

Ferrandon, S., Feinstein, T. N., Castro, M., Wang, B., Bouley, R., Potts, J. T., et al. (2009). Sustained cyclic AMP production by parathyroid hormone receptor endocytosis. *Nature Chemical Biology*, *5*(10), 734–742. https://doi.org/10.1038/nchembio.206.

Gaidarov, I., Krupnick, J. G., Falck, J. R., Benovic, J. L., & Keen, J. H. (1999). Arrestin function in G protein-coupled receptor endocytosis requires phosphoinositide binding. *The EMBO Journal*, *18*(4), 871–881. https://doi.org/10.1093/emboj/18.4.871.

Godbole, A., Lyga, S., Lohse, M. J., & Calebiro, D. (2017). Internalized TSH receptors en route to the TGN induce local Gs-protein signaling and gene transcription. *Nature Communications*, *8*(1), 443. https://doi.org/10.1038/s41467-017-00357-2.

Gurevich, V. V., & Gurevich, E. V. (2018). GPCRs and signal transducers: Interaction stoichiometry. *Trends in Pharmacological Sciences*, *39*(7), 672–684. https://doi.org/10.1016/j.tips.2018.04.002.

Hakak, Y., Shrestha, D., Goegel, M. C., Behan, D. P., & Chalmers, D. T. (2003). Global analysis of G-protein-coupled receptor signaling in human tissues. *FEBS Letters*, *550*(1–3), 11–17. https://doi.org/10.1016/S0014-5793(03)00762-2.

Hanyaloglu, A. C., & von Zastrow, M. (2008). Regulation of GPCRs by endocytic membrane trafficking and its potential implications. *Annual Review of Pharmacology and Toxicology*, *48*(1), 537–568. https://doi.org/10.1146/annurev.pharmtox.48.113006.094830.

Irannejad, R., Tomshine, J. C., Tomshine, J. R., Chevalier, M., Mahoney, J. P., Steyaert, J., et al. (2013). Conformational biosensors reveal GPCR signalling from endosomes. *Nature*, *495*(7442), 534–538. https://doi.org/10.1038/nature12000.

Irannejad, R., & von Zastrow, M. (2014). GPCR signaling along the endocytic pathway. *Current Opinion in Cell Biology*, *27*, 109–116. https://doi.org/10.1016/j.ceb.2013.10.003.

Jean-Alphonse, F. G., Wehbi, V. L., Chen, J., Noda, M., Taboas, J. M., Xiao, K., et al. (2017). β2-adrenergic receptor control of endosomal PTH receptor signaling via Gβγ. *Nature Chemical Biology*, *13*(3), 259–261. https://doi.org/10.1038/nchembio.2267.

Jonkman, J., Brown, C. M., Wright, G. D., Anderson, K. I., & North, A. J. (2020). Tutorial: guidance for quantitative confocal microscopy. *Nature Protocols*, *15*(5), 1585–1611. https://doi.org/10.1038/s41596-020-0313-9.

Kang, Y., Zhou, X. E., Gao, X., He, Y., Liu, W., Ishchenko, A., et al. (2015). Crystal structure of rhodopsin bound to arrestin by femtosecond X-ray laser. *Nature*, *523*(7562), 561–567. https://doi.org/10.1038/nature14656.

Kumari, P., Srivastava, A., Ghosh, E., Ranjan, R., Dogra, S., Yadav, P. N., et al. (2017). Core engagement with β-arrestin is dispensable for agonist-induced vasopressin receptor endocytosis and ERK activation. *Molecular Biology of the Cell*, *28*(8), 1003–1010. https://doi.org/10.1091/mbc.e16-12-0818.

Lally, C. C. M., Bauer, B., Selent, J., & Sommer, M. E. (2017). C-edge loops of arrestin function as a membrane anchor. *Nature Communications*, *8*(1), 14258. https://doi.org/10.1038/ncomms14258.

Latorraca, N. R., Wang, J. K., Bauer, B., Townshend, R. J. L., Hollingsworth, S. A., Olivieri, J. E., et al. (2018). Molecular mechanism of GPCR-mediated arrestin activation. *Nature*, *557*(7705), 452–456. https://doi.org/10.1038/s41586-018-0077-3.

Lee, M.-H., Appleton, K. M., Strungs, E. G., Kwon, J. Y., Morinelli, T. A., Peterson, Y. K., et al. (2016). The conformational signature of β-arrestin2 predicts its trafficking and signalling functions. *Nature*, *531*(7596), 665–668. https://doi.org/10.1038/nature17154.

Lee, J.-Y., & Kitaoka, M. (2018). A beginner's guide to rigor and reproducibility in fluorescence imaging experiments. *Molecular Biology of the Cell*, *29*(13), 1519–1525. https://doi.org/10.1091/mbc.E17-05-0276.

Lee, S., Park, S., Lee, H., Han, S., Song, J., Han, D., et al. (2019). Nedd4 E3 ligase and beta-arrestins regulate ubiquitination, trafficking, and stability of the mGlu7 receptor. *eLife*, *8*, e44502. https://doi.org/10.7554/eLife.44502.

Lin, F.-T., Chen, W., Shenoy, S., Cong, M., Exum, S. T., & Lefkowitz, R. J. (2002). Phosphorylation of β-arrestin2 regulates its function in internalization of β_2-adrenergic receptors. *Biochemistry*, *41*(34), 10692–10699. https://doi.org/10.1021/bi025705n.

Lin, F.-T., Krueger, K. M., Kendall, H. E., Daaka, Y., Fredericks, Z. L., Pitcher, J. A., et al. (1997). Clathrin-mediated endocytosis of the β-adrenergic receptor is regulated by phosphorylation/dephosphorylation of β-arrestin1. *Journal of Biological Chemistry*, *272*(49), 31051–31057. https://doi.org/10.1074/jbc.272.49.31051.

Lin, F.-T., Miller, W. E., Luttrell, L. M., & Lefkowitz, R. J. (1999). Feedback regulation of β-arrestin1 function by extracellular signal-regulated kinases. *Journal of Biological Chemistry*, *274*(23), 15971–15974. https://doi.org/10.1074/jbc.274.23.15971.

Lobingier, B. T., & von Zastrow, M. (2019). When trafficking and signaling mix: How subcellular location shapes G protein-coupled receptor activation of heterotrimeric G proteins. *Traffic*, *20*(2), 130–136. https://doi.org/10.1111/tra.12634.

Lohse, M., Benovic, J., Codina, J., Caron, M., & Lefkowitz, R. (1990). Beta-arrestin: A protein that regulates beta-adrenergic receptor function. *Science*, *248*(4962), 1547–1550. https://doi.org/10.1126/science.2163110.

Mattheyses, A. L., Simon, S. M., & Rappoport, J. Z. (2010). Imaging with total internal reflection fluorescence microscopy for the cell biologist. *Journal of Cell Science*, *123*(21), 3621–3628. https://doi.org/10.1242/jcs.056218.

Milano, S. K., Kim, Y.-M., Stefano, F. P., Benovic, J. L., & Brenner, C. (2006). Nonvisual arrestin oligomerization and cellular localization are regulated by inositol hexakisphosphate binding. *Journal of Biological Chemistry*, *281*(14), 9812–9823. https://doi.org/10.1074/jbc.M512703200.

Mundell, S. J., Luo, J., Benovic, J. L., Conley, P. B., & Poole, A. W. (2006). Distinct clathrin-coated pits sort different G protein-coupled receptor cargo. *Traffic*, *7*(10), 1420–1431. https://doi.org/10.1111/j.1600-0854.2006.00469.x.

Nguyen, A. H., Thomsen, A. R. B., Cahill, T. J., Huang, R., Huang, L.-Y., Marcink, T., et al. (2019). Structure of an endosomal signaling GPCR–G protein–β-arrestin megacomplex. *Nature Structural & Molecular Biology*, *26*(12), 1123–1131. https://doi.org/10.1038/s41594-019-0330-y.

Nobles, K. N., Xiao, K., Ahn, S., Shukla, A. K., Lam, C. M., Rajagopal, S., et al. (2011). Distinct phosphorylation sites on the β(2)-adrenergic receptor establish a barcode that encodes differential functions of β-arrestin. *Science Signaling*, *4*(185), ra51. https://doi.org/10.1126/scisignal.2001707.

Nuber, S., Zabel, U., Lorenz, K., Nuber, A., Milligan, G., Tobin, A. B., et al. (2016). β-Arrestin biosensors reveal a rapid, receptor-dependent activation/deactivation cycle. *Nature*, *531*(7596), 661–664. https://doi.org/10.1038/nature17198.

Oakley, R. H., Laporte, S. A., Holt, J. A., Caron, M. G., & Barak, L. S. (2000). Differential affinities of visual arrestin, βarrestin1, and βarrestin2 for G protein-coupled receptors delineate two major classes of receptors. *Journal of Biological Chemistry*, *275*(22), 17201–17210. https://doi.org/10.1074/jbc.M910348199.

Paradis, J. S., Ly, S., Blondel-Tepaz, É., Galan, J. A., Beautrait, A., Scott, M. G. H., et al. (2015). Receptor sequestration in response to β-arrestin-2 phosphorylation by ERK1/2 governs steady-state levels of GPCR cell-surface expression. *Proceedings of the National Academy of Sciences of the United States of America*, *112*(37), E5160–E5168. https://doi.org/10.1073/pnas.1508836112.

Pavlos, N. J., & Friedman, P. A. (2017). GPCR signaling and trafficking: The long and short of it. *Trends in Endocrinology and Metabolism*, *28*(3), 213–226. https://doi.org/10.1016/j.tem.2016.10.007.

Peterson, Y. K., & Luttrell, L. M. (2017). The diverse roles of arrestin scaffolds in G protein–coupled receptor signaling. *Pharmacological Reviews*, *69*(3), 256–297. https://doi.org/10.1124/pr.116.013367.

Shenoy, S. K., & Lefkowitz, R. J. (2003). Trafficking patterns of β-arrestin and G protein-coupled receptors determined by the kinetics of β-arrestin deubiquitination. *Journal of Biological Chemistry*, *278*(16), 14498–14506. https://doi.org/10.1074/jbc.M209626200.

Shenoy, S. K., McDonald, P. H., Kohout, T. A., & Lefkowitz, R. J. (2001). Regulation of receptor fate by ubiquitination of activated ^{N}L2-adrenergic receptor and ^{N}L-arrestin. *Science*, *294*, 8.

Shenoy, S. K., Xiao, K., Venkataramanan, V., Snyder, P. M., Freedman, N. J., & Weissman, A. M. (2008). Nedd4 mediates agonist-dependent ubiquitination, lysosomal targeting, and degradation of the β2-adrenergic receptor. *Journal of Biological Chemistry*, *283*(32), 22166–22176. https://doi.org/10.1074/jbc.M709668200.

Shukla, A. K., Westfield, G. H., Xiao, K., Reis, R. I., Huang, L.-Y., Tripathi-Shukla, P., et al. (2014). Visualization of arrestin recruitment by a G-protein-coupled receptor. *Nature*, *512*(7513), 218–222. https://doi.org/10.1038/nature13430.

Simaan, M., Bedardgoulet, S., Fessart, D., Gratton, J., & Laporte, S. (2005). Dissociation of β-arrestin from internalized bradykinin B2 receptor is necessary for receptor recycling and resensitization. *Cellular Signalling*, *17*(9), 1074–1083. https://doi.org/10.1016/j.cellsig.2004.12.001.

Smith, T. H., Coronel, L. J., Li, J. G., Dores, M. R., Nieman, M. T., & Trejo, J. (2016). Protease-activated receptor-4 signaling and trafficking is regulated by the clathrin adaptor protein complex-2 independent of β-arrestins. *Journal of Biological Chemistry*, *291*(35), 18453–18464. https://doi.org/10.1074/jbc.M116.729285.

Sochacki, K. A., Dickey, A. M., Strub, M.-P., & Taraska, J. W. (2017). Endocytic proteins are partitioned at the edge of the clathrin lattice in mammalian cells. *Nature Cell Biology*, *19*(4), 352–361. https://doi.org/10.1038/ncb3498.

Sposini, S., Jean-Alphonse, F. G., Ayoub, M. A., Oqua, A., West, C., Lavery, S., et al. (2017). Integration of GPCR signaling and sorting from very early endosomes via opposing

APPL1 mechanisms. *Cell Reports*, *21*(10), 2855–2867. https://doi.org/10.1016/j.celrep.2017.11.023.

Staus, D. P., Hu, H., Robertson, M. J., Kleinhenz, A. L. W., Wingler, L. M., Capel, W. D., et al. (2020). Structure of the M2 muscarinic receptor–β-arrestin complex in a lipid nanodisc. *Nature*, *579*(7798), 297–302. https://doi.org/10.1038/s41586-020-1954-0.

Thomsen, A. R. B., Plouffe, B., Iii, T. J. C., Shukla, A. K., Tarrasch, J. T., Dosey, A. M., et al. (2016). GPCR-G protein-b-arrestin super-complex mediates sustained G protein signaling. *Cell*, *14*, 907–919. 27499021.

Tobin, A. B., Butcher, A. J., & Kong, K. C. (2008). Location, location, location…site-specific GPCR phosphorylation offers a mechanism for cell-type-specific signalling. *Trends in Pharmacological Sciences*, *29*(8), 413–420. https://doi.org/10.1016/j.tips.2008.05.006.

Truan, Z., Tarancón Díez, L., Bönsch, C., Malkusch, S., Endesfelder, U., Munteanu, M., et al. (2013). Quantitative morphological analysis of arrestin2 clustering upon G protein-coupled receptor stimulation by super-resolution microscopy. *Journal of Structural Biology*, *184*(2), 329–334. https://doi.org/10.1016/j.jsb.2013.09.019.

van Koppen, C. J., & Jakobs, K. H. (2004). Arrestin-independent internalization of G protein-coupled receptors: Fig. 1. *Molecular Pharmacology*, *66*(3), 365–367. https://doi.org/10.1124/mol.104.003822.

Vassilatis, D. K., Hohmann, J. G., Zeng, H., Li, F., Ranchalis, J. E., Mortrud, M. T., et al. (2003). The G protein-coupled receptor repertoires of human and mouse. *Proceedings of the National Academy of Sciences of the United States of America*, *100*(8), 4903–4908. https://doi.org/10.1073/pnas.0230374100.

Vilardaga, J.-P., Jean-Alphonse, F. G., & Gardella, T. J. (2014). Endosomal generation of cAMP in GPCR signaling. *Nature Chemical Biology*, *10*(9), 700–706. https://doi.org/10.1038/nchembio.1611.

Wehbi, V. L., Stevenson, H. P., Feinstein, T. N., Calero, G., Romero, G., & Vilardaga, J.-P. (2013). Noncanonical GPCR signaling arising from a PTH receptor–arrestin–Gβγ complex. *Proceedings of the National Academy of Sciences of the United States of America*, *110*(4), 1530–1535. https://doi.org/10.1073/pnas.1205756110.

Weinberg, Z. Y., Zajac, A. S., Phan, T., Shiwarski, D. J., & Puthenveedu, M. A. (2017). Sequence-specific regulation of endocytic lifetimes modulates arrestin-mediated signaling at the μ opioid receptor. *Molecular Pharmacology*, *91*(4), 416–427. https://doi.org/10.1124/mol.116.106633.

Wilden, U., Wüst, E., Weyand, I., & Kühn, H. (1986). Rapid affinity purification of retinal arrestin (48 kDa protein) via its light-dependent binding to phosphorylated rhodopsin. *FEBS Letters*, *207*(2), 292–295. https://doi.org/10.1016/0014-5793(86)81507-1.

Xiao, K., McClatchy, D. B., Shukla, A. K., Zhao, Y., Chen, M., Shenoy, S. K., et al. (2007). Functional specialization of beta-arrestin interactions revealed by proteomic analysis. *Proceedings of the National Academy of Sciences of the United States of America*, *6*, 12011–12016. 17620599. https://doi.org/10.1073/pnas.0704849104.

Zhou, X. E., He, Y., de Waal, P. W., Gao, X., Kang, Y., Van Eps, N., et al. (2017). Identification of phosphorylation codes for arrestin recruitment by G protein-coupled receptors. *Cell*, *170*(3), 457–469.e13. https://doi.org/10.1016/j.cell.2017.07.002.

CHAPTER

Strategies for targeting cell surface proteins using multivalent conjugates and chemical biology

9

Shivani Sachdev, Chino C. Cabalteja, and Ross W. Cheloha*

National Institutes of Health, National Institute of Diabetes, Digestive, and Kidney Diseases (NIDDK), Laboratory of Bioorganic Chemistry, Bethesda, MD, United States

**Corresponding author: e-mail address: ross.cheloha@nih.gov*

Chapter outline

Abstract

Proper function of receptors on the cell surface is essential for homeostasis. Compounds that target cell surface receptors to address dysregulation have proven exceptionally successful as therapeutic agents; however, the development of compounds with the desired specificity for receptors, cells, and tissues of choice has proven difficult in some cases. The use of compounds that can engage more than one binding site at the cell surface offers a path toward improving biological specificity or pharmacological properties. In this chapter we summarize historical context for the development of such bivalent compounds. We focus on developments in chemical methods and biological engineering to provide bivalent compounds in which the high affinity and specificity of antibodies are leveraged to create multifunctional conjugates with new and useful properties. The development of methods to meld biological macromolecules with synthetic compounds will facilitate modulation of receptor biology in ways not previously possible.

Methods in Cell Biology, Volume 166, ISSN 0091-679X, https://doi.org/10.1016/bs.mcb.2021.06.004

1 Introduction

Proteins found on the cell surface are essential for proper responses to environmental cues. In multicellular organisms, cells communicate through the production and release of molecules that act on protein receptors found on the surface of neighbors, proximal and distal. This cell-to-cell communication is essential for coordinated biological responses in complex organisms. Dysfunction of these processes can have profound consequences and frequently results in disease. Addressing this dysregulation through the application of molecules that mimic or block the action of natural signaling molecules is a common and successful approach for therapeutic development. Efforts to target G protein-coupled receptors (GPCRs) and receptor tyrosine kinases (RTKs), two large and important families of cell surface proteins, exemplify these efforts.

GPCRs are the largest family of cell surface proteins, with over 800 family members in humans (Hauser, Attwood, Rask-Andersen, Schiöth, & Gloriam, 2017). They function through activation of guanine nucleotide-binding proteins (G-proteins), among other pathways, to regulate virtually every aspect of cell biology. Responses are induced through the binding of ligands such as small molecules, peptides, and full-size proteins to GPCRs. Over 30% of approved therapeutics target GPCRs (Hauser et al., 2017); however, several outstanding challenges remain (Wacker, Stevens, & Roth, 2017). There is substantial overlap in specificity among GPCRs and the ligands they recognize: approximately 75% of naturally occurring receptors and ligands interact with more than one receptor or ligand (Foster et al., 2019). Extraordinary efforts have been undertaken to identify synthetic small molecules that bind tightly and specifically to receptors of interest to induce desired biological responses (Griffith et al., 2020; Lyu et al., 2019; Moehle et al., 2020). Peptides and antibodies (Abs) have also been developed to modulate GPCRs, often proving effective in targeting receptors for which small molecules has proven insufficiently potent and specific (Davenport, Scully, de Graaf, Brown, & Maguire, 2020; Hutchings, Koglin, Olson, & Marshall, 2017). RTKs, like GPCRs, are a prime target in therapeutic development, with particular relevance in the development of cancer therapies (Pottier et al., 2020). Most RTKs are also activated by more than one ligand, complicating approaches to make selective and potent receptor modulators (Trenker & Jura, 2020). Efforts to induce desired biological outcomes by targeting either GPCRs or RTKs is further complicated by variation in receptor expression in different tissues, sometimes as different receptor isoforms (Marti-Solano et al., 2020). Furthermore, the consequences of receptor activation or blockade can vary based on the tissue in which targeting occurs (Cheloha, Gellman, Vilardaga, & Gardella, 2015; Hao & Tatonetti, 2016).

Given these considerations, improved methods to address receptor function would be valuable. This chapter focuses on the development of multivalent compounds to target cell surface receptors. A brief overview of the historical precedence for using small molecule bivalent compounds to modulate receptor function will be presented. Next, a framework for describing different varieties of multivalent compounds will be illustrated. This framework will be used to describe recent efforts to use biomolecules to construct multivalent conjugates for targeting cell surface proteins. Special emphasis will be placed on multivalent compounds that result from the combination of chemical and biological methods (chemical biology). The methodology used to construct these conjugates will then be described. Finally, the biological and pharmacological properties of these conjugates will be summarized followed by a perspective on advances that will propel this field forward.

2 Discussion

2.1 Categorization of conjugates

Multivalent ligands have been used to target cell surface receptors to provide specialized and useful properties for decades (Newman, Battiti, & Bonifazi, 2020). During this period, several varieties of multivalent ligands have been developed with overlapping and signature characteristics (Fig. 1). Here we define bivalent (and oligovalent) ligands as those that are comprised of two (or more) moieties that target the orthosteric site of cell surface proteins. These oligovalent ligands can consist of multiple copies of the same targeting moiety (homo-oligovalent, Fig. 1A) or more than one different targeting moiety (hetero-oligovalent, Fig. 1B). A variation on this approach relies on the use of targeting moieties that do not bind to the orthosteric site on a receptor, but rather some other site for which binding does not activate or block signaling, here named a secondary site. The linkage of moieties that differ in the type of site they bind (orthosteric vs secondary) provides bitopic (or oligotopic) ligands (Fig. 1C and D). Note that bitopic ligands differ from their bivalent counterparts in the use of a targeting moiety that binds to a secondary site. Bitopic ligands can bind to multiple, distinct sites on the same target protein (cis-bitopic, Fig. 1C) or to sites on neighboring proteins (trans-bitopic, Fig. 1D). These bivalent and bitopic compounds can also be used to engage targets expressed on different cells or cell populations (Fig. 1E). These classes of multivalent ligands have been deployed in a variety of contexts, with distinct advantages and drawbacks, as will be discussed below.

2.2 Historical precedent with small molecules

Some of the earliest efforts in this area focused on efforts to design bivalent ligands consisting of two copies of a pharmacophore connected by a linker, which bind to the orthosteric binding site of opioid receptors (Erez, Takemori, & Portoghese, 1982). This design was proposed as a way to use the binding of the first pharmacophore to reduce the entropic cost for the binding of the second ligand to its receptors. These early studies revealed that both the nature of the orthosteric site binding moieties and the length and composition of the linker connecting them were important parameters in determining bivalent ligand affinity and potency. Optimized homobivalent compounds showed potencies that exceeded those of monovalent ligands. Further studies revealed that such bivalent compounds could facilitate selective engagement of

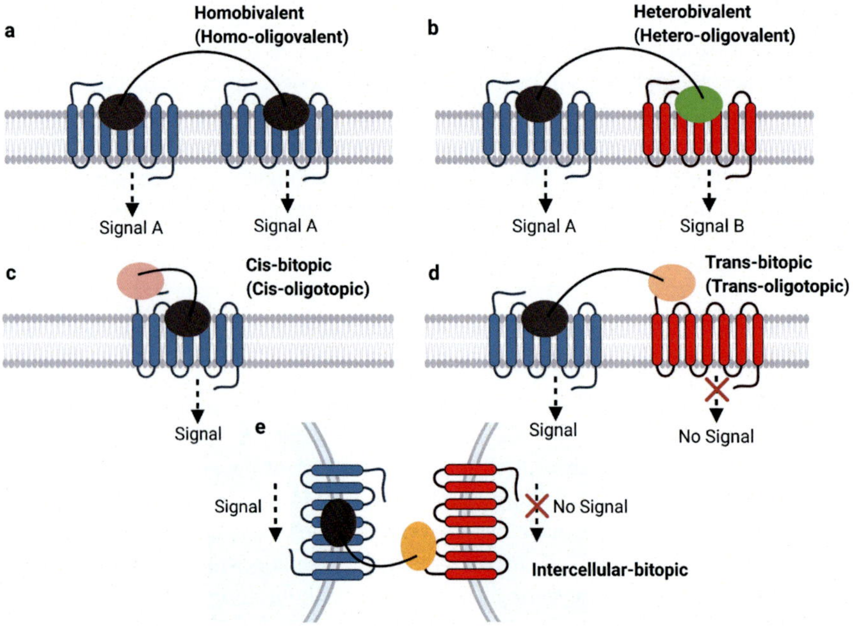

FIG. 1

Classes of multivalent conjugates used to target cell surface receptors. Schematic of (A and B) bivalent ligands selectively targeting two orthosteric binding site of the same receptor (A, homobivalent), or different receptor subtypes (B, heterobivalent); (C and D) bitopic ligands selectively targeting the primary orthosteric and a secondary binding site in a single receptor (C, cis-bitopic), or different receptor subtype (D, trans-bitopic). (E) Intercellular-bitopic to target receptor subtypes expressed on two different cells. This Figure was prepared using Biorender software.

specific opioid receptor subtypes, a feat which was difficult to achieve using monovalent compounds available at that time (Portoghese, Sultana, & Takemori, 1988). One contributing factor for the unique characteristics seen with bivalent ligands that target cell surface receptors is that many of these receptors oligomerize to form homodimers, heterodimers, or higher order complexes (Gomes et al., 2016). Selectively targeting cell surface protein oligomers using multivalent ligands offers the opportunity to induce biological responses only in cell types expressing each of the proteins needed to form the functional receptor oligomer. Engagement of receptor complexes has received increasing interest as a way to design new tool compounds and therapeutic candidates (Rosenbaum, Clemmensen, Bredt, Bettler, & Strømgaard, 2020).

GPCRs often dimerize to form assemblies with properties that differ from their constituent monomers (González-Maeso, 2011; Terrillon & Bouvier, 2004). Multivalent, small molecule compounds have been designed to target these assemblies.

Bivalent and bitopic compounds with high specificity toward serotonergic receptor subtypes and receptor assemblies have been reported (Soto et al., 2018; Zhang, Zhang, Branfman, Baldessarini, & Neumeyer, 2007). Bivalent compounds have also proven useful for targeting assemblies of cannabinoid receptors (Nimczick & Decker, 2015; Zhang et al., 2010), adrenergic receptors (Hague et al., 2006; Xu et al., 2003), adenosine receptors (Barlow, Baker, & Scammells, 2013; Soriano et al., 2009), dopamine receptors (Ågren et al., 2020; Kumar et al., 2016, 2017) and a variety of other GPCR heterodimers (Hübner et al., 2016; Toneatti et al., 2020). Bitopic compounds have also shown promise in stimulating signaling through a subset of the full repertoire of pathways engaged by standard ligands, or biased signaling (Bonifazi et al., 2019; Holze et al., 2020). There is an extensive literature on the unique and desirable properties exhibited by small molecule compounds designed to target GPCR dimers (Newman et al., 2020) and cell surface receptors more generally (Rosenbaum et al., 2020). This chapter will focus on examples in which targeting moieties besides small molecules have been applied.

2.3 Biomolecules for targeting receptors

Despite the extensive efforts invested in designing bivalent and bitopic compounds to target GPCRs and other cell surface receptors, there are many targets for which suitable small molecule compounds are not readily available. In some cases, alternative modalities such as peptides and proteins (Davenport et al., 2020), including Abs and their fragments (Heukers, De Groof, & Smit, 2019; Hutchings et al., 2017), facilitate uniquely powerful approaches for modulating target function. The utility of polypeptides for the selective targeting of receptors such as GPCRs is perhaps not surprising since a majority of GPCRs are activated by peptide ligands (Foster et al., 2019). However, natural ligands for receptors are often promiscuous, activating more than one receptor, leading to difficulties in their application as pharmacological tools or therapeutic candidates. In such cases, monoclonal Abs, or their fragments, widely applied for their high affinity and specific recognition of targets of interest, are useful. The examples discussed in detail below focus on Abs and their fragments; however, a variety of platforms relying on biomolecule oligomers for selective targeting of cell surface proteins of interest are under development (Richards, 2018).

Although Abs that recognize most proteins localized to the cell membrane are commercially available, many of these do not bind to the fully folded proteins found on live cells. Other Abs bind but do not affect the function of the bound target. Such Abs cannot be used directly to study receptor pharmacology or the corresponding biology. Approaches to specifically identify Abs that act as receptor antagonists have been published (McMahon et al., 2020; Ren et al., 2020) but they are sometimes difficult to generalize. Even rarer are examples in which Abs are converted into receptor agonists using design approaches (Liu et al., 2015; Ma et al., 2020). General methods to leverage the useful characteristics of Abs to directly modulate the function of cell surface receptors would be valuable.

2.4 Methodology for producing Ab conjugates

Chemical approaches and biological engineering offer a path toward equipping Abs with moieties that can directly impact the function of cell surface proteins. A comprehensive review of approaches used to modify Abs has recently been published (Walsh et al., 2021). Here we focus on examples in which conjugates are used to modulate cell surface protein function. One difficulty in making Ab conjugates to impact cell surface protein function relates to the methods used to connect receptor targeting moieties to Abs. In some cases, a bioactive polypeptide can be fused with an Ab (or Ab fragment) through recombinant expression. One complicating factor for this approach is that conventional monoclonal Abs (usually immunoglobulin-G or IgGs) are comprised of four separate polypeptide chains, two heavy chains (HCs) and two light chains (LCs), connected by disulfide bonds (Fig. 2). Efforts to produce genetic fusions between IgGs and bioactive polypeptides often results in a loss of stability or folding efficiency. Extensive empirical optimization was required to identify a construct and conditions that allowed for production of a bitopic fusion between an IgG that targeted the soluble protein PCSK9 and a polypeptide ligand agonist for the type-1 glucagon-like peptide receptor (GLP1R) (Chodorge et al., 2018). Many attempts at designing such fusions provided proteins that were prone to aggregation and insolubility, were produced in low yield, or contained truncations that abrogated GLP1R agonist activity (Chodorge et al., 2018). Similar issues can plague efforts to use the Ab fragments responsible for antigen binding (Fabs, Fig. 2), which are often produced by the enzymatic digestion of full size IgGs.

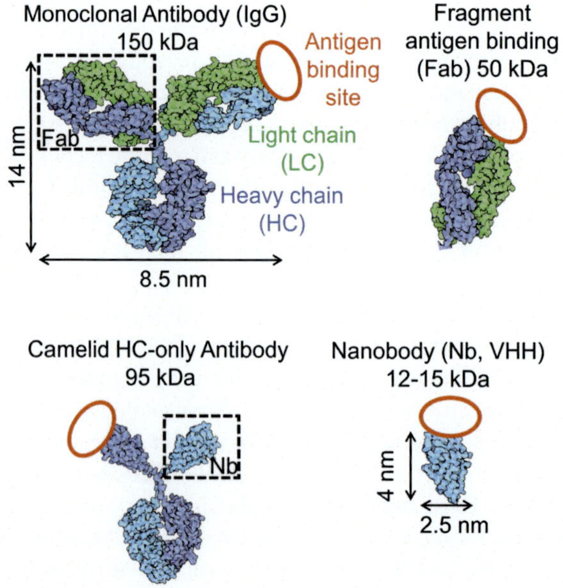

FIG. 2

Schematic of antibody composition, functional fragments, and sizes.

Issues of poor fusion protein behavior can be addressed to some extent by the use of single domain Abs (nanobodies, Nbs) in place of conventional Abs. Nbs are the antigen recognition domain of camelid heavy chain only Abs (Fig. 2) (Cheloha, Harmand, Wijne, Schwartz, & Ploegh, 2020). These single domain antibodies (Nbs) fold and function without HC-LC pairing and disulfide bond formation. Functionalized Nbs have been used to visualize GPCR distribution and trafficking (Heukers et al., 2019) and as a tool to assess GPCR conformation (Soave et al., 2020). A fusion between a GLP1R agonist peptide and two Nbs, one targeting GLP1R and the other targeting serum albumin to extend circulatory half-life, was readily produced by recombinant expression and effectively activated GLP1R in vitro and in mice (Pan et al., 2020) (Fig. 3A). Deploying a similar approach with conventional antibodies would be complicated by mispairing between HCs and LCs. Nbs can often be used for a variety of applications in which conventional antibodies are not optimal or fail (Cheloha, Harmand, et al., 2020).

Abs produced using conventional methods, such as recombinant expression, can be modified with a compound containing a functional group that spontaneously

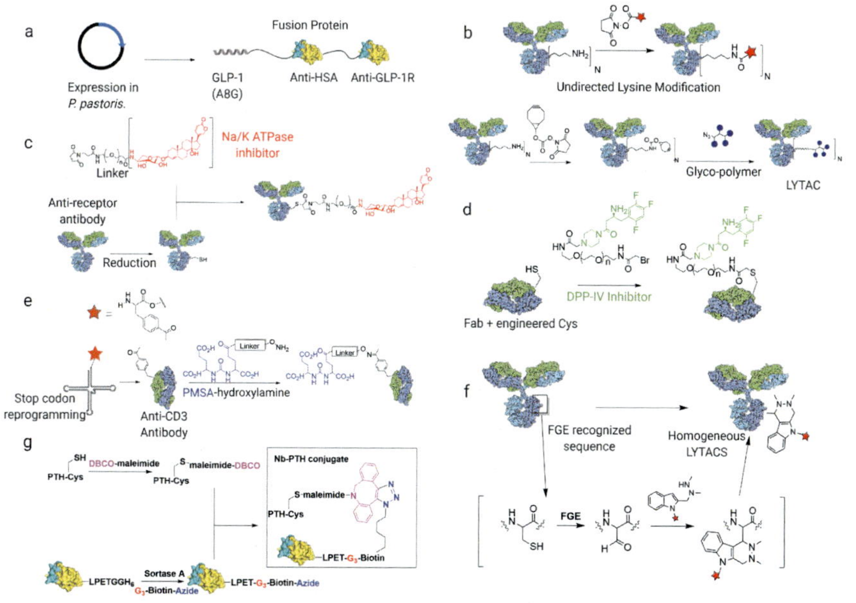

FIG. 3

Methods used in the production of Ab-ligand conjugates. (A) Recombinant expression of bioactive peptide-Nb fusions. (B) Undirected lysine side chain modification. (C) Disulfide bond reduction with subsequent cysteine side chain conjugation. (D) Unpaired cysteine residue side chain conjugation. (E) Stop codon reprogramming and incorporation of unnatural amino acid with ketone functional group. (F) Site-specific conversion of cysteine to formylglycine for subsequent conjugation. (G) Sortase A ligation.

reacts with an amino acid side chain group found in Abs. One common incarnation of this approach consists of modifying a small molecule of interest with an amine reactive functional group that can react with exposed lysine residues or the amine at the N-terminus (Fig. 3B). Lysine is a common amino acid in natural proteins so modification via amine reactivity can result in the appendage of several copies of the small molecule cargo and heterogeneous labeling. Heavy modification of Abs can sometime cause a loss of Ab stability and function (Vira, Mekhedov, Humphrey, & Blank, 2010). A lack of control of the degree of functionalization would be predicted to be especially problematic when appending large or hydrophobic compounds. Perhaps because of this complication, there are few examples of using undirected amine conjugation to create multivalent compounds to target the cell surface. In one example, trans-bitopic (or oligotopic) conjugates were prepared from Abs modified non-site-specifically with a glycans designed to promote lysosomal trafficking of the Ab bound target (Ahn et al., 2020; Banik et al., 2020) (Fig. 3B).

A more common labeling approach for creating multivalent conjugates relies on the unique reactivity of the thiol group found on the side chain of cystine (Cys) residues. Cys possesses unique reactivity among the natural amino acid side chains allowing for selective functionalization at Cys not involved in disulfide bonds. Common functional groups used for site-specific Cys labeling include maleimide- and α-halocarbonyl-functionalized compounds. Since reduced cysteine is much less common than lysine in natural proteins, tighter control of the site and degree of modification with Cys-reactive compounds is possible. For proteins that traverse the secretory pathway, such as Abs, Cys residues are often oxidized to participate in disulfide bonds, preventing their immediate functionalization. Thus, methods to reduce disulfide bonds to free the Cys side chain for functionalization have been developed. This approach was deployed to functionalize Abs with compounds that inhibit a cell surface ion pump to produce trans-bitopic conjugates that kill targeted cells (Marshall et al., 2016; Sweeny et al., 2013) (Fig. 3C). Alternatively, an unpaired Cys residue, which does not engage in disulfide bond formation, can be engineered into Abs, to allow for site-specific functionalization with an inhibitor of a cell surface enzyme to produce a cis-bitopic conjugate (Fig. 3D) (Cheng et al., 2018).

Site selective modification of Abs with compounds of interest can be precisely achieved through the introduction of non-natural amino acids with orthogonal reactivity through translational reprogramming or enzymatic modification of engineered peptide tags. These methods often produce homogeneous conjugates, unlike those that rely on modification of naturally occurring residues. DNA stop codon suppression methodology allows for incorporation of non-natural amino acids at specified positions in proteins of interest (Kim et al., 2013). Stop codon reprogramming was used to introduce into an Ab fragment an amino acid with a ketone functional group, which unlike canonical amino acids, readily reacts with aminoxy-containing compounds to form oxime linkages (Fig. 3E) (Kim et al., 2013). This approach provided an intercellular, trans-bitopic Ab fragment-small molecule conjugate (Fig. 1E) that bound to targets on T cells and tumor cells to promote T-cell-mediated tumor cell killing. Enzyme catalysis can also be used to append compounds of interest or reactive groups at short peptide tags engineered into Abs. Formylglycine generating enzyme can convert a Cys residue found within a recognition sequence into formylglycine, which can subsequently react with a variety of partners to allow for site-specific modification (Krüger, Dierks, & Sewald, 2019). This approach was used to functionalize Abs with glycans to produce homogeneous trans-bitopic conjugates that direct Ab bound targets for lysosomal degradation (Fig. 3F) (Ahn et al., 2020). Homogenous Ab-glycan conjugates prepared in this manner showed improved pharmacokinetic properties relative to those produced by undirected lysine modification. A different enzymatic labeling approach relies on the enzyme Sortase A, which ligates cargoes attached to peptides with N-terminal glycine residues to proteins with exposed LPETG motifs (Cheloha et al., 2019; Pishesha, Ingram, & Ploegh, 2018). This approach was used to create a series of homo-bitopic nanobody-ligand conjugates that activate parathyroid hormone receptor in cells and in mice (Fig. 3G) (Cheloha, Fishcer, et al., 2020; Cheloha, Harmand, et al., 2020).

2.5 Biological applications

Conjugates consisting of multiple components that target cell surface proteins can exhibit biological properties that differ from the individual components. The linkage of multiple recognition modules in which the different components bind to distinct targets on the cell surface, sometimes on different cells, has been applied in a variety of settings. This approach has been most extensively deployed in the field of bispecific Abs (Labrijn, Janmaat, Reichert, & Parren, 2019). Bispecific Abs are often used to improve specificity for targeting a cell type of interest or to simultaneously target multiple cell types. This well-documented approach will not be discussed in detail here (see Table 1 for selected references). A new incarnation of this approach, dubbed "RIPR" for receptor inhibition by phosphatase recruitment, was used to enforce proximity between a promiscuous cell membrane phosphatase (CD45) and RTKs by using trans bitopic fusions between Ab fragments (Fernandes et al., 2020). Ligand binding to RTKs induces phosphorylation of intracellular tyrosine residues, which can be reversed by enforced proximity with CD45. This approach was used to inhibit activation of a variety of RTKs relevant to immune cell function including PD-1, CTLA-4, TIM-3, and LAG3 (Table 1).

An alternative type of genetic fusion consists of Ab(s) (or Ab fragments) and cytokines to produce trans bitopic fusions called immunocytokines (Neri, 2019). Cytokines often act via activation of RTK signaling. Immunocytokines target tumor cells via Ab recognition and induce anti-tumor immune responses via cytokine receptor engagement. This approach has been used to deliver cytokines such as IL15 (Liu et al., 2018), IL2 (Dougan et al., 2018), and IL12 (Probst, Stringhini, Ritz, Fugmann, & Neri, 2019) to tumors or the surrounding vasculature using Abs or their fragments (Table 1). Immunocytokines sometimes suffer from off-target effects due to the high potency of cytokines relative to the targeting capacity of the Ab (Tzeng, Kwan, Opel, Navaratna, & Wittrup, 2015). This has been addressed

Table 1 Summary of biological properties for multivalent conjugates targeting cell surface receptors consisting of antibodies and fragments.

Multivalent conjugate	Category	Targeting moiety	Proteins targeted	Signaling or 2° targeting moiety	Signaling response	References
Established classes						
Bispecific antibodies	Trans-bitopic	Ab and fragments	Various (tumor)	Ab (various)	Various (none, immune activation)	Labrijn et al. (2019)
Immunocytokine	Trans-bitopic	Ab and fragments	Various (tumor)	Cytokine (IL15, IL2, IL12, IFNα, others)	Anti-tumor immune response, improved specificity	Neri (2019)
ADC (conventional)	Trans-bitopic	Ab and fragments	Various (tumor)	None (cytotoxic compound)	Induction of cell death, improved specificity	Lambert and Berkenblit (2018)
New and specific examples						
Nb-synthetic peptide ligand (CLAMP)	Cis-bitopic	Nb	PTHR1	PTHR1 synthetic peptide ligand	Activation of PTHR1, blood calcium spike	Cheloha, Fischer, et al. (2020)
Nb-peptide ligand genetic fusion	Cis-bitopic	Nb	GLP1R, serum albumin	GLP1 peptide ligand (genetic fusion)	Activation of GLP1R, suppression blood sugar	Pan et al. (2020)
Ab-small molecule ligand	Cis-bitopic	Fab	DPP-IV	DPP-IV small molecule inhibitor	Improved potency for DPP-IV inhibition	Cheng et al. (2018)
Nb-small molecule ligand	Cis-bitopic	Nb	mGluR4	Photo switchable small molecule ligand	Light controlled receptor activation	Farrants et al. (2018)
Ab-small molecule ligand	Trans-bitopic	Ab	Various (tumor)	Small molecule inhibitor Na+/K+ ATPase	Induction of cell death, improved specificity	Marshall et al. (2016) and Sweeny et al. (2013)
RIPR	Trans-bitopic	Ab and fragments	RTKs	Anti CD45 Ab or Ab fragment	Dephosphorylation of Ab-bound proximal targets	Fernandes et al. (2020)
LYTACs	Trans-bitopic	Ab, peptides	EGFR, HER2, PD-L1	Glycan (synthetic compound or polymer)	Trafficking of Ab bound target to lysosome, degradation	Ahn et al. (2020) and Banik et al. (2020)

through the use of an attenuated mutant versions of cytokines (interferon-α and leptin), which show strong activity only when directed to cell types of choice via fusion with nanobodies in trans-bitopic ligands, in an approach dubbed "activity-by-targeting," which diminishes off-target effects (Garcin et al., 2014; Pogue et al., 2016).

Genetic fusions between Ab fragments and peptidic GPCR ligands have also been developed. There are several examples of fusions between Abs and peptide GPCR ligands in which the Ab serves only to prolong the circulatory half-life of the peptide ligand, or to bind to a soluble extracellular protein (Chodorge et al., 2018). Rarer are examples in which the Ab component also binds to a cell surface protein. In one example, a peptide ligand for GLP1R was fused to two nanobodies: one nanobody that targets a non-orthosteric site on GLP1R and the other nanobody that targets serum albumin (Pan et al., 2020). These trans-bitopic (or tritopic) conjugates stimulated enduring suppression of blood glucose levels, demonstrating strong GLP1R activation (Table 1).

The linkage of Abs with cell surface protein ligands produced using synthetic chemistry offers an avenue to improve the properties of these ligands. These efforts benefit from the large body of literature on conventional antibody-drug conjugates (ADCs) (Lambert & Berkenblit, 2018). Conventional ADCs are typically comprised of an Ab that targets a cell surface marker expressed on pathogenic cells such as tumors and a cytotoxic compound that acts on targets found in the interior of the cell. The key difference between conventional ADCs and conjugates discussed here is the cellular localization of the target: conventional ADCs require delivery of the conjugated cargo to the interior of the cell, typically through endocytosis and proteolysis. An alternative approach emphasized here consists of using Abs to deliver ligands that induce cytotoxic responses through targeting a cell surface receptor. This approach has been adopted to deliver cytotoxic small molecule inhibitors of the Na^+/K^+ ATPase through conjugation to Abs that target proteins expressed on tumor cells (Table 1) (Marshall et al., 2016; Sweeny et al., 2013). This approach provided trans-bitopic compounds with potent cell killing activity for cells expressing the target bound by the Ab and weak toxicity for other cells.

A variation on the theme of ADCs is to prepare a fusion between an Ab and ligand that both target the receptor of interest, or a cis-bitopic conjugate. Such conjugates have been dubbed conjugates of ligands and antibodies for membrane proteins or CLAMPs (Cheloha, Fischer, et al., 2020). The linkage of weak, synthetic peptide ligands for type-1 parathyroid hormone receptor (PTH-receptor 1, PTHR1) and nanobodies that target the same receptor at a different site provided conjugates with improved receptor activation potency relative to the original ligand (Table 1) (Cheloha, Fischer, et al., 2020). Furthermore, PTH activates both type-1 and type-2 PTH-receptor (PTHR1 and PTHR2), whereas one nanobody characterized selectively bound PTHR1. The PTH-nanobody conjugate potently activated PTHR1 but not PTHR2, showcasing the capacity of Ab targeting to impart receptor subtype specificity. A related approach was used to link an Ab and a small molecule each of which inhibits the cell surface localized protease DPP-IV (Cheng et al., 2018).

DPP-IV is a clinically validated target for the treatment of type-2 diabetes. This cis-bitopic conjugate, structurally characterized in complex with DPP-IV, showed superior inhibition of DPP-IV relative to either the Ab or small molecule inhibitor alone (Table 1).

Conjugates between Abs and synthetic moieties can also be used to modulate receptor function in ways not possible using conventional methods. Glycans that target cation-independent mannose-6-phosphate receptor or asialoglycoprotein receptor and stimulate lysosomal trafficking can be conjugated with Abs that recognize other cell surface receptors to provide lysosomal targeting chimeras, or LYTACs. These conjugates stimulate lysosomal trafficking and degradation of Ab-bound targets (Table 1) (Ahn et al., 2020; Banik et al., 2020). This approach is applicable even in instances where agents that inhibit cell surface proteins are not available. LYTACs were used to stimulate degradation of a variety of targets including EGFR, HER2, and PD-L1 and to inhibit receptor-dependent cell survival. In a different approach, a small molecule agonist of the metabolic glutamate receptor (mGluR4), which undergoes isomerization across a double bond and changes agonist activity upon illumination was conjugated to a receptor binding nanobody. The resulting cis bitopic conjugate exhibits agonist activity at mGluR4 that can be toggled off and on by illumination (Farrants et al., 2018). This approach holds promise for imposing tight kinetic control of signaling, which is essential for interrogating biological processes on short time scales.

3 Conclusions and future directions

The wide-ranging approaches used to construct and apply multivalent conjugates have provided novel tools with useful and interesting properties. Despite this progress, there are shortcomings in chemical and biological methodology that restrict facile development and application of these conjugates, particularly those with Abs as a component. It has become increasingly apparent that monovalent and site-specific modification of Abs provides homogeneous conjugates that show more consistent and predictable behavior in vivo relative to heterogeneous conjugates produced by non-specific conjugation methods. Approaches to modify less common amino acids, such as methionine (Lin et al., 2017), may prove useful for these goals. Established enzymatic modification methods (Krüger et al., 2019; Pishesha et al., 2018) and newly developed alternatives (Rehm et al., 2019) will also likely be helpful in producing homogeneous conjugates.

Another limitation in developing multivalent conjugates with Abs relates to the availability of appropriately specific Abs in sufficient quantities to carry out modifications. Conventional monoclonal Abs against many targets are widely available commercially; however, they are often produced with specialized equipment and purchasing in bulk is often prohibitively expensive. Nanobodies offer a useful alternative because they can be produced from bacteria in high yield using standard research laboratory equipment (Cheloha, Harmand, et al., 2020). However, nanobodies for many cell surface proteins of interest have not yet been identified. Novel screening methods provide a pathway for identifying new nanobodies that bind to and impact the function of cell surface proteins of interest (Ma et al., 2020; McMahon et al., 2020; Ren et al., 2020). One intriguing possibility for future efforts to develop multivalent conjugates is to functionalize targeting moieties, such as Abs, with functional groups that will react to form a covalent bond when found in proximity to another targeting moiety equipped with a partner functional group. Enforced proximity can be achieved by simultaneous binding of Abs to different proteins on the cell surface to allow for conjugate production in situ (Komatsu et al., 2020). Multivalent Ab conjugates assembled in this way showed an improved ability to stimulate endocytosis relative to unlinked counterparts (Komatsu et al., 2020). This finding suggests that chemical methods could be harnessed to assemble multivalent conjugates at their site of action, avoiding the need to synthesize and purify conjugates prior to assay, which should enable accelerated identification of improved conjugates.

Acknowledgments

This work was supported by the Intramural Research Program of the NIH, The National Institute of Diabetes and Digestive and Kidney Diseases (NIDDK).

References

Ågren, R., Zeberg, H., Stępniewski, T. M., Free, R. B., Reilly, S. W., Luedtke, R. R., et al. (2020). Ligand with two modes of interaction with the dopamine D $_2$ receptor—An induced-fit mechanism of insurmountable antagonism. *ACS Chemical Neuroscience*, *11*(19), 3130–3143. https://doi.org/10.1021/acschemneuro.0c00477.

Ahn, G., Banik, S., Miller, C. L., Riley, N., Cochran, J. R., & Bertozzi, C. (2020). *Lysosome targeting chimeras (LYTACs) that engage a liver-specific asialoglycoprotein receptor for targeted protein degradation* (Preprint). https://doi.org/10.26434/chemrxiv.12736778.v1.

Banik, S. M., Pedram, K., Wisnovsky, S., Ahn, G., Riley, N. M., & Bertozzi, C. R. (2020). Lysosome-targeting chimaeras for degradation of extracellular proteins. *Nature*, *584*(7820), 291–297. https://doi.org/10.1038/s41586-020-2545-9.

Barlow, N., Baker, S. P., & Scammells, P. J. (2013). Effect of linker length and composition on heterobivalent ligand-mediated receptor cross-talk between the A1 adenosine and β2 adrenergic receptors. *ChemMedChem*, *8*(12), 2036–2046. https://doi.org/10.1002/cmdc.201300286.

Bonifazi, A., Yano, H., Guerrero, A. M., Kumar, V., Hoffman, A. F., Lupica, C. R., et al. (2019). Novel and potent dopamine D2 receptor go-protein biased agonists. *ACS Pharmacology & Translational Science*, *2*(1), 52–65. https://doi.org/10.1021/acsptsci.8b00060.

Cheloha, R. W., Fischer, F. A., Woodham, A. W., Daley, E., Suminski, N., Gardella, T. J., et al. (2020). Improved GPCR ligands from nanobody tethering. *Nature Communications*, *11*(1), 2087. https://doi.org/10.1038/s41467-020-15884-8.

Cheloha, R. W., Gellman, S. H., Vilardaga, J.-P., & Gardella, T. J. (2015). PTH receptor-1 signalling-mechanistic insights and therapeutic prospects. *Nature Reviews. Endocrinology*, *11*(12), 712–724. https://doi.org/10.1038/nrendo.2015.139.

Cheloha, R. W., Harmand, T. J., Wijne, C., Schwartz, T. U., & Ploegh, H. L. (2020). Exploring cellular biochemistry with nanobodies. *The Journal of Biological Chemistry*, *295*(45), 15307–15327. https://doi.org/10.1074/jbc.REV120.012960.

Cheloha, R. W., Li, Z., Bousbaine, D., Woodham, A. W., Perrin, P., Volarić, J., et al. (2019). Internalization of influenza virus and cell surface proteins monitored by site-specific conjugation of protease-sensitive probes. *ACS Chemical Biology*, *14*(8), 1836–1844. https://doi.org/10.1021/acschembio.9b00493.

Cheng, A. C., Doherty, E. M., Johnstone, S., DiMauro, E. F., Dao, J., Luthra, A., et al. (2018). Structure-guided discovery of dual-recognition chemibodies. *Scientific Reports*, *8*(1), 7570. https://doi.org/10.1038/s41598-018-25848-0.

Chodorge, M., Celeste, A. J., Grimsby, J., Konkar, A., Davidsson, P., Fairman, D., et al. (2018). Engineering of a GLP-1 analogue peptide/anti-PCSK9 antibody fusion for type 2 diabetes treatment. *Scientific Reports*, *8*(1), 17545. https://doi.org/10.1038/s41598-018-35869-4.

Davenport, A. P., Scully, C. C. G., de Graaf, C., Brown, A. J. H., & Maguire, J. J. (2020). Advances in therapeutic peptides targeting G protein-coupled receptors. *Nature Reviews. Drug Discovery*, *19*(6), 389–413. https://doi.org/10.1038/s41573-020-0062-z.

Dougan, M., Ingram, J. R., Jeong, H.-J., Mosaheb, M. M., Bruck, P. T., Ali, L., et al. (2018). Targeting cytokine therapy to the pancreatic tumor microenvironment using PD-L1-specific VHHs. *Cancer Immunology Research*, *6*(4), 389–401. https://doi.org/10.1158/2326-6066.CIR-17-0495.

Erez, M., Takemori, A. E., & Portoghese, P. S. (1982). Narcotic antagonistic potency of bivalent ligands which contain beta-naltrexamine. Evidence for bridging between proximal recognition sites. *Journal of Medicinal Chemistry*, *25*(7), 847–849. https://doi.org/10.1021/jm00349a016.

Farrants, H., Gutzeit, V. A., Acosta-Ruiz, A., Trauner, D., Johnsson, K., Levitz, J., et al. (2018). SNAP-tagged nanobodies enable reversible optical control of a G protein-coupled receptor via a remotely tethered photoswitchable ligand. *ACS Chemical Biology*, *13*(9), 2682–2688. https://doi.org/10.1021/acschembio.8b00628.

Fernandes, R. A., Su, L., Nishiga, Y., Ren, J., Bhuiyan, A. M., Cheng, N., et al. (2020). Immune receptor inhibition through enforced phosphatase recruitment. *Nature*, *586*(7831), 779–784. https://doi.org/10.1038/s41586-020-2851-2.

Foster, S. R., Hauser, A. S., Vedel, L., Strachan, R. T., Huang, X.-P., Gavin, A. C., et al. (2019). Discovery of human signaling systems: Pairing peptides to G protein-coupled receptors. *Cell*, *179*(4), 895–908.e21. https://doi.org/10.1016/j.cell.2019.10.010.

Garcin, G., Paul, F., Staufenbiel, M., Bordat, Y., Van der Heyden, J., Wilmes, S., et al. (2014). High efficiency cell-specific targeting of cytokine activity. *Nature Communications*, *5*(1), 3016. https://doi.org/10.1038/ncomms4016.

Gomes, I., Ayoub, M. A., Fujita, W., Jaeger, W. C., Pfleger, K. D. G., & Devi, L. A. (2016). G protein-coupled receptor heteromers. *Annual Review of Pharmacology and Toxicology*, *56*, 403–425. https://doi.org/10.1146/annurev-pharmtox-011613-135952.

González-Maeso, J. (2011). GPCR oligomers in pharmacology and signaling. *Molecular Brain*, *4*(1), 20. https://doi.org/10.1186/1756-6606-4-20.

Griffith, D. A., Edmonds, D. J., Fortin, J.-P., Kalgutkar, A. S., Kuzmiski, J. B., Loria, P. M., et al. (2020). *A small-molecule oral agonist of the human glucagon-like peptide-1 receptor*. [Preprint]. Pharmacology and Toxicology https://doi.org/10.1101/2020.09.29.319483.

Hague, C., Lee, S. E., Chen, Z., Prinster, S. C., Hall, R. A., & Minneman, K. P. (2006). Heterodimers of alpha1B- and alpha1D-adrenergic receptors form a single functional entity. *Molecular Pharmacology*, *69*(1), 45–55. https://doi.org/10.1124/mol.105.014985.

Hao, Y., & Tatonetti, N. P. (2016). Predicting G protein-coupled receptor downstream signaling by tissue expression. *Bioinformatics*, *32*(22), 3435–3443. https://doi.org/10.1093/bioinformatics/btw510.

Hauser, A. S., Attwood, M. M., Rask-Andersen, M., Schiöth, H. B., & Gloriam, D. E. (2017). Trends in GPCR drug discovery: New agents, targets and indications. *Nature Reviews. Drug Discovery*, *16*(12), 829–842. https://doi.org/10.1038/nrd.2017.178.

Heukers, R., De Groof, T. W. M., & Smit, M. J. (2019). Nanobodies detecting and modulating GPCRs outside in and inside out. *Current Opinion in Cell Biology*, *57*, 115–122. https://doi.org/10.1016/j.ceb.2019.01.003.

Holze, J., Bermudez, M., Pfeil, E. M., Kauk, M., Bödefeld, T., Irmen, M., et al. (2020). Ligand-specific allosteric coupling controls G-protein-coupled receptor signaling. *ACS Pharmacology & Translational Science*, *3*(5), 859–867. https://doi.org/10.1021/acsptsci.0c00069.

Hübner, H., Schellhorn, T., Gienger, M., Schaab, C., Kaindl, J., Leeb, L., et al. (2016). Structure-guided development of heterodimer-selective GPCR ligands. *Nature Communications*, *7*(1), 12298. https://doi.org/10.1038/ncomms12298.

Hutchings, C. J., Koglin, M., Olson, W. C., & Marshall, F. H. (2017). Opportunities for therapeutic antibodies directed at G-protein-coupled receptors. *Nature Reviews. Drug Discovery*, *16*(9), 787–810. https://doi.org/10.1038/nrd.2017.91.

Kim, C. H., Axup, J. Y., Lawson, B. R., Yun, H., Tardif, V., Choi, S. H., et al. (2013). Bispecific small molecule-antibody conjugate targeting prostate cancer. *Proceedings of the National Academy of Sciences of the United States of America*, *110*(44), 17796–17801. https://doi.org/10.1073/pnas.1316026110.

Komatsu, T., Kyo, E., Ishii, H., Tsuchikama, K., Yamaguchi, A., Ueno, T., et al. (2020). Antibody clicking as a strategy to modify antibody functionalities on the surface of targeted cells. *Journal of the American Chemical Society*, *142*(37), 15644–15648. https://doi.org/10.1021/jacs.0c05331.

Krüger, T., Dierks, T., & Sewald, N. (2019). Formylglycine-generating enzymes for site-specific bioconjugation. *Biological Chemistry*, *400*(3), 289–297. https://doi.org/10.1515/hsz-2018-0358.

Kumar, V., Bonifazi, A., Ellenberger, M. P., Keck, T. M., Pommier, E., Rais, R., et al. (2016). Highly selective dopamine D3 receptor (D3R) antagonists and partial agonists based on eticlopride and the D3R crystal structure: New leads for opioid dependence treatment. *Journal of Medicinal Chemistry*, *59*(16), 7634–7650. https://doi.org/10.1021/acs.jmedchem.6b00860.

Kumar, V., Moritz, A. E., Keck, T. M., Bonifazi, A., Ellenberger, M. P., Sibley, C. D., et al. (2017). Synthesis and pharmacological characterization of novel trans-cyclopropylmethyl-linked bivalent ligands that exhibit selectivity and allosteric pharmacology at the dopamine D3 receptor (D3R). *Journal of Medicinal Chemistry*, *60*(4), 1478–1494. https://doi.org/10.1021/acs.jmedchem.6b01688.

Labrijn, A. F., Janmaat, M. L., Reichert, J. M., & Parren, P. W. H. I. (2019). Bispecific antibodies: A mechanistic review of the pipeline. *Nature Reviews. Drug Discovery*, *18*(8), 585–608. https://doi.org/10.1038/s41573-019-0028-1.

Lambert, J. M., & Berkenblit, A. (2018). Antibody-drug conjugates for cancer treatment. *Annual Review of Medicine*, *69*, 191–207. https://doi.org/10.1146/annurev-med-061516-121357.

Lin, S., Yang, X., Jia, S., Weeks, A. M., Hornsby, M., Lee, P. S., et al. (2017). Redox-based reagents for chemoselective methionine bioconjugation. *Science (New York, N.Y.)*, *355*(6325), 597–602. https://doi.org/10.1126/science.aal3316.

Liu, Y., Wang, Y., Xing, J., Li, Y., Liu, J., & Wang, Z. (2018). A novel multifunctional anti-CEA-IL15 molecule displays potent antitumor activities. *Drug Design, Development and Therapy*, *12*, 2645–2654. https://doi.org/10.2147/DDDT.S166373.

Liu, T., Zhang, Y., Liu, Y., Wang, Y., Jia, H., Kang, M., et al. (2015). Functional human antibody CDR fusions as long-acting therapeutic endocrine agonists. *Proceedings of the National Academy of Sciences of the United States of America*, *112*(5), 1356–1361. https://doi.org/10.1073/pnas.1423668112.

Lyu, J., Wang, S., Balius, T. E., Singh, I., Levit, A., Moroz, Y. S., et al. (2019). Ultra-large library docking for discovering new chemotypes. *Nature*, *566*(7743), 224–229. https://doi.org/10.1038/s41586-019-0917-9.

Ma, Y., Ding, Y., Song, X., Ma, X., Li, X., Zhang, N., et al. (2020). Structure-guided discovery of a single-domain antibody agonist against human apelin receptor. *Science Advances*, *6*(3), eaax7379. https://doi.org/10.1126/sciadv.aax7379.

Marshall, D. J., Harried, S. S., Murphy, J. L., Hall, C. A., Shekhani, M. S., Pain, C., et al. (2016). Extracellular antibody drug conjugates exploiting the proximity of two proteins. *Molecular Therapy: The Journal of the American Society of Gene Therapy*, *24*(10), 1760–1770. https://doi.org/10.1038/mt.2016.119.

Marti-Solano, M., Crilly, S. E., Malinverni, D., Munk, C., Harris, M., Pearce, A., et al. (2020). Combinatorial expression of GPCR isoforms affects signalling and drug responses. *Nature*, *587*(7835), 650–656. https://doi.org/10.1038/s41586-020-2888-2.

McMahon, C., Staus, D. P., Wingler, L. M., Wang, J., Skiba, M. A., Elgeti, M., et al. (2020). Synthetic nanobodies as angiotensin receptor blockers. *Proceedings of the National Academy of Sciences of the United States of America*, *117*(33), 20284–20291. https://doi.org/10.1073/pnas.2009029117.

Moehle, M. S., Bender, A. M., Dickerson, J. W., Foster, D. J., Donsante, Y., Peng, W., et al. (2020). *Discovery of the first selective M_4 muscarinic acetylcholine receptor antagonists with in vivo anti-parkinsonian and anti-dystonic efficacy*. [Preprint]. Pharmacology and Toxicology https://doi.org/10.1101/2020.10.12.324152.

Neri, D. (2019). Antibody-cytokine fusions: Versatile products for the modulation of anticancer immunity. *Cancer Immunology Research*, *7*(3), 348–354. https://doi.org/10.1158/2326-6066.CIR-18-0622.

Newman, A. H., Battiti, F. O., & Bonifazi, A. (2020). 2016 Philip S. Portoghese medicinal chemistry lectureship: Designing bivalent or bitopic molecules for G-protein coupled receptors. The whole is greater than the sum of its parts. *Journal of Medicinal Chemistry*, *63*(5), 1779–1797. https://doi.org/10.1021/acs.jmedchem.9b01105.

Nimczick, M., & Decker, M. (2015). New approaches in the design and development of cannabinoid receptor ligands: Multifunctional and bivalent compounds. *ChemMedChem*, *10*(5), 773–786. https://doi.org/10.1002/cmdc.201500041.

Pan, H., Su, Y., Xie, Y., Wang, W., Qiu, W., Chen, W., et al. (2020). Everestmab, a novel long-acting GLP-1/anti GLP-1R nanobody fusion protein, exerts potent anti-diabetic effects. *Artificial Cells, Nanomedicine, and Biotechnology*, *48*(1), 854–866. https://doi.org/10.1080/21691401.2020.1770268.

Pishesha, N., Ingram, J. R., & Ploegh, H. L. (2018). Sortase A: A model for transpeptidation and its biological applications. *Annual Review of Cell and Developmental Biology*, *34*, 163–188. https://doi.org/10.1146/annurev-cellbio-100617-062527.

Pogue, S. L., Taura, T., Bi, M., Yun, Y., Sho, A., Mikesell, G., et al. (2016). Targeting attenuated interferon-α to myeloma cells with a CD38 antibody induces potent tumor regression with reduced off-target activity. *PLoS One*, *11*(9), e0162472. https://doi.org/10.1371/journal.pone.0162472.

Portoghese, P. S., Sultana, M., & Takemori, A. E. (1988). Naltrindole, a highly selective and potent non-peptide delta opioid receptor antagonist. *European Journal of Pharmacology*, *146*(1), 185–186. https://doi.org/10.1016/0014-2999(88)90502-x.

Pottier, C., Fresnais, M., Gilon, M., Jérusalem, G., Longuespée, R., & Sounni, N. E. (2020). Tyrosine kinase inhibitors in cancer: Breakthrough and challenges of targeted therapy. *Cancers*, *12*(3). https://doi.org/10.3390/cancers12030731.

Probst, P., Stringhini, M., Ritz, D., Fugmann, T., & Neri, D. (2019). Antibody-based delivery of TNF to the tumor neovasculature potentiates the therapeutic activity of a peptide anticancer vaccine. *Clinical Cancer Research: An Official Journal of the American Association for Cancer Research*, *25*(2), 698–709. https://doi.org/10.1158/1078-0432.CCR-18-1728.

Rehm, F. B. H., Harmand, T. J., Yap, K., Durek, T., Craik, D. J., & Ploegh, H. L. (2019). Site-specific sequential protein labeling catalyzed by a single recombinant ligase. *Journal of the American Chemical Society*, *141*(43), 17388–17393. https://doi.org/10.1021/jacs.9b09166.

Ren, H., Li, J., Zhang, N., Hu, L. A., Ma, Y., Tagari, P., et al. (2020). Function-based high-throughput screening for antibody antagonists and agonists against G protein-coupled receptors. *Communications Biology*, *3*(1), 146. https://doi.org/10.1038/s42003-020-0867-7.

Richards, D. A. (2018). Exploring alternative antibody scaffolds: Antibody fragments and antibody mimics for targeted drug delivery. *Drug Discovery Today: Technologies*, *30*, 35–46. https://doi.org/10.1016/j.ddtec.2018.10.005.

Rosenbaum, M. I., Clemmensen, L. S., Bredt, D. S., Bettler, B., & Strømgaard, K. (2020). Targeting receptor complexes: A new dimension in drug discovery. *Nature Reviews. Drug Discovery*, *19*(12), 884–901. https://doi.org/10.1038/s41573-020-0086-4.

Soave, M., Heukers, R., Kellam, B., Woolard, J., Smit, M. J., Briddon, S. J., et al. (2020). Monitoring allosteric interactions with CXCR4 using NanoBiT conjugated nanobodies. *Cell Chemical Biology*, *27*, 1250–1261.e5. S2451945620302294 https://doi.org/10.1016/j.chembiol.2020.06.006.

Soriano, A., Ventura, R., Molero, A., Hoen, R., Casadó, V., Cortés, A., et al. (2009). Adenosine A2A receptor-antagonist/dopamine D2 receptor-agonist bivalent ligands as pharmacological tools to detect A2A-D2 receptor heteromers. *Journal of Medicinal Chemistry*, *52*(18), 5590–5602. https://doi.org/10.1021/jm900298c.

Soto, C. A., Shashack, M. J., Fox, R. G., Bubar, M. J., Rice, K. C., Watson, C. S., et al. (2018). Novel bivalent 5-HT2A receptor antagonists exhibit high affinity and potency in vitro and efficacy in vivo. *ACS Chemical Neuroscience*, *9*(3), 514–521. https://doi.org/10.1021/acschemneuro.7b00309.

Sweeny, L., Hartman, Y. E., Zinn, K. R., Prudent, J. R., Marshall, D. J., Shekhani, M. S., et al. (2013). A novel extracellular drug conjugate significantly inhibits head and neck squamous cell carcinoma. *Oral Oncology*, *49*(10), 991–997. https://doi.org/10.1016/j.oraloncology.2013.07.006.

Terrillon, S., & Bouvier, M. (2004). Roles of G-protein-coupled receptor dimerization. *EMBO Reports*, *5*(1), 30–34. https://doi.org/10.1038/sj.embor.7400052.

Toneatti, R., Shin, J. M., Shah, U. H., Mayer, C. R., Saunders, J. M., Fribourg, M., et al. (2020). Interclass GPCR heteromerization affects localization and trafficking. *Science Signaling*, *13*(654). https://doi.org/10.1126/scisignal.aaw3122.

Trenker, R., & Jura, N. (2020). Receptor tyrosine kinase activation: From the ligand perspective. *Current Opinion in Cell Biology*, *63*, 174–185. https://doi.org/10.1016/j.ceb.2020.01.016.

Tzeng, A., Kwan, B. H., Opel, C. F., Navaratna, T., & Wittrup, K. D. (2015). Antigen specificity can be irrelevant to immunocytokine efficacy and biodistribution. *Proceedings of the National Academy of Sciences of the United States of America*, *112*(11), 3320–3325. https://doi.org/10.1073/pnas.1416159112.

Vira, S., Mekhedov, E., Humphrey, G., & Blank, P. S. (2010). Fluorescent-labeled antibodies: Balancing functionality and degree of labeling. *Analytical Biochemistry*, *402*(2), 146–150. https://doi.org/10.1016/j.ab.2010.03.036.

Wacker, D., Stevens, R. C., & Roth, B. L. (2017). How ligands illuminate GPCR molecular pharmacology. *Cell*, *170*(3), 414–427. https://doi.org/10.1016/j.cell.2017.07.009.

Walsh, S. J., Bargh, J. D., Dannheim, F. M., Hanby, A. R., Seki, H., Counsell, A. J., et al. (2021). Site-selective modification strategies in antibody–drug conjugates. *Chemical Society Reviews*, *50*, 1305–1353. https://doi.org/10.1039/D0CS00310G.

Xu, J., He, J., Castleberry, A. M., Balasubramanian, S., Lau, A. G., & Hall, R. A. (2003). Heterodimerization of alpha 2A- and beta 1-adrenergic receptors. *The Journal of Biological Chemistry*, *278*(12), 10770–10777. https://doi.org/10.1074/jbc.M207968200.

Zhang, Y., Gilliam, A., Maitra, R., Damaj, M. I., Tajuba, J. M., Seltzman, H. H., et al. (2010). Synthesis and biological evaluation of bivalent ligands for the cannabinoid 1 receptor. *Journal of Medicinal Chemistry*, *53*(19), 7048–7060. https://doi.org/10.1021/jm1006676.

Zhang, A., Zhang, Y., Branfman, A. R., Baldessarini, R. J., & Neumeyer, J. L. (2007). Advances in development of dopaminergic aporphinoids. *Journal of Medicinal Chemistry*, *50*(2), 171–181. https://doi.org/10.1021/jm060959i.

CHAPTER

10

Identifying *Plasmodium falciparum* receptor activation using bioluminescence resonance energy transfer (BRET)-based biosensors in HEK293 cells

Pedro H.S. Pereira[a,b], Celia R.S. Garcia[a,*], and Michel Bouvier[b]

[a]*Department of Clinical and Toxicological Analysis, School of Pharmaceutical Sciences, University of São Paulo, São Paulo, Brazil*

[b]*Department of Biochemistry and Molecular Medicine, Institute for Research in Immunology and Cancer, University of Montreal, Montreal, QC, Canada*

**Corresponding author: e-mail address: celiaregarcia@gmail.com*

Chapter outline

Methods in Cell Biology, Volume 166, ISSN 0091-679X, https://doi.org/10.1016/bs.mcb.2021.06.018

Abstract

Throughout evolution the need for unicellular organisms to associate and form a single cluster of cells had several evolutionary advantages. G protein coupled receptors (GPCRs) are responsible for a large part of the senses that allow this clustering to succeed, playing a fundamental role in the perception of cell's external environment, enabling the interaction and coordinated development between each cell of a multicellular organism. GPCRs are not exclusive to complex multicellular organisms. In single-celled organisms, GPCRs are also present and have a similar function of detecting changes in the external environment and transforming them into a biological response. There are no reports of GPCRs in parasitic protozoa, such as the *Plasmodium* genus, and the identification of a protein of this family in *P. falciparum* would have a significant impact both on the understanding of the basic biology of the parasite and on the history of the evolution of GPCRs. The protocol described here was successfully applied to study a GPCR candidate in *P. falciparum* for the first time, and we hope that it helps other groups to use the same approach to study this deadly parasite.

1 Introduction

Bioluminescent resonance energy transfer (BRET) is a natural phenomenon observed in marine animals such as *Renilla reniformis* and *Aequorea victoria*. The energy from a luminescent donor (e.g., *Renilla* Luciferase, RLuc) is transferred to a fluorescent receptor (GFP, YFP or RFP) when the proteins are physically close to each other. This natural event was transformed into a method for studying protein-protein interactions in bacteria (Xu, Piston, & Johnson, 2002) and mammals (Angers et al., 2000) (Fig. 1). Subsequently, the method became a powerful alternative for studying living cells in a non-invasive and highly sensitive way. The assays consist of the expression of a protein fused to a RLuc and another fused to a fluorescent protein (Hamdan, Percherancier, Breton, & Bouvier, 2006; Kobayashi, Picard, Schönegge, & Bouvier, 2019). Two of the currently used forms of BRET consist of the use of coelenterazine H (benzyl-coelenterazine) as a substrate (BRET1) that has an emission at 480 nm or bis-deoxy-coelenterazine (DeepBlueC™) that has an emission at 395 nm (BRET2). BRET1-based experiments result in strong and long-lasting signals, while BRET2 produces signals that are 300 times weaker and of shorter duration, but with an excellent signal-to-noise ratio (Kocan, See, Seeber, Eidne, & Pfleger, 2008). The main advantage of using BRET-based methods

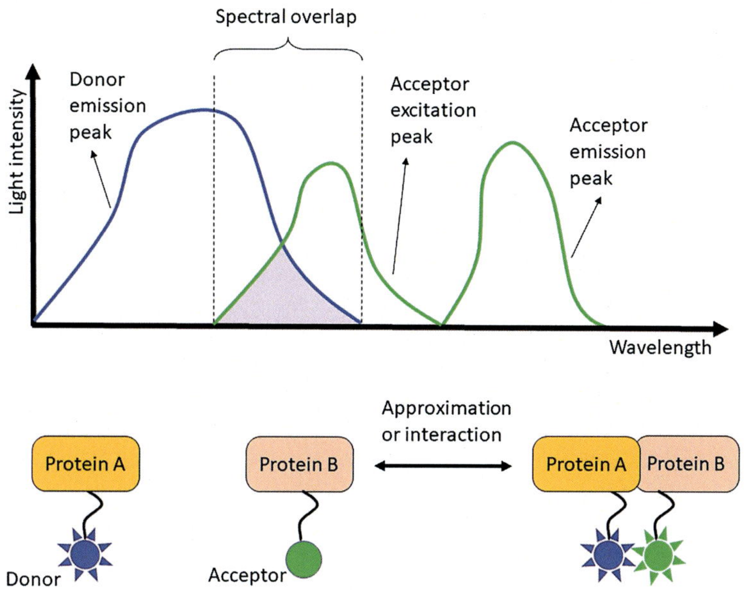

FIG. 1

BRET-based methods for studying protein-protein interaction. Two proteins of interest (A and B) are engineered to contain a donor and an acceptor sequence in their polypeptide chain. The transfer of energy is made when the luminescent donor protein, due to the proximity, excites the acceptor protein through a spectrum overlap and this, in turn, emits the energy in another wavelength. The source of excitation is a chemical substrate that is broken down by a bioluminescent enzyme (luciferase).

to detect the interaction of proteins is the ability to use living systems without the need to chemically modify the proteins or the cell under study, resulting in an interaction in a molecular environment closer to the natural one (Weibrecht et al., 2010). This method is also compatible with live cell microscopy in real time, making it even more robust (Kobayashi et al., 2019). In comparison with methods based on fluorescent resonance energy transfer (FRET), BRET-based methods have better resolution due to the lower background signal caused by the absence of a laser for excitation of the donor particle. BRET-based biosensors have already been used to study components of the signaling pathway in mammals, such as receptor multimerization (Ayoub et al., 2002; Mercier, Salahpour, Angers, Breit, & Bouvier, 2002), activation and coupling of G proteins (Galés et al., 2005, 2006), traffic (Kobayashi, Ogawa, Yao, Lichtarge, & Bouvier, 2009), recruitment and activation of accessory proteins (Charest & Bouvier, 2003; Terrillon, Barberis, & Bouvier, 2004), receptor activity modifying proteins (Héroux, Breton, Hogue, & Bouvier, 2007), ubiquitination (Perroy, Pontier, Charest, Aubry, & Bouvier, 2004) and assays for interaction with agonists (Stoddart et al., 2015).

2 Identifying *Plasmodium falciparum* receptors: BRET principles

To identify putative *Plasmodium falciparum* receptor, the candidate receptor cDNA is co-expressed with the appropriate components of the BRET-based sensors in HEK293 cells. Bioluminescence resonance energy transfer (BRET) is a natural phenomenon by which the energy of a luminescent donor is transferred to fluorescent acceptor that, in turn, emits light at a wave length that is different from that of the donor emission. Such transfer occurs only when the energy donor and acceptor are within 100 Å and the transfer efficiency is inversely proportional to 6th power of the distance between the donor and acceptor. This natural phenomenon has been exploited to develop proximity-based assays to monitor receptor activation and cell signaling events based on protein-protein interactions and conformational changes within proteins (Marullo & Bouvier, 2007). In most cases, luciferases and fluorescent proteins are used as donors and acceptors, respectively, and are genetically attached to components of the biosensors. Biosensors that can be used to detect the putative receptor activity include sensors detecting G protein activation (Galés et al., 2006) and second messenger generation (Leduc et al., 2009; Namkung et al., 2018).

3 Before you begin

Timing: 30–50 min

1. HEK293 cell line
2. Tyrode Buffer
   (a) 119 mM NaCl, 5 mM KCl, 25 mM HEPES buffer, 2 mM CaCl2, 2 mM MgCl2, 6 g/L glucose. pH 7.4
3. Phosphate-buffered saline (PBS)
   (a) 137 mM NaCl, 2.7 mM KCl, 4.3 mM Na2HPO4, 1.4 mM NaH2PO4, pH 7.4
4. Polyethyleneimine, linear, 25 kDa (PEI). For 1 g/L stock solution:
   (a) Dissolve 1 mg/mL PEI in PBS
   (b) Swirl for 1 h with stirring rod
   (c) Set pH to 6.5–7.5 with HCl
   (d) Adjust volume to 50 mL with PBS and let it swirl overnight
   (e) Check pH next morning
   (f) Sterilize by filtration (0.2 μm diameter)
   (g) Aliquot and store at −20 °C
5. PEI working solution
   (a) For 3:1 proportion of PEI:DNA, add 3 μL of PEI stock solution to 47 μL PBS

6. Receptor DNA
   (a) Prepare 100 ng/μL solution in PBS. cDNA should be codon optimized
7. Biosensor DNA
   (a) Prepare 10 ng/μL solution in PBS (or 100 ng/μL depending on the biosensor used)

4 Key resources table

Reagent or resource	Source	Identifier
Chemicals		
Coelenterazine 400a (DeepBlueC™)	NanoLight Technology	340
Trypsin-EDTA	Sigma-Aldrich	T4299
Polyethyleneimine, linear, 25 kDa	Sigma-Aldrich	9002-98-6
Salmon sperm DNA	ThermoFischer	15,632,011
Fetal bovine serum	Sigma-Aldrich	F2442
Software and algorithms		
GraphPad Prism 8	GraphPad	
Equipment		
Mithras LB 940	Berthold Technologies	

5 Materials and equipment

- HEK293SL cell line
- Dulbecco's modified Eagle's medium (DMEM)
- Fetal bovine serum (FBS)
- Tyrode Buffer
- Phosphate-buffered saline (PBS)
- Polyethyleneimine, linear, 25 kDa (PEI)
- Receptor DNA
- Biosensor DNA
- Salmon sperm DNA (ssDNA)
- Coelenterazine 400a (DeepBlueC™)
- Agonist solution
- 96-wells white plates, flat bottom
- Trypsin-EDTA solution
- BRET compatible plate reader
- Cell incubator at 37 °C with 5% CO_2

Alternatives: New Calf serum (NCS) can be used instead of FBS. Other transfection methods then PEI can be used, but need to be optimized. Methoxy e-Coelenterazine (Prolume purple) is an alternative to DeepBlueC that is more stable (longer reading periods), gives a higher signal, but a smaller window.

6 Step-by-step method details

6.1 Transfection mix preparation

Timing: 20–30 min

1. The transfection mix for one row (12 wells) of a 96 well cell plate must be prepared in a 2 mL Eppendorf tube, which will have a total of 200 μL of solution containing 1 μg of DNA. 12 wells is enough to test four different conditions for one receptor using one biosensor (including controls). If more conditions are needed, scale the transfection proportionally.
   - **(a)** Add 1–5 μL (100–500 ng) of receptor's DNA
   - **(b)** Add 1–10 μL (10–500 ng) of biosensor's DNA
   - **(c)** Add ssDNA to complete 1 μg of total DNA
   - **(d)** Complete volume to 100 μL using PBS
2. Add 100 μL of PEI working solution to the DNA mix.
3. Mix the solution thoroughly and incubate at room temperature for 15 min.

Note: PEI + DNA mixtures should not be incubated for long periods (hours) as this will result in damage to the DNA molecules.

Critical: Titration of receptor's DNA and biosensor's DNA is necessary to optimize the expression ratios that provide a robust response window.

6.2 Cell transfection

Timing: 10–20 min to prepare the cells and mix reagents. 48 h of incubation.

4. Considering a small cell culture flask (T25) of healthy and 70–90% confluent and mycoplasma free HEK293 cells.
   - **(a)** Remove media from the flask and gently wash with PBS.
   - **(b)** Add 500 μL of trypsin-EDTA solution and incubate for 5 min at 37 °C.
   - **(c)** Add 9,5 mL of complete DMEM, homogenize and transfer to a 15 mL falcon tube.
5. Transfer 1.3 mL of cells to the tube containing 200 μL of DNA + PEI transfection mix (from step 3). Mix by inverting the tube vigorously.

Critical: Be sure to mix thoroughly both cells and transfection mix before pipetting, and gently mix after adding cells to the DNA + PEI solution to ensure homogenous transfection and distribution to every well. This is even more important when using a biosensor with more than one component.

6. Transfer 100 µL to each well in one row of the 96 well plate.
7. Incubate the plate for 48 h at 37 °C with 5% CO_2.

6.3 Optional: Butyric Acid supplementation

Timing: 5 min

8. 24 h after transfection, supplement each well with butyric acid to a final concentration of 2 mM.
9. Return the plate to the incubator for 24 h at 37 °C with 5% CO_2.

Note: Butyric acid act as a molecular chaperone that stabilize the tertiary structures of proteins with complex folding, which can increase the amount of protein expressed and targeted to their proper site of action.

6.4 Reading BRET signals

Timing: 30–50 min

10. Remove cells from the incubator and wash every well once with 100 µL of PBS.
    (a) Remove PBS and add 80 µL of Tyrode buffer to all wells.
    (b) Incubate the plate at 37 °C for at least 20 min.

Note: Cells can remain in Tyrode buffer for up to 2 h before adding substrate and reading.

11. Add 10 µL of DeepBlueC solution and incubate for 5 min at 37 °C.
12. Add 10 µL of 10 × concentrated agonist solution. Proceed for reading immediately.
13. Read plates using a microplate reader equipped with a donor filter 485 nm and an acceptor filter 530 nm.

Critical: The plate reader selected need to be sensitive for the detection of luminescence signals emitted at different wave lengths. The reader must allow the detection of two different wave-length signal almost simultaneously to allow proper ratiometric measurements. This can be achieved through a light splitting device or rapid automatic exchange between filters.

7 Expected outcomes

The BRET signals reading can be done either at specific time points (30 s, 1 min, 5 min, etc. after stimulation), resulting in only one value per well, or over time, resulting in a time-dependent signal curve. In any of the methods used, the BRET signals are consistent and have great reproducibility between biological replicates. The expected result varies from a BRET signal around 0.005, when there is no interaction between donor and acceptor, to values between 0.03 and 2.0, when there is interaction. The amplitude of the signal depends on the biosensor used and the optimization of the experimental system. The sensitivity is sufficient to detect differences as small as 10% between conditions with statistical significance.

An example of Diacylglycerol (DAG) biosensor and its application for *P. falciparum* serpentine receptor 12 (PfSR12) can be accessed in the publication by Pereira et al. (2020). The DAG biosensor is used to detect the formation of diacylglycerol in the plasma membrane. It is composed of a protein chimera containing a domain that attaches the biosensor to the membrane through myristoylation/palmitoylation, a GFP (acceptor), a long flexible linker, RLuc (donor) and a DAG interaction domain c1b from PKCδ (Namkung et al., 2018). This experiment showed that when HEK293 cells are transfected with PfSR12, there is an increase in DAG formation when compared to cells transfected with the dopamine receptor (D2R). When cells were treated with Gq/11 inhibitor YM254890, the signal is reduced drastically, showing enrollment of Gq/11 coupled GPCR signaling pathway. The full set of data and other examples can be found in (Pereira et al., 2020).

8 Quantification and statistical analysis

The interpretation of BRET values is generally quite straightforward. A high value indicates that the donor and receptor molecules are close, and a low value means that they are separate. Before starting stimulation of the cells, the acceptor fluorescence (excited by an external source of light at the wave-length corresponding to its excitation peak) and luminescence can be estimated using a plate reader just to ensure that the transfection worked as expected. These measurements also allow to have a quantitative measurement of the biosensor components.

Statistical analysis can be done using Student's *t*-test to assess the difference between the means of two conditions or using two-dimensional ANOVA for comparisons between multiple conditions or samples. Be careful when analyzing BRET results as over time curves, as the area under the curve and the height of the response peak may be different.

9 Advantages

The choice of heterologous expression in a mammalian system is ultimately the best option for studying *P. falciparum* receptors because parasite knockout or conditional knockdown systems are complex, time consuming, and may be difficult to properly control. Also, the assays used to assess receptor activity in the parasite are indirect and do not report directly on the proximal effector of the putative receptors. As the culture of mammalian cells is simpler and the variety of BRET-based biosensors already validated in these cells is large, the application of this method for the study of *Plasmodium* receptors represents a promising avenue for the identification and characterization of *P. falciparum* receptors and other signaling molecules. Since *P. falciparum* parasites infect and express proteins inside human cells for a large portion of its lifecycle, it is rational that parasites proteins are stable and may associate to human proteins, which allows this method to function properly.

10 Limitations

The major limitation of the method for studying protein interactions and signaling in *Plasmodium* is the availability of BRET-based biosensors. Currently, there is no biosensor available for studying specifically *Plasmodium* proteins, only those that measure the downstream activity of a receptor (such as biosensors based on mammalian signaling cascades) are available. Therefore, *Plasmodium*-specific biosensors should be built for the study of these protein, if the use of biosensors already used for mammalian proteins is not applicable.

Another limitation of this method is the low abundance of *Plasmodium* protein expression in mammalian cells. In addition to the expression being much smaller than mammalian proteins, there is no guarantee that the protein will be expressed and sent to the correct destination in mammalian cells.

11 Optimization and troubleshooting

11.1 Problem: Low BRET signal, luminescence and/or fluorescence

When establishing a new BRET-based assay, the initial signals may be low and may not change upon activation of the candidate receptor protein. This may not be because the candidate protein is not a receptor but could result either from a receptor expression level that is too low or a ratio between the receptor and the biosensor components that is not optimal.

11.2 Potential solution to optimize the procedure

The titration of several components of the assay can solve the problems and allow to establish the optimized conditions of the procedure. The components that must be titrated to optimize the method are: DNA of the protein of interest, DNA of the biosensor or components of the biosensor, concentration of PEI, concentration of agonist/antagonist. In addition, an analysis of the signal strength over time should be done to establish the time-point of greatest response window. Sometimes the response can be so quick that a plate reader that automatically adds compounds is needed. Other methods of helping to solve these problems are the use of butyric acid to increase expression or transfection methods other than PEI that may lead to better transfection efficiency. Using low passage cells and DMEM supplemented with FBS instead of NCS can increase greatly the expression and signal of all components. In addition, to make sure that the biosensor components are functional under the experimental conditions, it is important to carry out control parallel experiment using *bona-fide* as positive controls.

References

Angers, S., Salahpour, A., Joly, E., Hilairet, S., Chelsky, D., Dennis, M., et al. (2000). Detection of β2-adrenergic receptor dimerization in living cells using bioluminescence resonance energy transfer (BRET). *Proceedings of the National Academy of Sciences of the United States of America*, *97*, 3684–3689. https://doi.org/10.1073/pnas.060590697.

Ayoub, M. A., Couturier, C., Lucas-Meunier, E., Angers, S., Fossier, P., Bouvier, M., et al. (2002). Monitoring of ligand-independent dimerization and ligand-induced conformational changes of melatonin receptors in living cells by bioluminescence resonance energy transfer. *Journal of Biological Chemistry*, *277*, 21522–21528. https://doi.org/10.1074/jbc.M200729200.

Charest, P. G., & Bouvier, M. (2003). Palmitoylation of the V2 vasopressin receptor carboxyl tail enhances β-arrestin recruitment leading to efficient receptor endocytosis and ERK1/2 activation. *Journal of Biological Chemistry*, *278*, 41541–41551. https://doi.org/10.1074/jbc.M306589200.

Galés, C., Rebois, R. V., Hogue, M., Trieu, P., Breit, A., Hébert, T. E., et al. (2005). Real-time monitoring of receptor and G-protein interactions in living cells. *Nature Methods*, *2*, 177–184. https://doi.org/10.1038/nmeth743.

Galés, C., Van Durm, J. J. J., Schaak, S., Pontier, S., Percherancier, Y., Audet, M., et al. (2006). Probing the activation-promoted structural rearrangements in preassembled receptor-G protein complexes. *Nature Structural and Molecular Biology*, *13*, 778–786. https://doi.org/10.1038/nsmb1134.

Hamdan, F. F., Percherancier, Y., Breton, B., & Bouvier, M. (2006). Monitoring protein-protein interactions in living cells by bioluminescence resonance energy transfer (BRET). *Current Protocols in Neuroscience*, *34*(1), 5.23.1–5.23.20. https://doi.org/10.1002/0471142301.ns0523s34.

Héroux, M., Breton, B., Hogue, M., & Bouvier, M. (2007). Assembly and signaling of CRLR and RAMP1 complexes assessed by BRET. *Biochemistry*, *46*, 7022–7033. https://doi.org/10.1021/bi0622470.

Kobayashi, H., Ogawa, K., Yao, R., Lichtarge, O., & Bouvier, M. (2009). Functional rescue of β1-adrenoceptor dimerization and trafficking by pharmacological chaperones. *Traffic*, *10*, 1019–1033. https://doi.org/10.1111/j.1600-0854.2009.00932.x.

Kobayashi, H., Picard, L.-P., Schönegge, A.-M., & Bouvier, M. (2019). Bioluminescence resonance energy transfer–based imaging of protein–protein interactions in living cells. *Nature Protocols*, *14*(4), 1084–1107. https://doi.org/10.1038/s41596-019-0129-7.

Kocan, M., See, H. B., Seeber, R. M., Eidne, K. A., & Pfleger, K. D. G. (2008). Demonstration of improvements to the bioluminescence resonance energy transfer (BRET) technology for the monitoring of g protein-coupled receptors in live cells. *Journal of Biomolecular Screening*, *13*, 888–898. https://doi.org/10.1177/1087057108324032.

Leduc, M., Breton, B., Gales, C., Le Gouill, C., Bouvier, M., Chemtob, S., et al. (2009). Functional selectivity of natural and synthetic prostaglandin EP4 receptor ligands. *Journal of Pharmacology and Experimental Therapeutics*, *331*, 297–307. https://doi.org/10.1124/jpet.109.156398.

Marullo, S., & Bouvier, M. (2007). Resonance energy transfer approaches in molecular pharmacology and beyond. *Trends in Pharmacological Sciences*, *28*, 362–365. https://doi.org/10.1016/j.tips.2007.06.007.

Mercier, J. F., Salahpour, A., Angers, S., Breit, A., & Bouvier, M. (2002). Quantitative assessment of β1- and β2-adrenergic receptor homo- and heterodimerization by bioluminescence resonance energy transfer. *Journal of Biological Chemistry*, *277*, 44925–44931. https://doi.org/10.1074/jbc.M205767200.

Namkung, Y., LeGouill, C., Kumar, S., Cao, Y., Teixeira, L. B., Lukasheva, V., et al. (2018). Functional selectivity profiling of the angiotensin II type 1 receptor using pathway-wide BRET signaling sensors. *Science Signaling*, *11*, eaat1631. https://doi.org/10.1126/scisignal.aat1631.

Pereira, P. H. S., Brito, G., Moraes, M., Kiyan, C. L., Avet, C., Bouvier, M., et al. (2020). BRET sensors unravel that Plasmodium falciparum serpentine receptor 12 (PfSR12) increases surface expression of mammalian GPCRs in HEK293 cells. *BioRxiv*. https://doi.org/10.1101/2020.04.17.047217.

Perroy, J., Pontier, S., Charest, P. G., Aubry, M., & Bouvier, M. (2004). Real-time monitoring of ubiquitination in living cells by BRET. *Nature Methods*, *1*, 203–208. https://doi.org/10.1038/nmeth722.

Stoddart, L. A., Johnstone, E. K. M., Wheal, A. J., Goulding, J., Robers, M. B., MacHleidt, T., et al. (2015). Application of BRET to monitor ligand binding to GPCRs. *Nature Methods*, *12*, 661–663. https://doi.org/10.1038/nmeth.3398.

Terrillon, S., Barberis, C., & Bouvier, M. (2004). Heterodimerization of V1a and V2 vasopressin receptors determines the interaction with -arrestin and their trafficking patterns. *Proceedings of the National Academy of Sciences of the United States of America*, *101*, 1548–1553. https://doi.org/10.1073/pnas.0305322101.

Weibrecht, I., Leuchowius, K. J., Clausson, C. M., Conze, T., Jarvius, M., Howell, W. M., et al. (2010). Proximity ligation assays: A recent addition to the proteomics toolbox. *Expert Review of Proteomics*, *7*, 401–409. https://doi.org/10.1586/epr.10.10.

Xu, Y., Piston, D. W., & Johnson, C. H. (2002). A bioluminescence resonance energy transfer (BRET) system: Application to interacting circadian clock proteins. *Proceedings of the National Academy of Sciences of the United States of America*, *96*, 151–156. https://doi.org/10.1073/pnas.96.1.151.

CHAPTER

Methods for binding analysis of small GTP-binding proteins with their effectors

11

Abhishek Sharma, Gaurav Kumar, Sheetal Sharma, Kshitiz Walia, Priya Chouhan, Bidisha Mandal, and Amit Tuli*

Division of Cell Biology and Immunology, CSIR-Institute of Microbial Technology (IMTECH), Chandigarh, India

**Corresponding author: e-mail address: atuli@imtech.res.in*

Chapter outline

Methods in Cell Biology, Volume 166, ISSN 0091-679X, https://doi.org/10.1016/bs.mcb.2021.06.003

Abstract

Proteins often do not function as a single biomolecular entity; instead, they frequently interact with other proteins and biomolecules forming complexes. There is increasing evidence depicting the essentiality of protein-protein interactions (PPIs) governing a wide array of cellular processes. Thus, it is crucial to understand PPIs. Commonly used approaches like genetic (e.g., Yeast Two-Hybrid, Y2H), optical (e.g., Surface Plasmon Resonance, SPR; Fluorescence Resonance Energy Transfer, FRET), and biochemical have rendered ease in developing interactive protein maps as freely available information in protein databases on the web. The underlying basis of traditional protein interaction analysis is the core of biochemical methodologies providing direct evidence of interactions. Co-Immunoprecipitation (Co-IP) is a powerful biochemical technique that facilitates identifying novel interacting partners of a protein of interest in vivo, allowing specific capture of their complexes on an immunoglobulin. Here, using Arf-like (Arl) GTPase-8b (Arl8b) and Pleckstrin Homology Domain-Containing Family M Member 1 (PLEKHM1) as an example of small GTPase-effector pair, we provide a detailed protocol for performing Y2H and Co-IP assays to confirm the interaction between a small GTPase and its effector protein.

1 Introduction

The analysis of PPIs not only is limited to determine the function of a protein within a cell but is also helpful in understanding a variety of cellular processes that they dictate, including metabolism signal transduction, cell cycle and many more (Braun & Gingras, 2012). Proteins are dynamic team players constantly interacting with peer proteins and other biomolecules, thus mediating biological processes in living cells.

PPI can be defined as a specific, physical, and purposeful associative event under biomolecular forces. PPI, based on persistence, can be both transient (for example, receptor-ligand interaction) and stable (forming macromolecular/subunit complexes, for example, ATP synthase; proteasome) (De Las Rivas & Fontanillo, 2010). Moreover, the establishment of PPIs depends on cell type, cell developmental stage, availability of co-factors, post-translational modifications (PTMs), environmental stimulus etc.

Fundamental processes, including protein folding, protein assembly formation and PPI, are mediated via non-covalent contacts between residue side chains. This is simultaneously influenced by differences in characteristics of interfaces between internal (within the same chain) and external (between different chains) interactions. Together, such interactions mediate PPI through the formation of homo-/hetero-oligomers and homo-/hetero-complexes, substantially based on differences in amino acid composition and residue-contact preferences (Ofran & Rost, 2003).

The complete map of protein interactions occurring in a living organism is called the interactome. It is estimated that there are approximately 25,000 protein-encoding genes in human cells, and 650,000 PPIs can occur that majorly drive cellular functionality. This reflects that the scope of regulation of interactome is vast (Stumpf, Thorne, et al., 2008). Thus it is both vital and challenging to identify interacting protein partners and their associative outcome. In the current era of biological research, there exists consistent and increasing effort in mapping interactome. Among this massive interactome, an actively-explored PPI responsible for tightly regulating a myriad of cellular pathways is between small GTPases and their interacting protein partners called effectors (Homma, Hiragi, et al., 2021; Takai, Sasaki, et al., 2001).

Generally, small GTPases consistently cycle between active [Guanosine triphosphate (GTP)-bound; membrane-bound] and inactive [Guanosine diphosphate (GDP)-bound; cytosolic] form (Fig. 1). In their active state, they recruit proteins (termed as effectors) to mediate various cellular processes such as signal transduction, cell cycle progression, membrane fusion and fission, vesicle trafficking, endocytosis, phagocytosis and autophagy (Homma et al., 2021). Arl8b is a ~20kDa, small GTPase localized primarily on lysosomes, where it recruits one of its effector PLEKHM1 by directly interacting with its N-terminal RUN domain. This PPI

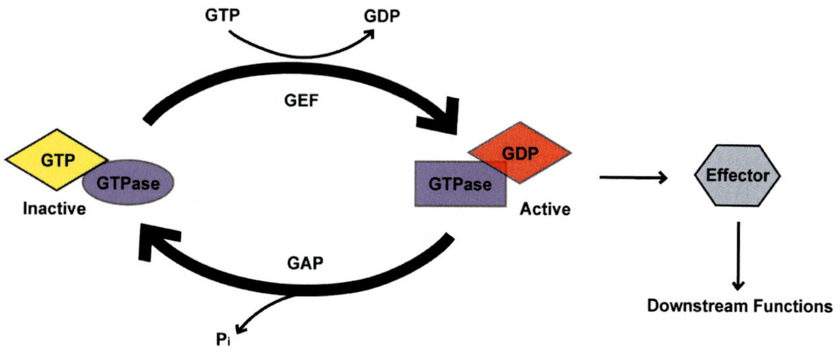

FIG. 1

The small GTPase cycle: Small GTPases or GTP-binding proteins possess a consensus sequence via which they interact with GTP/GDP and exhibit GTPase activity resulting in hydrolysis of GTP to GDP and Pi. These proteins have inter-convertible states, GDP-bound inactive and GTP-bound active. An upstream signal initiates the dissociation of GDP from the inactive form, followed by the GTP association. GTP binding causes a conformational change in the effector binding region of the GTPase, resulting in the recruitment of specific effectors. This association of GTPase and effector protein results in effector-specific downstream functions. The GDP/GTP exchange is a slow, rate-limiting step and is stimulated by guanine nucleotide releasing factors (GEFs). GEF interacts with the GDP-bound GTPase, ultimately displacing GDP. This results in the formation of a binary complex followed by GTP displacing the GEF. Furthermore, GTPases are intrinsically able to hydrolyze GTP to GDP. Since this conversion is slow, GTPase activating Protein (GAPs) catalyzes the reaction by interacting with the GTPase. This ultimately releases the bound downstream effector. In this manner, a small GTPase cycle between inactive and active forms and functions as critical molecular switches in an array of biological processes.

subsequently promotes the fusion of autophagic and endocytic vesicles with lysosomes marked for cargo degradation (Khatter, Sindhwani, et al., 2015; Marwaha, Arya, et al., 2017).

Although GTPases regularly interact with numerous proteins in the cell, for a protein to be designated as its effector, it preferentially needs to interact with the active form of GTPase. To affirm the Arl8b-PLEKHM1 interaction, we exploit previously identified GTP/GDP-locked point mutants of Arl8b, namely Q75L and T34N. Arl8b-Q75L mutant mimics the GTP-bound active state of Arl8b and localizes on lysosomes, while Arl8b-T34N mutant imitates GDP-bound inactive state residing in the cytosol (Marwaha et al., 2017). Below, we describe assays, namely Y2H and Co-IP, to demonstrate the binding of the small GTPase Arl8b with its newly identified effector PLEKHM1.

2 Yeast two-hybrid (Y2H) assay

Y2H is an in vivo, high-throughput and widely used technique based on the reconstitution of a functional transcription factor (TF) in response to the interaction of two proteins or polypeptides of interest. This takes place in genetically modified yeast strains, in which the transcription of a reporter gene leads to a specific phenotype. The most popular reporter genes, namely His3 (encodes for imidazole glycerol-phosphate dehydratase involved in histidine biosynthesis) and LacZ (encodes for β-galactosidase), facilitates selection of yeast on medium lacking Histidine and to screen yeast in a colorimetric assay, respectively. Each of the protein of interest is fused with either of the two domains of a yeast transcription factor (like Gal4), i.e., Gal4-DNA-binding domain (DBD) and Gal4-activation domain (AD). The protein fused to the DBD is referred to as the "bait," while the protein fused to the AD is referred to as "prey." Upon interaction between the bait and prey, the DBD and AD of Gal4 are brought in close proximity such that the functional TF is reconstituted, ultimately transcribing the reporter gene (Fields & Song, 1989; Ito, Chiba, et al., 2001).

Y2H offers a sensitive and cost-effective means to test direct interaction between two query proteins in their native conformations or use a protein of interest as bait to screen a library of proteins or protein fragments prepared from different cell types or tissues. However, the Y2H assay can generate a high number of false-positive interactions. Among possible reasons resulting in false-positives can be high prey and bait expressions, and their localization in a cellular compartment is not corresponding to their natural environment. Moreover, incorrectly folded proteins exhibit non-specific interactions, further adding to the generation of false-positives. Another real limitation of the Y2H assay is the generation of false-negatives. The false-negative interactions when analyzing higher eukaryotes proteins arise due to differences in the status of PTM of proteins of interests in the yeast system. Furthermore, the query proteins must be localized to the nucleus since proteins, which are less likely to be present or fail to enter into the nucleus due to their large size, cannot transcribe reporter genes. Despite these shortcomings, the Y2H assay remains a powerful genetic tool to map protein-protein interactions. In the past two decades, many Y2H-based technologies, coupled with different screening approaches, have been developed to circumvent the issues mentioned above, increasing access to almost the entire cellular proteome (Bruckner, Polge, et al., 2009).

Here, we provide a protocol for carrying out a Y2H assay to detect the interaction between two targeted proteins using Arl8b and PLEKHM1 as an example. For this, human Arl8b gene—wild type (WT), GTP-bound form (Q75L mutant) and GDP-bound (T34N mutant) used in the assay were cloned in Gal4-DBD Y2H bait vector, pGBKT7 and human PLEKHM1 gene was cloned in Gal4-AD Y2H prey vector-pGADT7.

2.1 Materials

2.1.1 Yeast transformation

1. Bait plasmid: plasmid encoding Gal4-DBD in fusion with the bait protein (*see Note 1*)
2. Prey plasmid: plasmid encoding Gal4-AD in fusion with the prey protein (*see Note 1*)
3. Yeast cells: Y2H Gold yeast strain (Clontech, #ST0030)
4. Yeast Peptone Dextrose (YPD) broth (BD Difco, #242820)
5. Yeast Nitrogen Base (YNB) without Amino acids (BD Difco, #291940)
6. Dextrose (Merck, #1.94925.0521)
7. Bacteriological agar (HiMedia, #GRM026)
8. Adenine (Sigma-Aldrich, #A8626)
9. Yeast synthetic drop-out media supplements without Leucine and Tryptophan (Double drop-out -Leu/-Trp) (Sigma-Aldrich, #Y0750)
10. Lithium Acetate (LiAc) Dihydrate (Sigma-Aldrich, #L6883)
11. Polyethylene Glycol (PEG) (Sigma-Aldrich, #P4338)
12. Carrier DNA (Deoxyribonucleic acid sodium salt from salmon testes) (Sigma-Aldrich, #D1626)
13. Dimethyl Sulfoxide (DMSO) (Sigma-Aldrich, #D8418)
14. Autoclaved distilled water (DW)
15. Dry bath for incubation
16. Incubator
17. Incubator/shaker

2.1.2 Yeast spotting

1. Yeast transformants
2. YNB without amino acids
3. Dextrose

4. Bacteriological agar
5. Double drop-out (-Leu/-Trp) agar plates
6. Triple drop-out (-His/-Leu/-Trp) agar plates

2.1.3 Preparation of stock reagents and media

- *All reagents need to be sterilized by autoclaving unless otherwise stated.*
  1. YPAD (Yeast Peptone Adenine Dextrose) agar plates: Dissolve YPD broth (5 g), Adenine (3 mg) and Bacteriological agar (2 g) in 100 mL of DW. After autoclaving, pour the media in sterile petri-dishes.
  2. YPAD broth: Dissolve YPD broth (5 g) and Adenine (3 mg) in 100 mL of DW.
  3. 10 × LiAc stock solution, pH 7.5: Dissolve Lithium Acetate (10 g) in 100 mL of DW.
  4. 10 × TE stock solution, pH 7.5: Add 1 M Tris pH 7.5 (10 mL) and 0.5 M EDTA pH 8.0 (2 mL) and make up the volume to 100 mL using DW.
  5. Carrier DNA stock (10 mg/mL): Add Carrier DNA (100 mg) to 10 mL of autoclaved 1 × TE buffer. Heat the tube at 99 °C with intermittent vortexing until a clear solution is obtained.
  6. 50% PEG solution: Dissolve PEG (50 g) in 100 mL of DW.
  7. Yeast competent solution (for a 20 mL yeast culture): Mix 2.4 mL DW, 300 µL 10 × LiAc (pH 7.5), 300 µL of 10 × TE pH 7.5 and 300 µL of Carrier DNA (10 mg/mL).
  8. PEG competent solution (for a 20 mL yeast culture): Mix 50% PEG solution (20 mL), 10 × LiAc (2.3 mL) and 10 × TE (2.3 mL).
  9. Double drop-out (-Leu/-Trp) agar plates: Dissolve Bacteriological agar (2 g), Dextrose (2 g), YNB (670 mg) and Double drop-out supplement (154 mg) in 100 mL of DW. After autoclaving, pour the media in sterile petri-dishes.
  10. Triple drop-out (-His/-Leu/-Trp) agar plates: Dissolve Bacteriological agar (2 g), Dextrose (2 g), YNB (670 mg) and Triple drop-out supplement (146 mg) in 100 mL of DW. After autoclaving, pour the media in sterile petri-dishes.

2.2 Protocol

- *Perform the protocol under sterile conditions inside the laminar airflow cabinet unless otherwise stated.*

2.2.1 Setting up yeast culture

1. Streak the Y2H yeast Gold strain on YPAD agar plate and incubate at 30 °C for 2–3 days. Store the plate having colonies at 4 °C until further use.
2. Inoculate three colonies of the Y2H yeast Gold strain in 20 mL YPAD broth and incubate at 30 °C in a shaking incubator (200–220 rpm) for 20–24 h (*see Note 2*).

2.2.2 Yeast transformation

1. Spin down the culture at 1000 × *g* for 5 min at room temperature (RT).
2. Discard the supernatant and wash the yeast cell pellet with 20 mL DW. Centrifuge the suspension at 1000 × *g* for 5 min at RT.

Table 1 Tabular representation of yeast two-hybrid assay grid to show the setup of the required number of reactions to be performed.

AD BD	pGADT7-PLEKHM1	pGADT7
pGBKT7-Arl8b (WT)	1	2
pGBKT7-Arl8b (Q75L)	3	4
pGBKT7-Arl8b (T34N)	5	6
pGBKT7	7	8

The pGBKT7 and pGADT7 are yeast two-hybrid "bait" and "prey" vectors, respectively.

3. Discard the supernatant and resuspend the pellet in 3.3 mL of yeast competent solution (*see Note 2*).
4. Label sterile microcentrifuge tubes (MCTs) as per the number of required reactions to be carried out (see Table 1).
5. Add 0.5 μg of bait- and prey-protein expressing plasmid in their respective MCTs.
6. Add 100 μL of the resuspended yeast cells to each reaction tube and vortex the tubes.
7. Add 500 μL PEG competent solution to each MCT and vortex. Incubate the MCTs at 30 °C (dry bath) for 30 min.
8. Add 70 μL DMSO to each reaction and mix the components by gently inverting tubes twice. Incubate the tubes in the dry bath at 42 °C for 45 min (*see Note 3*).
9. Place the tubes on ice for a minute and spin down at 16,600 × *g* for 3 min at 4 °C.
10. Discard the supernatant and resuspend each pellet in 40 μL DW. Spread 30 μL of each suspension in their respective section on the -Leu/-Trp agar plate shown in Fig. 2A (*see Note 4*).
11. Incubate the plates at 30 °C for 72 h for the transformants to appear (Fig. 2B).

2.2.3 Yeast spotting for detecting interactions

1. For spotting, take two MCTs per reaction and label them as set "A" and "B." To each MCT, add 600 and 500 μL DW, respectively.
2. For each transformant, pick 2–3 colonies (of similar size) and dissolve them in their respective set "A" MCT to get yeast suspension. Transfer 500 μL yeast solution from set "A" to set "B" MCT.
3. Vortex the contents of each MCT and record the absorbance (O.D. at 600 nm) of set "B" MCTs. Normalize the absorbance reading of each reaction to 0.1 by adding an appropriate amount of DW to the MCTs of set "A" (*see Notes 5 and 6*).
4. Spot the required volume of each reaction on drop-out agar plates, namely -Leu/-Trp and -Leu/-Trp/-His (*see Notes 7–9*).
5. Allow the plates to dry inside the laminar airflow cabinet with their lids open and incubate at 30 °C for 72 h.
6. Examine the plates and score the positive interactions by observing the growth of spotted yeast on the reporter plate, i.e., -Leu/-Trp/-His. As shown in Fig. 3,

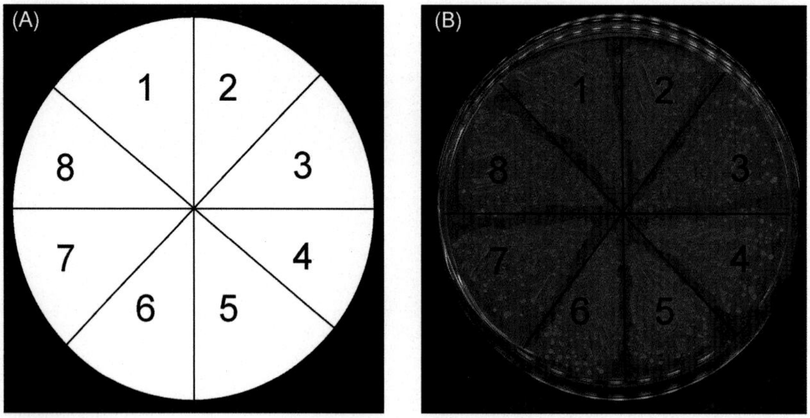

FIG. 2

(A) Schematic showing arrangement to plate transformed yeast cells on a petri-dish.
(B) Image showing growth of yeast cells on a -Leu/-Trp agar plate after co-transformation with bait and prey Y2H plasmids.

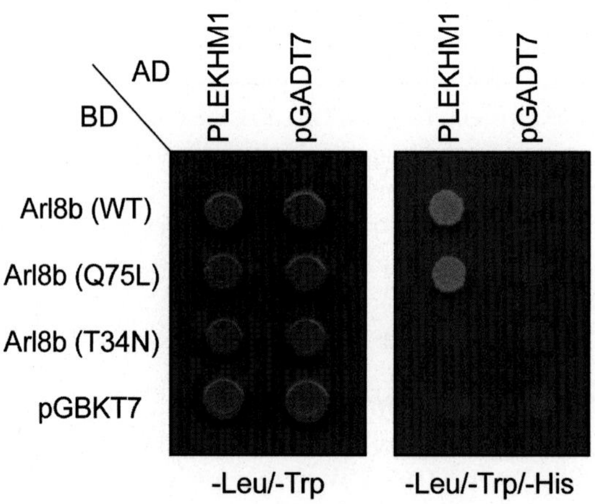

FIG. 3

Yeast-two hybrid assay to show direct interaction of PLEKHM1 with Arl8b. Plasmids encoding GAL4-Activation Domain (AD) fused to PLEKHM1 were co-transformed with Arl8b-WT, -Q75L (GTP-bound) and -T34N (GDP-bound) forms fused to GAL4-DNA Binding Domain (BD) in *Saccharomyces cerevisiae* to examine the interactions. The co-transformants were spotted on non-selective medium (-Leu/-Trp) to confirm viability and on selective medium (-Leu/-Trp/-His) to detect interactions.

PLEKHM1 interacted with the WT (wild-type) and Q75L (constitutively GTP-bound) forms of Arl8b, but not with the T34N (constitutively GDP-bound) form, indicating that PLEKHM1 directly interacts with the GTP-bound conformation of Arl8b.

3 Co-immunoprecipitation (co-IP)

Immunoprecipitation (IP), as the name suggests, is based on the principle of antigen-antibody interaction. It is a sensitive method that allows isolation of a protein/antigen of interest using a specific antibody to precipitate it (Kapingidza, Kowal, et al., 2020). Antigen-antibody interaction is a highly specific and irreversible association based on four short-distance non-covalent forces: (a) hydrogen bonding, in which two electronegative atoms share a hydrogen atom; (b) ionic bonding, in which two opposite-charged moieties interact; (c) hydrophobic interaction, in which water molecules force hydrophobic groups to come together; and (d) van der Waals interactions, between outer electron clouds of two or more atoms. In an aqueous environment, these interactions (generally called bonding affinity) are weak (compared to covalent interaction) and depend upon the complementarities between the hypervariable region of the antibody and the epitope of the protein.

The antigen source can be unlabeled cells or tissues, metabolically or extrinsically labeled cells or tissues, subcellular fractions, in vitro translated proteins, etc. A standard IP protocol has several stages; stage one includes lysing cellular compartments to extract soluble and membrane proteins using; (a) non-denaturing conditions usually employing non-ionic detergent, like Triton-X/NP-40, preserving weak PPIs; (b) denaturing condition (for an inaccessible epitope of the protein or if the protein cannot be extracted under non-denaturing conditions) employing ionic detergents like SDS or Sarkosyl, and; (c) mechanical disruption via repeated cell passaging and centrifugation under detergent-free conditions to extract highly solubilized cytosolic and luminal-organellar proteins.

Stage two requires immobilizing an antibody on to a sedimentable solid-phase matrix-like derivatized agarose-attached protein or magnetic beads. Genetically-engineered recognition tags provide a fast way to capture PPIs or purify proteins. The basis of using recognition tags is to attach it to the target protein by recombinant DNA methodologies. The tag itself functions as an antigenic determinant or epitope for an antibody, facilitating retention of the protein of interest and its involvement in protein complexes. Commercially, pre-conjugated reagents are directly available to isolate protein fused with frequently-used peptide motifs, notably myc-, HA-, FLAG-tag and others (Young, Britton, et al., 2012).

Stage three includes incubating solubilized antigens obtained from stage one with immobilized antibody such that target antigen is captured, ultimately followed by extensive washing to remove non-specifically bound proteins.

Co-IP is an extended arm to IP; however, it aims to identify physiologically relevant interacting partner(s) of a certain immunoprecipitated protein, allowing capture of protein-protein interaction on an antibody (Kaboord & Perr, 2008; Lin & Lai, 2017). Co-IP enables the determination of PPI in their native state with little or no external influence. Also, it is cost-saving, easy to handle, and can be applied to a variety of cell types like plant cells, mammalian cells, yeast etc. Co-IP ambiently reveals transient protein-protein interactions, and thus it is conditionally suggested to perform cross-linking of cells before lysis to fix or stabilize weak PPIs. The pre-clearing of cell lysates on the beads slurry is recommended to remove proteins that bind non-specifically to the beads slurry and insoluble material that could interfere with target protein interaction with the beads-bound antibody. Co-immunoprecipitated complexes are resolved on SDS-PAGE and subjected to western blotting and/or mass spectrometry to characterize protein partners, interacting ligands, post-translational modifications, etc.

Here, we describe a protocol to perform Co-IP using lysates of HEK293T cells expressing FLAG-tagged-PLEKHM1 with different forms of HA-tagged-Arl8b (WT, QL and TN).

3.1 Materials

3.1.1 Transfection

1. Human Embryonic Kidney (HEK) 293T cells (ATCC, #CRL-3216)
2. 1 × Dulbecco's Minimal Essential Media (DMEM) [Lonza-12, #604F] supplemented with 10% Fetal Bovine Serum (Gibco, #10082-147), 1 mM Sodium Pyruvate (Gibco, #11360-070), 1 × Antibiotic Antimycotic solution (Gibco, #15240-062), 1 × HEPES buffer (Gibco, #15630-080), 1 × Non-Essential Amino Acids (Gibco, #11140-050) and 1 × GlutaMax (Gibco, #35050-061)
3. 1 × Opti-MEM (Gibco, #11058-021)
4. Expression Plasmids
5. X-treme GENE HP DNA Transfection Reagent (Roche, #06366546001)
6. 35 mm cell culture dishes (Falcon, #353001)
7. Humidified cell culture incubator (Thermo Fisher Scientific)

3.1.2 Preparation of cell lysates for Co-IP

1. Cell lysis buffer: 20 mM Tris-Cl (Invitrogen, # AM9856) pH 8.00, 150 mM NaCl (Invitrogen AM, #9759), 0.5% NP-40 (USB Corporation, #19628), 1 mM $MgCl_2$ (Invitrogen, #AM9530G), 1 mM Na_3VO_4 (Sigma-Aldrich, #S6508), 1 mM NaF (Sigma, #215309), 1 mM PMSF (Sigma, #P7626) and protease inhibitor cocktail (Sigma, #P8340).
2. Wash buffer: 20 mM Tris-Cl pH 8, 150 mM NaCl, 0.1% NP-40, 1 mM $MgCl_2$, 1 mM Na_3PO_4 and1 mM NaF.
3. 1 × Phosphate Buffer Saline (PBS) pH 7.4: 137 mM NaCl (Sigma, #S3014), 2.7 mM KCl (Sigma, #P9541), 10 mM Na_2HPO_4 (Sigma, #S3264), 2 mM KH_2PO_4 (Sigma, # P9791).

4. Monoclonal anti-HA antibody-conjugated agarose beads (Sigma, #82095).
5. HulaMixer (Life Technologies, #15920D).
6. 4 × Laemmli sample buffer: 250 mM Tris-Cl pH 6.8, 8% SDS (Sigma, #L3771), 40% Glycerol (Sigma, #G5516), 0.04% bromophenol blue (Sigma, #114391) and 20% β-mercaptoethanol (Sigma, #M3148).
7. SDS-PAGE reagents and apparatus.
8. Western blot transfer reagents and apparatus.

3.1.3 Immunoblotting

1. Primary antibodies against the desired antigens/epitopes to be probed
2. HRP-conjugated secondary antibodies
3. 0.05% Phosphate Buffer Saline containing Tween-20 (PBST): 1 × PBS, 0.05% Tween-20 (Sigma, #P1379)
4. 0.3% PBST: 1 × PBS, 0.3% Tween-20 (Sigma, #P1379)
5. Blocking buffer: 5% Skim milk (Difco, #232100) in 0.05% PBST
6. ECL Plus western blotting substrate (Thermo Scientific, #32132)
7. Rocking platform (Tarson)
8. X-ray films (Carestream)
9. Automatic X-Ray Film Processor (OptiMax)

3.2 Protocol

- *Follow the standard cell culture practice for maintaining and passaging HEK293T cells.*
- *Perform all steps of Co-IP in the cold room or on ice unless otherwise stated.*

3.2.1 Transfection of HEK293T cells with expression plasmids

(1) For each co-expression pair of plasmids, seed 0.5 million HEK293T cells in a 35 mm culture dish. Perform the transfection at 60–70% cellular confluency, i.e., 18–24 h after seeding the cells.

(2) For a 200 µL transfection mix per dish, add the following reagents in an MCT:
   - 0.5 µg of a plasmid encoding effector protein, i.e., FLAG-PLEKHM1: x µL
   - 0.5 µg of a plasmid encoding small GTPase, i.e., Arl8b-HA (WT/Q75L/T34N): y µL
   - Add 1 × Opti-MEM: 200 µL—(x + y) µL
   - Add X-tremeGENE HP DNA transfection reagent: 1 µL (1:1 ratio, i.e., 1 µL for 1 µg DNA)

(3) Mix well the transfection mix using a micro-pipette and incubate at RT for 30 min.

(4) During the incubation, replace the existing media from cells with fresh media (*see Note 9*).

(5) Add the transfection mix to the cells in a drop-wise manner (*see Note 10*).

(6) Incubate the cells in a humidified chamber at 37 °C/5%CO_2.

3.2.2 Preparation of cell lysates for Co-IP

1. 18–24 h post-transfection, collect the cells in 1 × cold PBS using a micro-pipette. Spin down the suspension at 700 × *g* for 2 min to pellet the cells.
2. Discard the supernatant and re-suspend each cell pellet in 350 μL of ice-cold cell lysis buffer. Collect each cell lysate in a separate 1.5 mL MCT and incubate on a HulaMixer for 30 min at 4 °C. Finally, spin the lysates at a high-speed of 16,500 × *g* for 5 min at 4 °C. Now transfer the supernatant to sample labeled fresh MCTs (*see Note 11 and 12*).
3. Meanwhile, to equilibrate anti-HA antibody-conjugated-agarose beads (50% slurry in preservative solution), aliquot 10–15 μL beads in ice-cold 500 μL of cell lysis buffer and incubate on ice for 2 min. Spin down the beads at 700 × *g* for 1 min at 4 °C and remove the supernatant (*see Note 13*).
4. From step 2, save 1% of each supernatant separately as inputs containing the protein of interest. Load each of the remaining supernatants on to the equilibrated beads. Incubate the mix for 3–4 h at 4 °C on a Hulamixer for sufficient binding to take place.
5. Spin down the mix at 700 × *g* for 2 min. Discard the supernatant and wash the beads thrice with 500 μL wash buffer (*see Note 14*).
6. After the final wash, discard the supernatant and prepare samples (~40 μL each) for gel electrophoresis by adding 30 μL wash buffer and 10 μL 4 × Laemmli sample buffer. Also, prepare input samples as well in a similar fashion. Boil all the samples at 100 °C for 5–7 min.
7. Separate the samples on SDS-PAGE using appropriate gel percentages based on the sizes of proteins of interest.
8. Follow the standard protocol for the transfer of resolved proteins onto the PVDF membrane.

3.2.3 Immunoblotting

1. After the Western blot transfer is over, incubate the PVDF membrane in blocking buffer for 2 h at RT on a rocking platform. This ensures saturation of free binding sites on the membrane, preventing non-specific binding of primary and secondary antibodies in downstream steps.
2. Wash the membrane three times for 10 min each by using 0.3% PBST buffer to remove the blocking buffer. Meanwhile, prepare a sufficient volume of the primary antibody solution of recommended dilution in 0.05% PBST. As shown in Fig. 4, to immunoblot for FLAG-tagged-PLEKHM1, we have used a mouse anti-FLAG antibody. Similarly, to check for direct IP of HA-tagged forms of Arl8b, we have used a rabbit anti-HA antibody.
3. Incubate the membrane with primary antibody solution for 2 h at RT on a rocking platform.

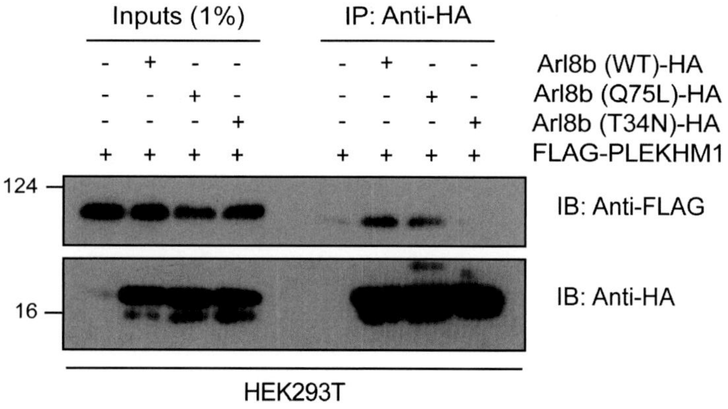

FIG. 4

Co-Immunoprecipitation of PLEKHM1 with the GTP-bound form of Arl8b. FLAG tagged-PLEKHM1 was co-transfected with different forms of C-terminal HA-tagged Arl8b in HEK293T cells and the lysates were immunoprecipitated (IP) with anti-HA antibodies conjugated-agarose beads and the precipitates were immublotted (IB) with the indicated antibodies.

4. Wash the membrane three times for 10 min each by using 0.3% PBST buffer. Incubate the membrane with HRP-conjugated secondary antibody solution for 1 h at RT on a rocking platform. The secondary antibody solution of recommended dilution should be prepared in 0.05% PBST.
5. Wash the membrane three times for 10 min each by using 0.05% PBST.
6. To develop signal, incubate the membrane with the luminescence detection reagent, expose the membrane to X-ray film for different time intervals, and develop in an automatic X-ray developer machine placed in a darkroom. As shown in Fig. 4, PLEKHM1 was Co-IP with the WT (wild-type) and Q75L (constitutively GTP-bound) forms of Arl8b, but not with the T34N (constitutively GDP-bound) form, indicating that PLEKHM1 interacts with the GTP-bound conformation of Arl8b under in vivo conditions.

4 Notes

(1) Quality and concentration of plasmids must be optimized to ensure the success of yeast transformation.

(2) Typically with 20 mL of yeast culture, 30 co-transformation reactions can be carried out. If the number of reactions is less, accordingly inoculate the volume of yeast culture. Also, use the appropriate volume of yeast competent solution based on the volume of yeast culture.

(3) Avoid vortexing or shaking MCTs vigorously after adding DMSO as it may damage yeast cells.

O.D.$_{600}$	Volume of water to be added to normalize the OD_{600} to 0.1	Dilution Factor (D.F.)	Volume to be spotted = $\frac{0.1*5\ \mu L}{x}$ where, $x = OD_{600}/D.F.$
0.2	100 μL	2	5 μL

(4) The supernatant containing the PEG should be completely removed without losing any yeast cells as PEG hinders the growth of transformants.

(5) If the absorbance of the yeast suspension solution is less than 0.1 at 600 nm then use more number of colonies for making the suspension and perform the absorbance measurement again.

(6) To calculate the volume of the yeast suspension to be spotted on dropout plates, follow the calculation example shown below.

(7) The spotting of transformed yeast cells should be done both on -Leu/-Trp agar plate (non-selective media) and -Leu/-Trp/-His agar plate (selective/reporter media) to ensure uniformity of yeast spotted for each reaction.

(8) Sometimes fusion of bait protein to GAL4-BD or fusion of prey protein at GAL4-AD can result in self-activation. In such cases, spotting of yeast cells should be done on reporter plates containing 3-AT, a competitive inhibitor of the HIS3 gene product.

(9) Based on the genotype of the yeast strain employed for performing Y2H assay, alternative methods, such as β-galactosidase assays, can also be carried out for further confirmation of Y2H interactions. For more information refer to (Galletta & Rusan, 2015).

(10) Several factors can significantly affect the transfection efficiency. Therefore, an appropriate type of transfection reagent must be selected depending on the type of cells, cytotoxic effects and cell culture conditions.

(11) An important step Co-IP assay is the choice of lysis buffer. This depends on the protein of interest under study. Usually, the use of non-ionic detergents likes NP-40 or Triton X-100 disorders lipid-protein interactions while keeping PPIs intact. However, the ideal concentration of the non-ionic detergent must be optimized to minimize conformational change in the protein of interest.

(12) PMSF should be added freshly to the lysis buffer just before use.

(13) The antibody-conjugated-agarose beads should be given gentle spin to avoid damage to the beads during centrifugation step. Also, the agarose beads should never dry out during the protocol.

(14) Selecting the correct composition of wash buffer is equally essential as choosing an optimum lysis buffer. The wash buffer must keep interactions with the protein of interest intact and efficiently reduce its non-specific binding or background. Thus, one can make adjustments in the composition of wash buffer to optimize the washing condition.

Acknowledgments

The authors would like to acknowledge funding support from DBT-India Alliance/Wellcome Trust, DST-SERB and infrastructural support from CSIR-IMTECH. IMTECH communication No. 015/2021.

Funding

The authors acknowledge support from Council of Scientific & Industrial Research (CSIR). A.T. acknowledges financial support from the Department of Biotechnology (DBT) India Alliance/Wellcome Trust [IA/I/14/2/501543] and DST-SERB [CVD/2020/000733]. A.T. is a recipient of the DBT-India Alliance/Wellcome Trust Intermediate Fellowship Award.

Contributions

A.S. designed and wrote the protocol. A.S., G.K. and S.S. performed the experiments and analyzed the results. G.K. prepared the figures. K.W., P.C. and B.M. assisted in carrying out the experiments and contributed in overall design of the protocol. A.T. supervised the project.

References

Braun, P., & Gingras, A. C. (2012). History of protein-protein interactions: From egg-white to complex networks. *Proteomics*, *12*(10), 1478–1498.

Bruckner, A., Polge, C., et al. (2009). Yeast two-hybrid, a powerful tool for systems biology. *International Journal of Molecular Sciences*, *10*(6), 2763–2788.

De Las Rivas, J., & Fontanillo, C. (2010). Protein-protein interactions essentials: Key concepts to building and analyzing interactome networks. *PLoS Computational Biology*, *6*(6), e1000807.

Fields, S., & Song, O. (1989). A novel genetic system to detect protein-protein interactions. *Nature*, *340*(6230), 245–246.

Galletta, B. J., & Rusan, N. M. (2015). A yeast two-hybrid approach for probing protein-protein interactions at the centrosome. *Methods in Cell Biology*, *129*, 251–277.

Homma, Y., Hiragi, S., et al. (2021). Rab family of small GTPases: An updated view on their regulation and functions. *The FEBS Journal*, *288*(1), 36–55.

Ito, T., Chiba, T., et al. (2001). A comprehensive two-hybrid analysis to explore the yeast protein interactome. *Proceedings of the National Academy of Sciences of the United States of America*, *98*(8), 4569–4574.

Kaboord, B., & Perr, M. (2008). Isolation of proteins and protein complexes by immunoprecipitation. *Methods in Molecular Biology*, *424*, 349–364.

Kapingidza, A. B., Kowal, K., et al. (2020). Antigen-antibody complexes. *Sub-Cellular Biochemistry*, *94*, 465–497.

Khatter, D., Sindhwani, A., et al. (2015). Arf-like GTPase Arl8: Moving from the periphery to the center of lysosomal biology. *Cellular Logistics*, *5*(3), e1086501.

Lin, J. S., & Lai, E. M. (2017). Protein-protein interactions: Co-immunoprecipitation. *Methods in Molecular Biology*, *1615*, 211–219.

Marwaha, R., Arya, S. B., et al. (2017). The Rab7 effector PLEKHM1 binds Arl8b to promote cargo traffic to lysosomes. *The Journal of Cell Biology*, *216*(4), 1051–1070.

Ofran, Y., & Rost, B. (2003). Analysing six types of protein-protein interfaces. *Journal of Molecular Biology*, *325*(2), 377–387.

Stumpf, M. P., Thorne, T., et al. (2008). Estimating the size of the human interactome. *Proceedings of the National Academy of Sciences of the United States of America*, *105*(19), 6959–6964.

Takai, Y., Sasaki, T., et al. (2001). Small GTP-binding proteins. *Physiological Reviews*, *81*(1), 153–208.

Young, C. L., Britton, Z. T., et al. (2012). Recombinant protein expression and purification: A comprehensive review of affinity tags and microbial applications. *Biotechnology Journal*, *7*(5), 620–634.

CHAPTER

Investigating protein expression, modifications and interactions in the brain: Protocol for preparing rodent brain tissue for mass spectrometry-based quantitative- and phospho-proteomics analysis

12

Louis Dwomoh*

The Centre for Translational Pharmacology, Institute of Molecular, Cell and Systems Biology, College of Medical, Veterinary and Life Sciences, University of Glasgow, Glasgow, United Kingdom

**Corresponding author: e-mail address: louis.dwomoh@glasgow.ac.uk*

Chapter outline

Methods in Cell Biology, Volume 166, ISSN 0091-679X, https://doi.org/10.1016/bs.mcb.2021.06.010

Abstract

Neurodegenerative diseases remain a major global health challenge, affecting millions of individuals worldwide. Despite significant research efforts, the etiology and pathological mechanisms of most neurodegenerative disorders remain uncertain. This indicates that new disease models and research strategies need to be developed to enhance biomarker discovery and drug development processes.

Studies have shown that dysregulation of protein-protein interactions in the brain may be key to the pathophysiological processes in a number of neurodegenerative diseases. Consequently, a deeper understanding of protein-protein interactions in the brain in both healthy and neurodegenerative states may provide greater insights into disease mechanisms and progress, and aid the development of preventive, management and treatment strategies.

Efforts to investigate protein-protein interactions in the brain have largely depended on purifying the protein, together with its interacting partner proteins (interactome) and then analyzing these interactions by mass spectrometry. However, factors including abundance and phosphorylation status of the protein of interest and its interactome can lead to identification of false protein-protein interactions. Moreover, studies have shown that combining quantitative proteomics data with data from affinity-purification-mass spectrometry data can reduce these false positives and also provide more insights into protein-protein interactions. Thus, we have developed a protocol for preparing rodent brain tissue for quantitative- and phospho-proteomics analysis, combining in-house and commercially available kits. Data from this protocol will enhance the analysis and interpretation of protein-protein interactions in both physiological and pathological states.

1 Introduction

1.1 Neurodegenerative disorders

The brain is the central component of the central nervous system (CNS), and controls almost all the key processes in mammals, including movement, coordination, memory and behaviors. For these reasons, abnormalities associated with the brain and in extension, the CNS have significant effects on the organism. In humans, the CNS is vulnerable to a number of disorders and injuries including infections, trauma, degeneration, tumors, autoimmune attacks, ischemia and/or hypoxia due to disrupted blood flow, as well as structural defects (Ashraf et al., 2019; Mackay et al., 2019).

Neurodegenerative disorders are a group of diseases that are characterized by progressive loss of neurons in the brain, leading to significant deficits in memory and learning, cognition, movement, behaviors and other debilitating abnormalities. Neurodegenerative disorders include Alzheimer's disease (AD), Parkinson's disease (PD), Huntington's disease, amyotrophic lateral sclerosis (ALS) and others. With the increasing aged population worldwide, neurodegenerative disorders present a major threat to global health and despite continuing research efforts, these disorders remain incurable and irreversible (Deuschl et al., 2020; Feigin et al., 2019).

Efforts at developing drugs to prevent or treat neurodegenerative diseases have not been particularly successful, largely due to a number of factors. Many drugs developed to stop progression or treat symptoms of AD—which is the most common cause of dementia worldwide—have failed at different stages of trials. A recent example is the failure of Biogen's Aducanumab to receive approval from the United States Food and Drug Administration. Aducanumab was developed to target and reduce amyloid plaques present in the brain to slow cognitive and functional impairment in individuals with early AD (Knopman, Jones, & Greicius, 2021). These setbacks, together with other developments, indicate the importance of using newer techniques to test existing and new hypotheses underpinning etiology of many neurodegenerative diseases. In addition, these techniques must applied to improve biomarker discovery and drug development.

1.2 Animal models of neurodegenerative disorders

To understand the key concepts of the CNS and to enhance drug discovery for CNS diseases, almost 400 rodent (mouse and rat) models of neurodegeneration have been developed for use across laboratories worldwide (Dawson, Golde, & Lagier-Tourenne, 2018). These models recapitulate the human disorders to an extent and provide important ways to study the disease in the context of biomarker discovery, understanding disease processes as well as investigating the effects of candidate ligands on disease pathology in the pre-clinical phases of drug development and discovery.

2 Mass spectrometry-based proteomics

One of the recent techniques applied to the study of neurodegenerative diseases is mass spectrometry-based proteomics. Proteomics involves the large-scale study of the proteome—the entire set of protein complement that are expressed and/or modified by a cell, tissue or organism—at a particular point under specific conditions (Ahmad & Lamond, 2014). This method has several advantages over the conventional methods of studying protein abundance and modifications such as ELISA and Western blotting. Compared to ELISA and Western blotting, proteomics gives information about abundance and modifications of thousands of proteins simultaneously. Considering the importance of fully understanding the biochemical and molecular basis of neurodegeneration, application of mass spectrometry-based proteomics presents an important step in several areas of neurodegenerative disease research.

2.1 Significance of mass spectrometry-based proteomics in investigation of neurodegenerative disorders

Mass spectrometry-based proteomics, when combined with other "omics" approaches (genomics and transcriptomics) present a powerful tool for studying neurodegenerative disorders. When validated and optimized for biological relevance, proteomics can unravel changes in protein abundance and post-translational modifications (PTMs) such phosphorylation, ubiquitination and acetylation (Ayyadevara et al., 2019; Hosp & Mann, 2017; Rayner et al., 2019). Thus, this method has significant potential in identifying changes in expression and PTMs that may occur in the brain and cerebrospinal fluid (CSF) as result of neurodegeneration. This has significant potential for biomarker discovery, which is important for early and accurate screening and diagnosis of neurodegenerative diseases, and to design therapies or strategies to modify disease progression or deal with early symptoms of the disease. Several studies have applied proteomics for biomarker discovery in neurodegenerative diseases. In human studies, global proteomics analysis of CSF from AD patients and cognitively normal healthy controls revealed that over 200 proteins were significantly decreased in abundance in AD relative to controls, including VGF, CD74 and AHSG, while just over 300 proteins were significantly increased in abundance in AD relative to controls, including NEFL, MAPT and GAD43 (Higginbotham et al., 2020). Most of these protein abundance changes were "validated" or correlated to independent assessments with ELISA ($r=0.78$), further supporting the argument that mass spectrometry-based proteomics can give accurate measurements when validated.

In rodent models, it is possible to investigate changes in protein abundance and PTMs in both brain tissue and CSF, and several studies have looked at these using mass spectrometry-based proteomics (Johnson et al., 2018; Ping et al., 2018; Swarup et al., 2020; Wingo et al., 2020).

In our group, using a mouse prion neurodegeneration model, we have recently used mass spectrometry-based proteomics to investigate changes in protein abundance in the hippocampus. We have shown that several proteins are reduced in abundance compared to controls. These proteins are involved in neurotransmitter synthesis and secretion, memory and synaptic plasticity. On the other hand, key proteins that drive neuroinflammation, aberrant complement activation and proteolysis are increased in abundance relative to controls (Dwomoh et al., 2021). Interestingly, most of the changes in protein abundance (increased or decreased) were "reversed" or "normalized" after chronic administration of a ligand that enhances acetylcholine-based cholinergic neurotransmission in the brain. This study underlies the scope of proteomics applications in neurodegenerative diseases in the context of biomarker discovery, unraveling new disease mechanisms and drug discovery development.

Furthermore, proteomics has a potential in the investigation of the effect of neurodegeneration on protein-protein interactions (PPIs), and how this knowledge could be used in target validation and drug discovery. This will be discussed in the next section.

3 Protein-protein interactions in neurodegenerative disorders

Protein-protein interactions (PPIs) refer to the transient or stable interactions between proteins, whether the interaction is direct or indirect. It is estimated that almost 80% of all proteins in living organisms are involved or function as part of a protein complex or network (Berggard, Linse, & James, 2007). In the brain, PPIs modulate key processes such as neurotransmission, synaptogenesis and cell-to-cell communication (Peng, Wang, Peng, Wu, & Pan, 2017). This suggests that a greater understanding of PPIs in the brain has the potential to unravel disease pathways, identify new protein complexes and predict the role of these complexes in health and disease.

Moreover, studies have shown that dysregulation of PPIs may be part of the pathological mechanisms in neurodegenerative disorders. Using computational tools to predict PPIs in the Parkinson's disease brain, a study observed that dysregulated interactions between α-synuclein and it's interacting proteins could be central to protein misfolding and aggregation that occur during the disease process (Hernandez, Tikhonova, & Karamyshev, 2020). The α-synuclein interactome included enzymes, chaperones and translational targeting factors.

Further studies could be central to our understanding of the mechanisms of neurological disorders, and the processes of identifying and validating targets for drug development.

3.1 Investigating protein-protein interactions (PPIs) in neurodegenerative diseases

PPIs in the brain can be investigated using two main approaches. The first involves computational prediction of PPIs by superimposing whole proteome datasets onto PPI databases. The second approach involves experimental procedures where protein(s) of interest, together with their interacting proteins are captured by affinity-purification or other methods, and then analyzed by mass spectrometry to unravel the PPIs (Peng et al., 2017).

Combining affinity-purification with mass spectrometry has significantly improved the analysis of PPIs, however, a number of factors can affect this approach and lead to identification of false PPIs in the brain. For instance, protein abundance and post-translational modifications such as phosphorylation have been shown to significantly alter the affinity of proteins for other proteins, thereby creating false PPI networks (Ivanic, Yu, Wallqvist, & Reifman, 2009; Nishi, Hashimoto, & Panchenko, 2011).

Considering the effect of abundance and phosphorylation on PPIs, results from quantitative and phosphoproteomics analyses can significantly increase confidence in the PPI data and also provide further insights for the data from affinity-purification and mass spectrometry-based interactome analysis. This suggestion is supported by studies that have combined quantitative proteomics with affinity-purification and mass spectrometry to define PPIs in neurodegenerative diseases. In one study, a combination of quantitative proteomics and affinity-purification-mass spectrometry analysis identified the phosphorylated tau interactome in post-mortem neurofibrillary tangles (NFTs) that had been micro-dissected from patients with advanced AD (Drummond et al., 2020). The study showed 542 proteins were identified from the total proteomics of the NFTs, and the affinity-purified mass spectrometry showed that 75 of these proteins interacted with phosphorylated tau.

This chapter will discuss protocols for preparing rodent brain tissue for mass spectrometry-based quantitative proteomics and phosphoproteomics analysis. The data generated from these experiments can provide important information to complement PPI network data from both experimental and computational analyses.

4 Overview of the protocol

The protocol below describes the preparation of any dissected part of the rodent brain for mass spectrometry-based proteomics analysis. The protocol describes in detail the processes of tissue lysis, protein extraction and tryptic digestion into peptides. The processes of isobaric tandem mass tag (TMT) labeling of tryptic peptides, off-line sample fractionation and phosphoenrichment of the samples will be also be discussed. Of important note is that this process can be used for CSF (after immunodepletion steps) and tissues from other parts of the rodent. However, the process of protein extraction must be optimized for the tissue.

Any process submitting the prepared samples for LC-MS/MS will not be described, but links and references for some of the analysis suites used for proteomics analysis are suggested at the end of the protocol.

5 Step-by-step protocol

5.1 Day 1

5.1.1 Tissue lysis, protein extraction and digestion

The brain tissue is homogenized to extract protein for the downstream processes. This protocol has been adapted to be compatible with the suspension-trapping filter-based approach (S-Trap™) of protein extraction and digestion. Other methods of extraction and tryptic digestion are available.

5.1.1.1 Materials and reagents

- Hand-operated tissue grinder (Sigma, Z359971)
- Pellet pestles (Sigma, Z359971-1EA)
- Triethylammonium bicarbonate buffer (TEAB) (Sigma, T7408)
- S-Trap™ midi columns (Protifi)
- Sodium dodecyl sulfate (SDS, 20% solution in water) (Sigma, 05030)
- CHAPS hydrate (Sigma, C9426)
- NP-40 alternative (Millipore, 492016)
- Low protein binding microcentrifuge tubes (2.0mL) (ThermoScientific, 88380)

5.1.1.2 Protocol

a. Transfer freshly dissected or frozen brain tissue into a clean 1.5mL centrifuge tube.

b. Add 300 μL of 1 × SDS lysis buffer (50mM TEAB, 5% SDS; pH 7.5) to the tissue and homogenize using a hand-held cordless motor. Change the pestle between samples.

Note 1: The volume of lysis buffer can be changed depending on the size of the tissue, but the final volume prior to reduction and alkylation must not exceed 500 μL.

Critical: Avoid using amine-containing buffers in sample preparation. These buffers interfere with downstream processes such as TMT labeling. TEAB is an amine-free buffer.

c. Add 20% (w/v, dissolved in deionized water) CHAPS and 10% NP-40 (v/v in deionized water) to final concentrations of 2% (v/v) and 1% (v/v), respectively.

d. Vortex briefly and incubate at 4°C for 10min with end-to-end rotation.

e. Sonicate the lysate at 10–15 amplitude for 15s on ice. Repeat the process twice, with 10-s between sonication sessions.

f. Centrifuge at 10,000 × *g* for 10min.

g. Transfer the supernatant (protein lysate) into new pre-chilled microcentrifuge tubes.

5.1.2 BCA protein quantification

The Pierce BCA Protein Assay Kit was used to measure the protein concentration, following the manufacturer's instructions for the microplate procedure.

5.1.2.1 Materials and reagents

- Pierce BCA protein assay kit (ThermoScientific, 23225)
- 96-well microplate (Sigma, CLS3358)
- Microplate reader

5.1.2.2 Protocol

a. Prepare albumin standards for the microplate procedure as outlined in the data sheet.
b. Add 20 μL of each albumin standard and protein lysate in replicates into 96-well microplate wells.

Note: Dilute a small volume of protein lysate 5–20 × in water for the BCA analysis.

c. Add 200 μL of the BCA working reagent to each well, mix and incubate at 37 °C for 30 min.
d. Measure absorbance at or near 562 nm (540–590 nm).
e. Estimate protein concentration using the standard curve.
f. Normalize all samples to the same starting protein concentration and volume (500 μL) using 1 × lysis buffer.

Note: The concentration of starting protein material depends on many factors, including the downstream procedures. We have used starting concentrations of 300–2000 μg and above in our lab.

5.1.3 S-Trap™ trypsin digestion

This part describes trypsin digestion of the reduced and alkylated proteins using the suspension-trapping filter-based (S-Trap™) method.

5.1.3.1 Materials and reagents

- 1,4-Dithiothreitol (Sigma, 10197777001)
- Iodoacetamide (Sigma, I1149)
- Orthophosphoric acid (FisherScientific, A260)
- Trypsin/Lys-C mix (Promega, V5071)
- Water for HPLC (Sigma, 270733)
- Triethylammonium bicarbonate buffer (TEAB) (Sigma, T7408)
- Methanol for HPLC (Sigma, 34860)
- S-Trap™ midi columns (Protifi, C02-midi)

5.1.3.2 Protocol

a. Reduce proteins by adding 1 M DTT (dissolved in HPLC-grade water) to final concentration of 20 mM. Vortex briefly and incubate at 95 °C for 10 min.

b. Allow samples to cool to room temperature.
c. Alkylate proteins by adding 1 M iodoacetamide (dissolved in HPLC-grade water) to a final concentration of 100 mM. Vortex briefly and incubate at room temperature for 30 min in the dark.
d. Add phosphoric acid to a final concentration of 1.2% (v/v) to acidify proteins.
e. Add six volumes of S-Trap™ buffer (100 mM TEAB, 90% methanol, pH 7.1) to the acidified sample. Mix gently. The solution will appear translucent if there is sufficient protein.
f. Transfer sample into S-Trap™ column, insert column into 15 mL collection tube and centrifuge at 4000 × *g* for 30 s.
g. Reapply the flow-through to the column and centrifuge at 4000 × *g* for 30 s.
h. Wash the captured protein three times with 3 mL of S-Trap™ buffer at 4000 × *g* for 30 s.
i. Transfer the column into a new 15 mL collection tube and add 350 μL of S-Trap™ digestion buffer (50 mM TEAB) containing trypsin/lys-c mix or trypsin at protease to protein ratio of 1:25.

Note: Optimize the protease to protein ratio. We have used ratios between 1:20 and 1:50.

j. Centrifuge the solution into the column at 4000 × *g* for 30 s and reapply the solution to the top of the column.
k. Apply the cap of the tube, ensuring not to screw in to too tight. Wrap parafilm around the cap to reduce evaporation.
l. Incubate at 37 °C overnight in a water bath, making sure that the part of the tube containing the column is in contact with the water.

5.2 Day 2

5.2.1 Elution of tryptic peptides

5.2.1.1 Materials and reagents

- Formic acid (Sigma, F0507)
- Acetonitrile (ThermoScientific, 51101)
- Trifluoroacetic acid (ThermoScientific, 85183)
- Low protein binding microcentrifuge tubes (2.0 mL) (ThermoScientific, 88380)
- SpeedVac concentrator

5.2.1.2 Protocol

a. Add 500 μL of S-Trap™ digestion buffer to the column containing the protease solution and centrifuge at 4000 × *g* for 1 min. This flow-through contains the majority of the digested peptides.
b. Add 500 μL 0.2% (v/v) aqueous formic acid to the column, centrifuge at 4000 × *g* for 1 min and collect flow-through into the same collection tube.

c. Add 500 μL of 50% (v/v) aqueous acetonitrile containing 0.2% (v/v) formic acid to the column, centrifuge at 4000 × *g* for 1 min and collect flow-through.
d. Pool the elutions together and dry the tryptic peptides using a SpeedVac concentrator.
e. Resuspend peptides in 300 μL of 0.1% (v/v) TFA in water.

5.2.2 Peptide desalting

The Pierce™ Peptide Desalting Spin Columns kit was used for desalting, following the manufacturer's instructions. Use low protein binding microcentrifuge tubes throughout this process.

5.2.2.1 Materials and reagents

- The Pierce™ Peptide Desalting Spin Columns (ThermoScientific, 89851)
- Low protein binding microcentrifuge tubes (2.0 mL) (ThermoScientific, 88380)
- pH indicator strips (Sigma, P4786)
- Trifluoroacetic acid (ThermoScientific, 85183)
- Acetonitrile (ThermoScientific, 51101)
- Triethylammonium bicarbonate buffer (TEAB) (Sigma, T7408)
- SpeedVac concentrator
- Sodium hydroxide (Sigma, 221465)

5.2.2.2 Protocol

a. Check the pH of the peptide solution (dissolved in 300 μL of 0.1% TFA) with pH indicator strips. Adjust to 2.5–3.5 using dilute TFA or 1 M NaOH.
b. Remove the white tip from the bottom of the desalting column and insert column into a 2 mL microcentrifuge tube. Centrifuge at 5000 × *g* for 1 min and discard the solution.
c. Wash column twice with 300 μL of acetonitrile at 5000 × *g* for 1 min.
d. Wash column twice with 300 μL of 0.1% TFA (v/v) at 5000 × *g* for 1 min.
e. Insert the spin column into a new microcentrifuge tube and load the peptide solution into the column. Make sure peptides are completely dissolved before loading into column.
f. Centrifuge at 3000 × *g* for 1 min. Discard the flow-through.
g. Wash the column three times with 300 μL of 0.1% TFA (v/v) at 3000 × *g* for 1 min.
h. Transfer the column into a new microcentrifuge tube and add 300 μL of elution solution (50% acetonitrile, 0.1% TFA). Centrifuge at 3000 × *g* for 1 min and keep the eluate.
i. Add another 300 μL of elution solution and centrifuge at 3000 × *g* for 1 min into the same microcentrifuge tube.
j. Dry the peptides using a SpeedVac concentrator.
k. Dissolve dried peptides in 100 μL of 50 mM TEAB.

5.2.3 TMT isobaric labeling of peptides

This protocol has been optimized for the TMT labeling kit from ThermoScientific. The type of TMT (duplex, sixplex, 10plex or 16plex) to use depends on the number of samples being prepared. This protocol describes using 5 × 0.8 mg TMTSixplex labels. The label to peptide ratio, incubation times and other processes have been optimized in our lab. It is important to optimize these steps to save cost and improve upon the peptide labeling.

Prior to labeling, the concentration of peptides must be determined using any peptide quantification assay kit.

5.2.3.1 Material and reagents

- TMTSixplex™ Isobaric Label Reagent Set, 5 × 0.8 mg
- Triethylammonium bicarbonate buffer (TEAB) (Sigma, T7408)
- pH indicator strips (Sigma, P4786)
- Acetonitrile (ThermoScientific, 51101)
- Hydroxylamine solution (Sigma, 467804)
- Sodium hydroxide (Sigma, 221465)
- Trifluoroacetic acid (ThermoScientific, 85183)

5.2.3.2 Protocol

a. Normalize all peptide concentrations using 50 mM TEAB in a final volume of 100 μL.
b. Adjust the pH to 8–8.5 using dilute TFA or 1 M NaOH.
c. Equilibrate TMT reagent to room temperature. Spin briefly to collect reagent at bottom of tube.
d. Add 41 μL of acetonitrile to each of the 0.8 mg vials, vortex briefly and spin.
e. Allow reagent to dissolve for 5 min with occasional vortexing.
f. Add the TMT reagent to the respective samples.

Note: The TMT:peptide ratio must be optimized. ThermoScientific recommends a ratio of 8:1 (800 μg TMT to 100 μg peptide). We have optimized and used ratios ranging from 8:1 to 2:1 and have shown very similar labeling efficiency.

Note: The final concentration of acetonitrile in the TMT-peptide mixture should be around 30% (v/v).

g. Vortex and spin briefly. Incubate at 25 °C for 1 h.

Note: As an optional step, a TMT labeling check can be performed by analyzing a small volume of the sample on a mass spectrometer to determine the percentage of peptides that are TMT labeled. This step must be done prior to quenching the labeling process.

h. Add 5% hydroxylamine to a final concentration of 0.4% (v/v) and incubate at 25 °C for 15 min to quench the labeling.
i. Pool all the samples into one tube and store at −20 °C.

5.3 Day 3

The steps described so far are applicable to preparing samples for analysis of protein expression or post-translational modification of proteins. The next steps will describe the protocols for preparing samples to analyze protein abundance. A separate protocol for phosphoproteomics will be discussed.

On day 3, defrost the frozen TMT-labeled peptides on ice, dry and resuspend in 300 μL of 0.1% TFA and desalt (as described previously) to remove unbound TMT reagent. Dry the peptides using a SpeedVac concentrator.

5.3.1 Procedure: High-pH reverse-phase fractionation

The Pierce™ high pH reversed-phase peptide fractionation kit was used, following the manufacturer's instructions.

5.3.1.1 Materials and reagents

- Pierce™ high pH reversed-phase peptide fractionation kit (ThermoScientific, 84868)
- Acetonitrile (ThermoScientific, 51101)
- Trifluoroacetic acid (ThermoScientific, 85183)
- Water for HPLC (Sigma, 270733)
- Low protein binding microcentrifuge tubes (2.0 mL) (ThermoScientific, 88380)
- SpeedVac concentrator

5.3.1.2 Protocol

a. Prepare peptide elution buffers as shown described in the manufacturer's instructions. Adjust the volumes according to the number of samples being processed.

b. Remove the white tip from the bottom of the fractionation column and insert column into a 2 mL microcentrifuge tube. Centrifuge at 5000 × *g* for 2 min and discard the solution.

c. Wash column twice with 300 μL of acetonitrile at 5000 × *g* for 2 min.

d. Wash column twice with 300 μL of 0.1% (v/v) TFA at 5000 × *g* for 2 min.

e. Dissolve the peptides in 300 μL of 0.1% (v/v) TFA.

f. Place the fractionation column into a new 2 mL low protein binding microcentrifuge tube and load the 300 μL peptide sample into the column.

g. Replace the cap and centrifuge at 3000 × *g* for 2 min. Retain the eluate and label as "flow-through" fraction.

h. Place the column into a new microcentrifuge tube and add 300 μL of HPLC-grade water to the column. Centrifuge at 3000 × *g* for 2 min. Retain the eluate and label as "wash" fraction.

i. Place the column into a new microcentrifuge tube and add 300 μL of the "water wash" solution to the column. Centrifuge at 3000 × *g* for 2 min. Retain the eluate and label as "TMT wash" fraction.

j. Place the column into a new microcentrifuge tube and repeat step (i) using the appropriate elution solution.
k. Dry eluates and resuspend in 20 μL 0.1% (v/v) formic acid.

The sample is ready for liquid chromatography tandem mass spectrometry (LC-MS/MS) analysis.

If the sample is for phosphoproteomics analysis, the peptides must be enriched before fractionation.

5.3.2 Enrichment of phosphorylated peptides for phosphoproteomics analysis

Two methods of phosphopeptide enrichment will be described: The High-select™ titanium oxide (TiO_2) phosphopeptide enrichment using a kit from Thermo-Scientific, and our in-house immobilized metal affinity chromatography (IMAC)/ TiO_2 enrichment protocol.

Note: There are several other methods of enriching phosphopeptides, with varying degrees of efficiency. It is important to select and optimize a method for your work, taking into account factors such as cost, availability of equipment and efficiency.

5.3.2.1 Phosphopeptide enrichment alternative 1: High-select™ titanium oxide (TiO_2) phosphopeptide enrichment (ThermoScientific)

This protocol follows the manufacturer's instructions.

5.3.2.1.1 Materials and reagents.

- High-select™ TiO_2 phosphopeptide enrichment kit (ThermoScientific, A32993)
- pH indicator strips (Sigma, P4786)
- Low protein binding microcentrifuge tubes (2.0 mL) (ThermoScientific, 88380)
- Water for HPLC (Sigma, 270733)
- Trifluoroacetic acid (ThermoScientific, 85183)
- Formic acid (Sigma, F0507)
- Sodium hydroxide (Sigma, 221465)

5.3.2.1.2 Protocol.

a. Completely dissolve the dried, desalted TMT-labeled peptides in 150 μL of binding/equilibration buffer.
b. Check pH of the sample is <3.0. Adjust with dilute TFA or 1 M NaOH. This is very critical to the enrichment protocol.
c. Insert a centrifuge column adaptor into a new 2 mL microcentrifuge tube and insert the TiO_2 spin tip into the centrifuge column adaptor.
d. Load 20 μL of wash buffer into the spin tip and centrifuge at $3000 \times g$ for 2 min.
e. Load 20 μL of binding/equilibration buffer into the spin tip and centrifuge at $3000 \times g$ for 2 min.

f. Transfer the spin tip and the column adaptor into a new low protein binding microcentrifuge tube and load the 150 μL peptide sample into the spin tip.
g. Centrifuge at 1000 × *g* for 5 min.
h. Reapply the sample from the collection tube into the spin tip column and centrifuge at 1000 × *g* for 5 min.

Note: The flow-through, which contains majority of non-phosphorylated peptides can be retained for analysis.

i. Transfer the spin tip and the column adaptor into a new microcentrifuge tube, add 20 μL of binding/equilibration buffer to the column and centrifuge at 3000 × *g* for 2 min.
j. Repeat step (i) with 20 μL of wash buffer.
k. Repeat steps (i) and (j).
l. Wash the column with 20 μL of HPLC-grade water as described in step (i).
m. Transfer the spin tip and column adapter into a new low protein binding microcentrifuge tube.
n. Load 50 μL of phosphopeptide elution buffer, centrifuge at 1000 × *g* for 5 min to elute the phosphopeptides.
o. Repeat step (n) and elute into the same tube.
p. Dry the eluted phosphopeptides immediately and resuspend in 0.1% TFA.

5.3.2.2 Phosphopeptide enrichment alternative 2: IMAC/TiO_2 phosphopeptide enrichment

This protocol has been optimized "in-house." It involves two steps of phosphopeptide enrichment; starting with PHOS-Select™ Iron Affinity Gel, followed by TiO_2 enrichment.

5.3.2.2.1 Materials and reagents.

- PHOS-Select™ Iron Affinity Gel (Sigma, P9740)
- Trifluoroacetic acid (ThermoScientific, 85183)
- Acetic acid, glacial (FisherScientific, 10365020)
- Acetonitrile (ThermoScientific, 51101)
- Water for HPLC (Sigma, 270733)
- Sodium hydroxide (Sigma, 221465)
- Filter (small) 10 μm pore size (2B Scientific, M2110)
- "Classic" with screw cap and plug, without filters spin columns (2B Scientific, M1003)
- pH indicator strips (Sigma, P4786)
- Low protein binding microcentrifuge tubes (2.0 mL) (ThermoScientific, 88380)
- Ammonia solution (FisherScientific, 1336-21-6)
- 2,5-Dihydroxybenzoic acid (DHB) (Sigma, 149357)
- Titansphere TiO_2 5 μM (5020-75000)

5.3.2.2.2 Protocol: Immobilized metal affinity chromatography (IMAC).

a. Completely dissolve the dried, desalted TMT-labeled peptides in 1 mL of 0.1% (v/v) TFA.
b. Adjust pH to 2.5–3.0 using dilute TFA or 1 M NaOH.
c. Transfer IMAC beads into 2 mL microcentrifuge tube.
d. Wash beads three times with 1 mL wash/load buffer (0.25 M glacial acetic acid, 30% acetonitrile).
e. After final wash, add wash/load buffer to prepare a 50% IMAC slurry.
f. Add 50 μL of IMAC slurry to each sample.
g. Incubate at room temperature for 90 min with end-to-end rotation.
h. Insert the 10 μm pore size filters into the Mobicol spin columns.
i. Insert spin columns into 2 mL microcentrifuge tubes.
j. Add half (~525) of the sample-IMAC mix to the column.
k. Centrifuge at 1000 × *g* for 30 s at room temperature. Keep the flow-through. This will be used for TiO_2 enrichment.
l. Transfer the column into a new 2 mL microcentrifuge tube and add the remaining sample-IMAC mix.
m. Centrifuge at 1000 × *g* for 30 s at room temperature. Keep the flow-through.
n. Combine the two flow-throughs and keep for TiO_2 enrichment.
o. Transfer column into a new 2 mL microcentrifuge tube and add 200 μL of wash/load buffer.
p. Resuspend the sample in the buffer with a pipette and centrifuge at 1000 × *g* for 30 s. Discard flow-through and repeat the wash once.
q. Repeat the wash once with HPLC water.
r. Insert column in a new low protein binding microcentrifuge tube.
s. Add 100 μL of elution buffer (30% acetonitrile, 0.7% v/v ammonia), resuspend beads and incubate at for 2 min at room temperature.
t. Centrifuge at 1000 × *g* for 30 s. Keep flow-through.
u. Repeat steps (r) and (s).
v. Combine flow-throughs and keep at −20 °C.

5.3.2.2.3 Protocol: TiO_2 phosphopeptide enrichment. The flow-throughs from the PHOS-select-treated samples (IMAC) were used for further phosphopeptide enrichment with TiO_2.

a. Weigh the required amount of TiO_2 (5 mg per sample) and dissolve in 1 mL of wash buffer 1 (30% acetonitrile, 0.5% TFA).
b. Vortex briefly and centrifuge at 3000 × *g* for 1 min to wash. Repeat wash.
c. Weigh the required amount of DHB and resuspend in wash buffer 1 (30 mg/150 μL per sample). Vortex briefly to mix.
d. Remove supernatant and resuspend TiO_2 in 150 μL of DHB. Vortex briefly to mix.

e. Add 150 μL of the TiO_2-DHB mix to sample. Vortex briefly to mix.
f. Incubate at room temperature for 90min with end-to-end rotation.
g. Insert the 10μm pore size filters into the Mobicol spin columns.
h. Insert spin columns into 2mL microcentrifuge tubes.
i. Add half TiO_2-DHB-sample mix to the column.
j. Centrifuge at $1000 \times g$ for 30s at room temperature. Discard flow-through.
k. Add the remaining TiO_2-DHB-sample mix to the column.
l. Centrifuge at $1000 \times g$ for 30s at room temperature. Discard the flow-through.
m. Transfer column into a new 2mL microcentrifuge tube and add 200 μL wash buffer 1.
n. Resuspend the sample in the buffer with a pipette and centrifuge at $1000 \times g$ for 30s. Discard flow-through.
o. Repeat the wash once with wash buffer 2 (80% acetonitrile, 0.1% TFA).
p. Insert column in a new low protein binding microcentrifuge tube.
q. Add 100 μL of elution buffer (15% v/v ammonia) to the column, resuspend and incubate at for 2min at room temperature.
r. Centrifuge at $1000 \times g$ for 30s. Keep flow-through.
s. Repeat steps (q) and (r) and elute into same tube.
t. Combine eluates.
u. Combine the eluted phosphopeptides from the IMAC and TiO_2.
v. Dry and resuspend in 0.1% (v/v) TFA.

After enrichment of the phosphopeptides, carry out high-pH reverse-phase fractionation (described in Section 5.3.1), dry eluted fractions and resuspend in 20 μL of 0.1% formic acid. The sample is ready for LC-MS/MS for phosphoproteomics analysis.

A summary of the protocol is illustrated in Fig. 1.

5.3.3 Analysis of mass spectrometry-based proteomics data for protein abundance and post-translational modification

A number of analytical suites can be used to analyze the quantitative proteomics and phosphoproteomics data. The Perseus bioinformatics analysis platform (Tyanova & Cox, 2018; Tyanova et al., 2016) is a free, comprehensive platform for analysis of the brain proteome in the context of abundance and PTMs. Other bioinformatics platforms are available, and users can access them online.

To analyze, predict and visualize protein-protein interactions (PPIs), computational platforms such as STRING (Szklarczyk et al., 2019) and Ingenuity Pathway Analysis (IPA; Qiagen) are available for use. The STRING platform combines PPI data obtained from primary studies with those obtained from computational predictions to generate a wide scope of PPI data.

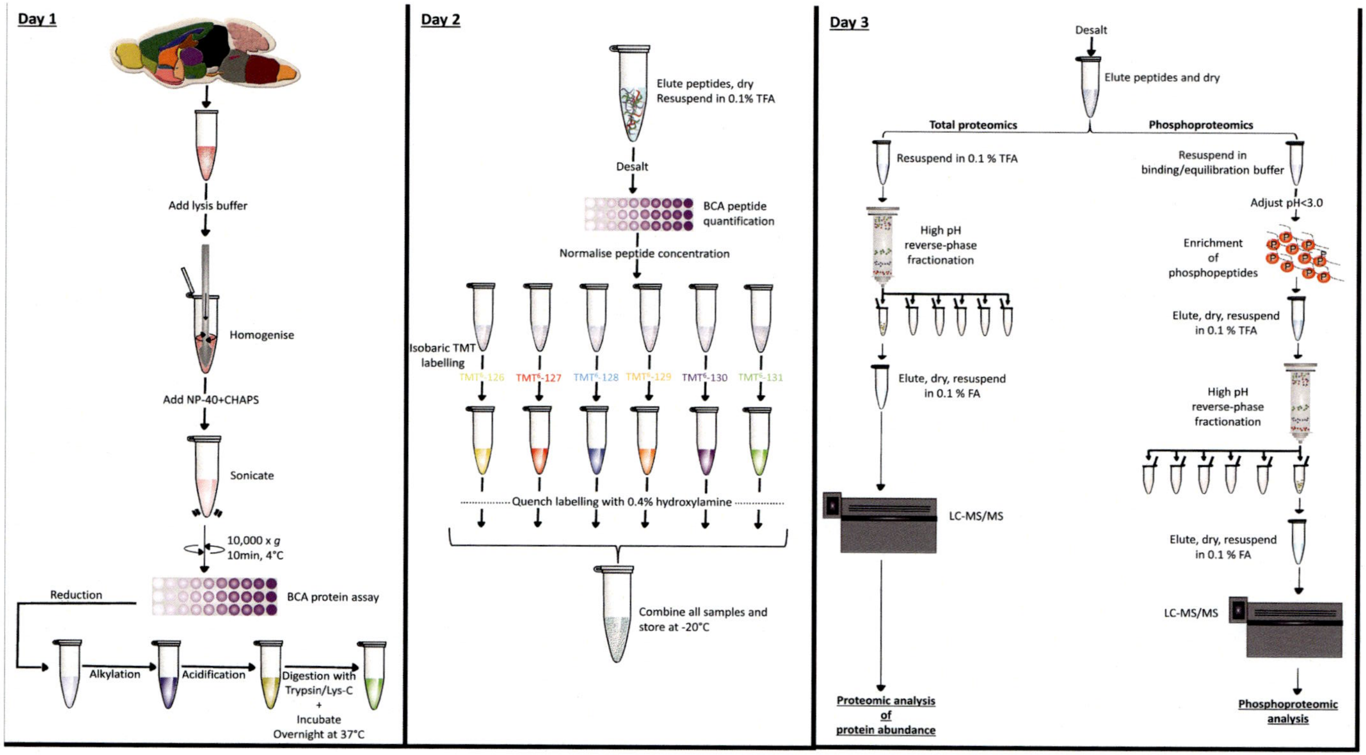

FIG. 1

Preparation of rodent brain tissue for LC-MS/MS-based proteomics analysis. This figure summarizes the protocols carried out over three days. The protocols can be used to prepare samples for proteomics-based analysis of proteins expression profiling (abundance) and/or post-translational modifications (phosphorylation).

6 Summary

Mass spectrometry-based quantitative proteomics and phosphoproteomics remain an attractive approach to understanding the pathophysiological mechanisms of neurodegeneration and other neurological disorders. In this chapter, optimized protocols for the preparation of rodent brain tissue for mass spectrometry-based proteomics have been discussed. Data obtained from these protocols can provide information about changes in protein expression and modification as a result of disease or drug treatment. The data can also be combined with affinity-purification mass spectrometry data to unravel protein-protein interactions in brain, how these interactions are affected by neurodegenerative disorders and the effect of drugs on these changes.

Acknowledgments

This work was supported by a Wellcome Trust Collaborative Award (201529/Z/16/Z) and an Alzheimer's Research UK Junior Small Grant. I am sincerely grateful to my colleague, Dr. Natasja Barki for her help with the summary figure and Dr. Samantha Jumbe for reading through the manuscript. I also thank Dr. Andrew Bottrill (RTP Proteomics Manager, University of Warwick) and Dr. Lev Solyakov for their support in developing these protocols.

References

Ahmad, Y., & Lamond, A. I. (2014). A perspective on proteomics in cell biology. *Trends in Cell Biology*, *24*(4), 257–264.

Ashraf, G. M., Tarasov, V. V., Makhmutova, A., Chubarev, V. N., Avila-Rodriguez, M., Bachurin, S. O., et al. (2019). The possibility of an infectious etiology of Alzheimer disease. *Molecular Neurobiology*, *56*(6), 4479–4491.

Ayyadevara, S., Ganne, A., Hendrix, R. D., Balasubramaniam, M., Shmookler Reis, R. J., & Barger, S. W. (2019). Functional assessments through novel proteomics approaches: Application to insulin/IGF signaling in neurodegenerative disease. *Journal of Neuroscience Methods*, *319*, 40–46.

Berggard, T., Linse, S., & James, P. (2007). Methods for the detection and analysis of protein-protein interactions. *Proteomics*, *7*(16), 2833–2842.

Dawson, T. M., Golde, T. E., & Lagier-Tourenne, C. (2018). Animal models of neurodegenerative diseases. *Nature Neuroscience*, *21*(10), 1370–1379.

Deuschl, G., Beghi, E., Fazekas, F., Varga, T., Christoforidi, K. A., Sipido, E., et al. (2020). The burden of neurological diseases in Europe: An analysis for the global burden of disease study 2017. *The Lancet Public Health*, *5*(10), e551–e567.

Drummond, E., Pires, G., MacMurray, C., Askenazi, M., Nayak, S., Bourdon, M., et al. (2020). Phosphorylated tau interactome in the human Alzheimer's disease brain. *Brain*, *143*(9), 2803–2817.

Dwomoh, L., Rossi, M., Scarpa, M., Khajehali, E., Molloy, C., Herzyk, P., et al. (2021). Activation of M_1 muscarinic receptors reduce pathology and slow progression of neurodegenerative disease. *Proceedings of the National Academy of Sciences of the United States of America*. Manuscript under review.

Feigin, V. L., Nichols, E., Alam, T., Bannick, M. S., Beghi, E., Blake, N., et al. (2019). Global, regional, and national burden of neurological disorders, 1990–2016: A systematic analysis for the Global Burden of Disease Study 2016. *The Lancet Neurology*, *18*(5), 459–480.

Hernandez, S. M., Tikhonova, E. B., & Karamyshev, A. L. (2020). Protein-protein interactions in alpha-synuclein biogenesis: New potential targets in Parkinson's disease. *Frontiers in Aging Neuroscience*, *12*, 72.

Higginbotham, L., Ping, L., Dammer, E. B., Duong, D. M., Zhou, M., Gearing, M., et al. (2020). Integrated proteomics reveals brain-based cerebrospinal fluid biomarkers in asymptomatic and symptomatic Alzheimer's disease. *Science Advances*, *6*(43), eaaz9360.

Hosp, F., & Mann, M. (2017). A primer on concepts and applications of proteomics in neuroscience. *Neuron*, *96*(3), 558–571.

Ivanic, J., Yu, X., Wallqvist, A., & Reifman, J. (2009). Influence of protein abundance on high-throughput protein-protein interaction detection. *PLoS One*, *4*(6), e5815.

Johnson, E. C. B., Dammer, E. B., Duong, D. M., Yin, L., Thambisetty, M., Troncoso, J. C., et al. (2018). Deep proteomic network analysis of Alzheimer's disease brain reveals alterations in RNA binding proteins and RNA splicing associated with disease. *Molecular Neurodegeneration*, *13*(1), 52.

Knopman, D. S., Jones, D. T., & Greicius, M. D. (2021). Failure to demonstrate efficacy of Aducanumab: An analysis of the EMERGE and ENGAGE trials as reported by Biogen, December 2019. *Alzheimer's & Dementia*, *17*, 696–701.

Mackay, D. F., Russell, E. R., Stewart, K., MacLean, J. A., Pell, J. P., & Stewart, W. (2019). Neurodegenerative disease mortality among former professional soccer players. *The New England Journal of Medicine*, *381*(19), 1801–1808.

Nishi, H., Hashimoto, K., & Panchenko, A. R. (2011). Phosphorylation in protein-protein binding: Effect on stability and function. *Structure*, *19*(12), 1807–1815.

Peng, X., Wang, J., Peng, W., Wu, F. X., & Pan, Y. (2017). Protein-protein interactions: Detection, reliability assessment and applications. *Briefings in Bioinformatics*, *18*(5), 798–819.

Ping, L., Duong, D. M., Yin, L., Gearing, M., Lah, J. J., Levey, A. I., et al. (2018). Global quantitative analysis of the human brain proteome in Alzheimer's and Parkinson's disease. *Scientific Data*, *5*(1), 180036.

Rayner, S. L., Morsch, M., Molloy, M. P., Shi, B., Chung, R., & Lee, A. (2019). Using proteomics to identify ubiquitin ligase–substrate pairs: How novel methods may unveil therapeutic targets for neurodegenerative diseases. *Cellular and Molecular Life Sciences*, *76*(13), 2499–2510.

Swarup, V., Chang, T. S., Duong, D. M., Dammer, E. B., Dai, J., Lah, J. J., et al. (2020). Identification of conserved proteomic networks in neurodegenerative dementia. *Cell Reports*, *31*(12), 107807.

Szklarczyk, D., Gable, A. L., Lyon, D., Junge, A., Wyder, S., Huerta-Cepas, J., et al. (2019). STRING v11: Protein–protein association networks with increased coverage, supporting functional discovery in genome-wide experimental datasets. *Nucleic Acids Research*, *47*, D607–D613.

Tyanova, S., & Cox, J. (2018). Perseus: A bioinformatics platform for integrative analysis of proteomics data in cancer research. *Methods in Molecular Biology (Clifton, N.J.)*, *1711*, 133–148.

Tyanova, S., Temu, T., Sinitcyn, P., Carlson, A., Hein, M. Y., Geiger, T., et al. (2016). The Perseus computational platform for comprehensive analysis of (prote)omics data. *Nature Methods*, *13*(9), 731–740. https://doi.org/10.1038/nmeth.3901.

Wingo, A. P., Fan, W., Duong, D. M., Gerasimov, E. S., Dammer, E. B., Liu, Y., et al. (2020). Shared proteomic effects of cerebral atherosclerosis and Alzheimer's disease on the human brain. *Nature Neuroscience*, *23*(6), 696–700. https://doi.org/10.1038/s41593-020-0635-5.

CHAPTER

Protein-protein interactions at a glance: Protocols for the visualization of biomolecular interactions

13

Mariangela Agamennone[a], Alessandro Nicoli[b], Sebastian Bayer[b], Verena Weber[b], Luca Borro[c], Shailendra Gupta[d], Marialuigia Fantacuzzi[a], and Antonella Di Pizio[b,]*

[a]*Department of Pharmacy, University "G. d'Annunzio" of Chieti-Pescara, Chieti, Italy*

[b]*Leibniz-Institute for Food Systems Biology at the Technical University of Munich, Freising, Germany*

[c]*Department of Imaging, Advanced Cardiovascular Imaging Unit, Bambino Gesù Children's Hospital, IRCCS, Rome, Italy*

[d]*Department of Systems Biology and Bioinformatics, University of Rostock, Rostock, Germany*

**Corresponding author: e-mail address: a.dipizio.leibniz-lsb@tum.de*

Chapter outline

Methods in Cell Biology, Volume 166, ISSN 0091-679X, https://doi.org/10.1016/bs.mcb.2021.06.012

Abstract

Protein-protein interactions (PPIs) play a key role in many biological processes and are intriguing targets for drug discovery campaigns. Advancements in experimental and computational techniques are leading to a growth of data accessibility, and, with it, an increased need for the analysis of PPIs. In this respect, visualization tools are essential instruments to represent and analyze biomolecular interactions. In this chapter, we reviewed some of the available tools, highlighting their features, and describing their functions with practical information on their usage.

1 Introduction

Protein-protein interactions (PPIs) are essential for many biological processes within and between cells and are therefore the target of research investigations for a wide range of biomedical applications. The first example of interacting biomolecules was reported in the literature at the end of the 19th century when Paul Ehrlich identified an antibody produced in mice counteracting the action of ricin, a toxic substance from plants (Ehrlich, 1891). At the beginning of the 20th century, the inhibiting interaction of trypsin and albumin was discovered (Northrop, 1922). Since then, the study of interacting proteins has moved hand in hand with breakthroughs in both experimental and computational technologies (Braun & Gingras, 2012). At the beginning of the 2000s, with the awareness that "no protein is an island entire of itself" (Kumar & Snyder, 2002), the interest toward PPIs increased, and nowadays interacting proteins are regarded as the fundament of most life processes. A bibliographic search carried out in SciFinder (https://scifinder.cas.org, consulted on February 7th, 2021) using the keywords "protein-protein interactions" retrieved almost 130,000 references, with an exponential increase from 2005, leading to a plateau of around 8000 references per year with a maximum of 10,000 items in 2013, and reflecting the consolidated interest in this subject. With the growth of the research interest, PPI data accessibility and analysis are becoming more and more relevant. The visualization of research outcomes impacts the way they will be perceived and used for further developments across the scientific community. In the case of proteins and PPIs, graphical representations are not only relevant for efficient scientific communication but are also functional to the analysis workflow. This prompted the development of numerous and increasingly sophisticated protein visualization tools (O'Donoghue et al., 2010; Olson, 2018). In this chapter, we briefly describe protein-protein interactions, starting from the interaction network till to the single molecules, and then we provide examples of PPIs and visualization protocols.

2 Protein-protein interaction network

In 1999, the term "interactome" was introduced to specify all molecular interactions occurring in a cell, and represents the intermediate and necessary step to connect the genome to phenotypic events (Sanchez et al., 1999). Indeed, the proteome adapts to different cells and conditions, therefore, the effective role of a protein is highly affected by its biological environment and partner macromolecules. Furthermore, most pathological conditions can be related to dysfunctional or altered PPIs. As a consequence, understanding the protein-protein interaction networks (PPINs) represents a gold mine for biomedical research. Different macromolecules, such as DNA, RNA, and proteins, can interact with each other, but the vast majority of contacts happens between pair of proteins, and the terms interactome and PPIN are often used indifferently.

Numerous databases have been established to collect and analyze current information about PPINs. They can be classified based on the technique used for their generation, the accurateness of reported data, the number and types of considered organisms, the functional classification of interacting molecules, etc. Some of them collect data through accurate literature mining, such as BioGrid (Oughtred et al., 2019), IntACT (Hermjakob et al., 2004), or MINT (Licata et al., 2012); other tools integrate literature mining along with computational prediction and/or with experimental structural data, such as STRING (Szklarczyk et al., 2019), Interactome3D (Mosca, Pons, Céol, Valencia, & Aloy, 2013), or APID (Alonso-López et al., 2019). High-throughput yeast two-hybrid (HT-Y2H), affinity purification followed by mass spectrometry (AP-MS), FRET and co-immunoprecipitation are among the experimental tools mostly used to identify interacting proteins (Iacobucci, Monaco, Cozzolino, & Monti, 2021). The last version of BioPlex 3.0 provides a database of protein-protein contacts detected through the AP-MS technique (Huttlin et al., 2020): it encompasses almost 70% of the human proteome including around 120,000 interactions from 15,000 proteins in two cell lines (human colon cancer HCT116 and human embryonic kidney HEK293 cells) and represents, to date, the largest database with experimentally derived contacts. Furthermore, the picture of connected proteins can be completed by localizing them in cell compartments, providing a more detailed description of the interactome and its topological distribution. As an example, the Cell Atlas was built by combining data from antibody-based, immunofluorescence and confocal microscopy studies to localize 12,003 proteins in 30 different cellular compartments, and can be used for subcellular localization (Thul et al., 2017).

Networks of interacting biomolecules can be visualized in schematic maps, where proteins are represented as nodes and their connections as edges. Generally, highly connected proteins are master regulators and are considered as *hubs*. Examples of network representations are reported in Fig. 1, where we generated a protein interaction network around the human Matrix Metalloproteinase 2 (MMP2), a zinc-dependent endoprotease known to exert regulatory roles in tumor growth, inflammation and cardiovascular diseases (Bauvois, 2012).

The size of PPINs largely depends on the text mining algorithms or the experimental methods selected for the screening of interacting protein partners, thus may

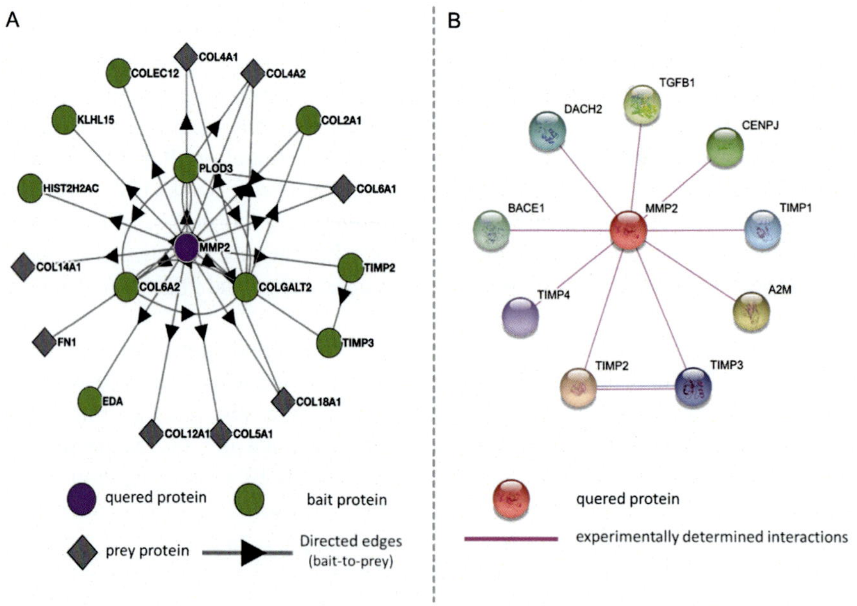

FIG. 1

Representation of experimentally validated protein interaction network around MMP2 from *Homo sapiens*. (A) Network generated using BioPlex explorer (Huttlin et al., 2020). The network contains 18 proteins interacting with MMP2 in the HEK293T cell line, identified using AP-MS experiments in the BioPlex project. Proteins used as baits in the AP-MS method are shown as green circles, while proteins identified only as preys are shown as gray diamonds. (B) Experimentally validated protein-protein interaction network generated with STRING (Szklarczyk et al., 2019). Only nine interacting partners were identified with an interaction score more than the medium confidence score (>0.400), an MMP2 PPIN with a low confidence score (>0.150) contains 58 nodes and 161 edges (network not shown).

vary according to the databases. Databases, such as STRING, use their own interaction score functions, which largely depend on the low and/or high throughput experiments for the detection of interacting proteins, to generate high, medium or low confidence PPINs around the protein under investigation. A large number of computational workflows and tools are available for prioritizing key PPIs from PPINs by combining network topological parameters with experimental data (Khan et al., 2017; Khan, Sadeghi, Gupta, & Wolkenhauer, 2018). PPIN analyses retrieve often partially overlapping records, fragmentation and/or incompleteness of information, making difficult the discrimination between structural and functional interactions. Methods and applications used to depict structurally interacting proteins at a molecular level will be discussed in the following paragraphs.

3 Protein structures and protein complexes

Because of their clinical significance, PPIs are attractive targets for drug discovery (Lu et al., 2020). In order to develop modulators of protein-protein contacts, moving toward a PPI description at atomic resolution is an essential requirement.

X-ray crystallography, nuclear magnetic resonance (NMR), and cryo-electron microscopy (cryo-EM) are the main techniques used to experimentally characterize the 3D structure of proteins. All solved macromolecule structures are deposited in the Brookhaven Protein Data Bank (PDB), which celebrates the 50th anniversary this year (2021) (Berman et al., 2000; Burley et al., 2021). As of January 21st, 2021, most of the structures were obtained by X-ray (153,464), followed by NMR (13,178) and cryo-EM (6660) (PDB-stats, 2021). The advancements in cryo-EM techniques were celebrated with the Nobel Prize in Chemistry in 2017 (Cressey & Callaway, 2017). Cryo-EM is based on the flash-freezing of solutions containing proteins that are analyzed by an electron beam. This way, it is possible to determine the 3D coordinates of macromolecules coming from the physiological environment (Callaway, 2020b; Cianfrocco & Kellogg, 2020). Rapid instrumental implementation is overcoming cryo-EM main limitations, and nowadays atomic resolutions can be reached (Nakane et al., 2020). Through NMR, molecules are studied in solution, better representing their dynamic behavior and capturing also transient contacts between proteins. NMR determines the structures through multidimensional nuclear magnetic resonance experiments to labeled proteins. However, this method is not suitable for very large macromolecules (Arthanari, Takeuchi, Dubey, & Wagner, 2019). X-ray crystallography is a well-established technique based on the collision of an X-ray beam with atomic nuclei present in a crystal and producing a diffraction pattern that depends on the 3D coordinates of macromolecule atoms. The application of Fourier transform algorithms provides a tridimensional map of electron density. This technique can only be applied to proteins that can crystallize in a non-physiological state, and produces only a static representation of macromolecules, introducing sometimes crystallization artifacts (Niedzialkowska et al., 2016). This procedure, in fact, entails that some experimentally retrieved macromolecular contacts could not be biologically relevant. The number of protein-protein contacts in the PDB is estimated to be 28,230 for heteromeric and 124,050 for homomeric complexes, but not all are functional. To circumvent this issue several tools were developed to study and refine the interacting surfaces (Elez, Bonvin, & Vangone, 2020), including EPPIC (Evolutionary Protein-Protein Interface Classifier), a web-based service that classifies the interacting surfaces to discriminate between biologically relevant from crystal-induced contacts (Duarte, Srebniak, Scharer, & Capitani, 2012), or ClusPro-DC that provides the same classification by using protein-protein docking calculations (Yueh et al., 2017).

When structural data about protein complexes are not available, PPIs can be simulated through computational tools. In silico techniques can also be used to predict 3D coordinates of single proteins. Such methods range from template-based to ab initio modeling. Protein structure prediction is highly challenging and the computational community is called every two years to participate in the CASP (Critical Assessment of Techniques for Protein Structure Prediction) competition, established in 1994 to compare performances of different structure prediction methods (Moult, Pedersen, Judson, & Fidelis, 1995). During the last CASP round, CASP14, AI-based protocols emerged as the most predictive tools for protein folding. AlphaFold 2, a program developed by DeepMind, the arm of Google AI, based in London, was able to determine protein structures with an accuracy level previously obtained only with experimental methodologies (Callaway, 2020a; Service, 2020). This "folding revolution" can speed the structural exploration of the human proteome and, consequently, PPIs.

We currently have structural data for 13% of proteins in the human interactome and for only 5% of PPIs (Interactome3d-stats, 2021), therefore, computer simulations are in standard use to fill the gap, predicting interaction modes and binding affinities (Kaczor, Bartuzi, Stepniewski, Matosiuk, & Selent, 2018; Siebenmorgen & Zacharias, 2019a). The first attempt to simulate a protein-protein association was made by Wodak and Janin (1978). Since then, many protein-protein docking strategies have been developed (Barradas-Bautista, Rosell, Pallara, & Fernandez-Recio, 2018; Huang, 2014). Generally, all available tools have an exploration algorithm that generates multiple solutions, and then apply a scoring function to rank resulting complexes. Protein-protein docking software can be classified on the basis of the protein treatment, rigid or flexible, and the degree of exploration, global or local. Rigid body docking explores the six degrees of freedom (three rotational and three translational) of partner proteins: one protein is fixed in the 3D space, while the other explores the surface of the fixed protein attempting to find the best complementarity. The application of Fast Fourier transform (FFT) and, more recently fast manifold Fourier transform (FMFT), greatly improves the speed of calculations (Padhorny et al., 2016). Examples of FFT-based software are ZDOCK (Pierce et al., 2014), pyDOCK (Jimenez-Garcia, Pons, & Fernandez-Recio, 2013), PIPER (Kozakov et al., 2017). Some of the mentioned programs are able to refine obtained complexes by minimization or sampling of side-chain conformations. A rigid docking approach can easily fail when conformational rearrangements occur in the binding process.

Attempts have been made to include flexibility consideration into protein-protein docking protocols (Harmalkar & Gray, 2020), usually by generating protein conformations before the docking steps, using randomized methods (genetic algorithm, simulated annealing, Monte Carlo sampling, etc.) and coarse-grained models to speed the calculations. Examples of such flexible docking software are SwarmDOCK (Torchala, Moal, Chaleil, Fernandez-Recio, & Bates, 2013), RosettaDock 4.0 (Marze, Roy Burman, Sheffler, & Gray, 2018), ATTRACT (Glashagen et al., 2020), HADDOCK (Van Zundert et al., 2016). In most cases, these programs can perform both global and local searches, and some of them (RosettaDock and HADDOCK) can account for experimental data such as site-directed mutagenesis data or NOE constraints to drive the docking calculations with improved reliability of resulting complexes. In addition to the exploration of conformational flexibility (Rakers, Bermudez, Keller, Mortier, & Wolber, 2015), computational efforts in docking protocols currently rely on the development of new scoring methods. Machine-learning based scoring functions are now being developed (Das & Chakrabarti, 2021; Wang, Terashi, Christoffer, Zhu, & Kihara, 2020): IRaPPA (Integrative Ranking of Protein–Protein Assemblies) uses the physicochemical descriptors of interacting surfaces to generate a ranked super-vector machine model (Moal et al., 2017). Calculations of the complex binding free energy can also be used to refine and score protein-protein complexes. A recently developed replica-exchange-based scheme with different levels of a repulsive biasing between partners showed good correlations with experimental data (Siebenmorgen, Engelhard, & Zacharias, 2020; Siebenmorgen & Zacharias, 2019b).

The performance of protein-protein docking protocols is assessed by CAPRI (Critical Assessment of PRedicted Interactions) competitions (Wodak, Velankar, & Sternberg, 2020). In September 2020, the 51st edition was launched, with 55 participants. As an expansion, the joint CAPRI–CASP initiative aims to assess the ability of computational tools to produce homo- or heteromeric protein structures starting from their sequence (Lensink et al., 2019).

4 Protein-protein interface: shape and chemical complementarity

Analyses of experimentally determined PPIs indicate that intermolecular recognition is determined by numerous, weak, non-covalent interactions at protein-protein interfaces, which are characterized by high complementarity of the shape and the chemical properties of the interacting surfaces (Jones & Thornton, 1996, 1997). Interacting surfaces are usually enriched in hydrophobic residues, especially in the most buried regions, and hydrophobic contacts are the major thermodynamic driving force for protein binding (Chandler, 2005; Levy, 2010). Electrostatic interactions have a relevant role to binding specificity, however, their contribution to the free energy is more complex and debated: they can positively affect protein stability

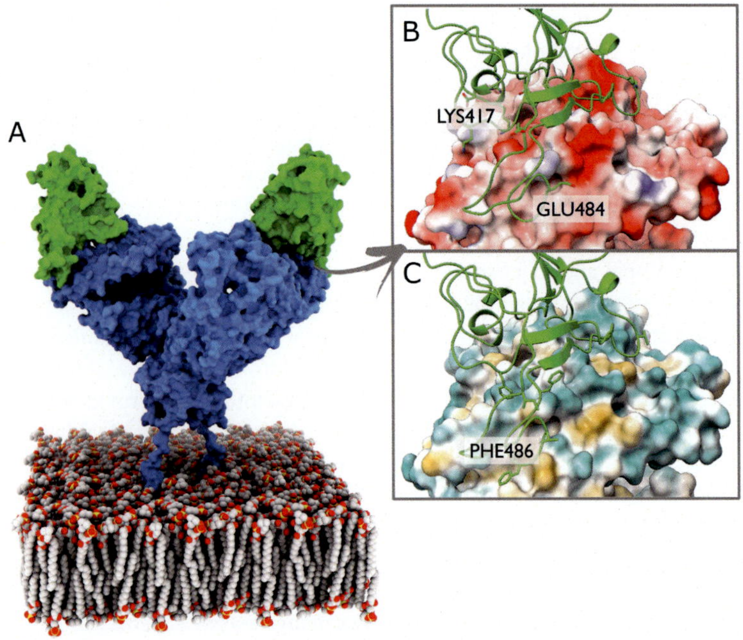

FIG. 2

Full (A) and zoomed view (B and C) of the interaction between the spike protein RBD and the ACE2 receptor. (A) RBDs and ACE2 are represented as green and blue molecular surfaces, embedded in a 1-palmitoyl-2-oleyl-sn-glycerol-3-phosphocholine (POPC) lipid bilayer. POPC residues are depicted as spheres, where lipid tail carbons are colored in white, lipid head phosphorus atoms are colored in yellow, and oxygen atoms in red. (B) RBD is shown in a green cartoon representation, positively and negatively charged residues as sticks. ACE2 is represented as surface colored by the Coulomb electrostatic potential. (C) RBD is shown as green cartoons and hydrophobic residues as sticks. ACE2 is represented as surface colored according to levels of lipophilicity. Only residues discussed in the main text are labeled.

but this effect is often counterbalanced by unfavorable desolvation (Hendsch & Tidor, 1994; Kundrotas & Alexov, 2006; Sheinerman & Honig, 2002; Sheinerman, Norel, & Honig, 2000).

The electrostatic and hydrophobic character of interacting surfaces can be visualized and analyzed with most of the developed 3D visualization tools. We provide an example in Fig. 2, where the complex between the receptor binding domain (RBD) of the SARS-CoV-2 spike (S) glycoprotein with the Angiotensin I Converting Enzyme 2 (ACE2) receptor is reported.

As of February 2021, more than 100 million people worldwide were affected by the Coronavirus Disease 2019 (COVID-19), caused by the Severe Acute Respiratory Syndrome SARS-CoV-2 coronavirus strain (WHO-coronavirus2019, 2021). The response of the scientific community to the pandemic was immediate and many aspects of the molecular mechanisms involved in the SARS-CoV-2 infection were rapidly unraveled.

Similar to SARS-CoV, SARS-CoV-2 was found to infect cells through the interaction between its spike glycoprotein and the ACE2 receptor on target cells (Hoffmann et al., 2020; Lan et al., 2020; Li et al., 2003; Walls et al., 2020; Yan et al., 2020).

The S protein functions as a homotrimer, where each monomer is formed by S1 and S2 subunits (Huang, Yang, Xu, Xu, & Liu, 2020). The host-virus interaction is mediated by the receptor binding domain (RBD) of the S1 subunit in the "up" (host cell receptor-accessible) conformation: binding to the host target destabilizes the pre-fusion homotrimer, which sheds off the S1 subunit, and allows for the transition of the S2 subunit to a highly stable post-fusion conformation (Cai et al., 2020; Casalino, Gaieb, et al., 2020; Wrapp et al., 2020).

Numerous experimental and structural studies targeted the spike interaction with ACE2, providing relevant insights into the binding process (Cao et al., 2020; Deganutti, Prischi, & Reynolds, 2021; Douangamath et al., 2020; Joshi, Joshi, & Degani, 2020; Kalathiya et al., 2020; Tiwari, Beer, Sankaranarayanan, Swanson-Mungerson, & Desai, 2020; Wrapp et al., 2020; Yang et al., 2020). In Fig. 2, we show the RBD-ACE2 interaction as determined by Yan et al. (2020).

The three representations shown in Fig. 2 were prepared using UCSF ChimeraX v1.1. ChimeraX is an open-source software for the analysis and visualization of molecular structures, released to the public in 2017 (Goddard et al., 2018; Pettersen et al., 2021). It was developed by the Resource for Biocomputing, Visualization, and Informatics (RBVI) at the University of California, San Francisco (UCSF), as a follow-up of Chimera (Pettersen et al., 2004).

ChimeraX does not have all features present in Chimera, a software of high success and use (the original paper has almost 20,000 citations): the idea behind ChimeraX was to implement Chimera's most highly used tools in order "to meet the increasing demands of today's visualization requirements" (Pettersen et al., 2021). Compared to Chimera, ChimeraX has a virtual reality mode to prepare structures for 3D printing, new tools to work with DICOM medical imaging data, and a new Toolbar that allows the user to easily set lighting modes and styles.

ChimeraX presents a user-friendly all-in-one window interface with panels: the *main window* for molecule visualization and 3D graphics, a *title bar* at the top, the *Toolbar* right below the title bar, a *command line* at the bottom. The user can do all operations and analyses using either the Graphical User Interface (GUI) or the command line. Different release versions of ChimeraX can be downloaded for different operating systems. A detailed User Guide is released with the download, but more dedicated tutorials are available on the website (ChimeraX-website, 2021).

The preparation of Fig. 2A is detailed below.

4.1 System preparation—Panel A

The RBD-ACE2 complex (PDB ID: 6M17) was downloaded from the PDB. We deleted chains A and C and zinc atoms that are not of our interest. The complex was embedded into a 1-palmitoyl-2-oleyl-sn-glycerol-3-phosphocholine (POPC) bilayer of 120×120 Å using VMD Membrane Builder Plugin Tool (Humphrey, Dalke, & Schulten, 1996). The orientation of the RBD-ACE2 structure within the membrane bilayer was obtained from the Orientations of Proteins in Membranes (OPM) database. The resulting PDB file was imported in UCSF ChimeraX (GUI: File > Open). We then set the background color to white (GUI: Toolbar > Home > Background > white) and saved the ChimeraX session (Home > Save > Files of type > .cxs). We recommend to the users to frequently save the Chimera session because only one undo is allowed.

4.2 Protein visualization—Panel A

We selected chains B and D corresponding to ACE2 (GUI: select > Chains > B and select > Chains > D; or command line: select/B/D). Then, we showed the surfaces of the selected chains (GUI: Toolbar > Molecule Display > Surface > Show), and colored them in blue (GUI: Action > Color > All Options… > Show all colors > Dodger Blue). The same procedure was then applied to color chains E and F, corresponding to spike protein RBDs, in green (GUI: Action > Color > All Options… > Show all colors > Lime Green).

4.3 Membrane visualization—Panel A

In our PDB file, the lipid bilayer corresponds to chain CL. Individual elements were selected and colored: phosphorus atoms in yellow (command line: select/CL@P; GUI: Action > Color > Yellow); carbon atoms in white (command line: select/CL@C*; GUI: Action > Color > White), and oxygen atoms in red (command line: select/CL@O*; GUI: Action > Color > Red). To obtain a clear visualization, all hydrogen atoms were hidden (command line: hide/CL@H*).

4.4 Lighting—Panel A

Once we prepared all components, we set the lighting to soft (GUI: Toolbar > Graphics > Lighting & Effects > Soft), and we hid the silhouettes mode that is active by default (GUI: Toolbar > Graphics > Lighting & Effects > Silhouettes).

For a better understanding of the interactions and the complementarity between ACE2 and RBD, we focused our attention on the interface of one of the two RBD present in the crystal structures and calculated the electrostatic and hydrophobic surfaces of ACE2 (Fig. 2B and C).

4.5 Panel B

As described above, we created a new ChimeraX session, imported the original PDB file (GUI: File > Open > 6M17.pdb), and set the background in white. ACE2 protein was selected (command line: select/B/D), and its molecular surface was visualized by the Coulombic Electrostatic Potential (ESP) (GUI: Toolbar > Molecule Display > electrostatic icon). The ESP coloring is red-white-blue over the value range − 10 to 10.

The spike RBD was selected (command line: select/B/D) and depicted as green cartoons (GUI: Action > Color > All Options... > Show all colors > Lime Green). To visualize the RBD residues corresponding to different electrostatic potential zones in ACE2, the RBD charged residues (ARG403, ASP405, ARG408, LYS417, LYS458, GLU471, GLU484) were displayed as sticks in the zoomed area (command line: show/A:403:405:417:458:471:484). We set the lighting as Full (GUI: Toolbar > Graphics > Lighting & Effects > Full) and we adjusted the side-view (GUI: Toolbar > Graphics > Camera > Side view) to get a white background.

In Fig. 2B, we labeled two residues that were found to play a relevant role in the spike-ACE2 interaction: $LYS417_{RBD}$ is involved in a salt bridge with the $ASP30_{ACE2}$ (Lan et al., 2020), whereas $GLU484_{RBD}$ forms an electrostatic interaction with the $LYS31_{ACE2}$ (Wrobel et al., 2020). The ESP visualization implemented in ChimeraX offers an easy and quick way to identify these interactions: $LYS417_{RBD}$ is indeed located on a red surface spot of ACE2 and $GLU484_{RBD}$ points to a blue surface spot of ACE2.

4.6 Panel C

The ChimeraX session was duplicated to prepare the protein representation in panel C with the same orientation and zoomed area as in panel B. We hid all the electrostatic residues (command line: hide). Then, ACE2 was selected (command line: select/B/D), and its molecular surface visualized by the Molecular Lipophilicity Potential (MLP) (Ghose, Viswanadhan, & Wendoloski, 1998) (GUI: Toolbar > Molecule Display > hydrophobic icon). The surface coloring ranges from dark goldenrod, for the most hydrophobic potentials, through white, to dark cyan, for the most hydrophilic. RBD residues VAL445, PHE456, ILE472, VAL483, PHE486, PHE490 were shown as sticks (command line: show/A:445:456:472:483:486:490).

In Fig. 2C, to provide examples of hydrophobic interactions that can be visualized with this visualization modality, we labeled one residue accommodated in a hydrophobic (yellowish surface) subpocket, e.g., $PHE486_{RBD}$ that was also highlighted in previous structural biology studies on the spike-ACE2 interaction (Lan et al., 2020; Wrobel et al., 2020).

5 Protein complexes in motion

Proteins are flexible entities, local adjustments such as small side-chains repositioning, or more global movements such as domain rearrangements affect and sometimes are induced by the interactions with other proteins. Molecular Dynamics (MD) simulations can be used to investigate at an atomistic level the interaction between protein partners in motion. Considering the number of atoms involved, the simulation time is higher than that required for small molecule-protein complexes; however,

MD simulations are successfully used for assessing the stability of the interactions identified by docking or experimental determination, and the conformational rearrangements associated with the PPI (Pan et al., 2019; Perricone et al., 2018; Rakers et al., 2015). Moreover, this technique allows to identify transient pockets on the PPI surface or to define the conformational transition of disordered proteins that fold in contact with their partner protein (Rakers et al., 2015).

Combined structural biology and MD studies revealed the high flexibility of the spike protein (Casalino, Dommer, et al., 2020; Casalino, Gaieb, et al., 2020; Turonova et al., 2020), and numerous MD studies are now directed to investigate the interaction of spike with ACE2 (Barros et al., 2021; Mugnai, Templeton, Elber, & Thirumalai, 2020; Raghuvamsi et al., 2021). The S protein was simulated within a complete viral envelope (305 million atoms) (Casalino, Dommer, et al., 2020). The computational effort was awarded the Special Prize of the ACM Gordon Bell Prize 2020 for High Performance Computing-Based COVID-19 Research. Most simulations are made available to the scientific community (Amaro & Mulholland, 2020; Merz et al., 2020). We downloaded the MD trajectory of the spike RBD-ACE2 complex from the COVID-19 Data Sets of the Amaro Lab (University of California, San Diego) (Amaro & Mulholland, 2020) to prepare Fig. 3 and therefore provide an example of the information that can be retrieved and visualized from an MD trajectory. Specifically, we show the occupancy of hydrogen bonds during the simulations, as a measure of the stability of these contacts, using Visual Molecular Dynamics (VMD) v1.9.3.

VMD is a molecular visualization open-source program for the modeling, analysis, animation, and visualization of biological systems. It was developed by Klaus Schulten (University of Illinois) in 1996 (Humphrey et al., 1996) and, since then, the VMD original paper was cited more than 28,000 times. VMD can be downloaded from the UIUC website, where also dedicated tutorials are available. VMD is principally used to analyze and visualize MD trajectories and is also integrated with NAMD (Nanoscale Molecular Dynamics) for MD simulations (Phillips et al., 2020). VMD provides numerous functionalities and can be launched from the GUI (from the VMD Main) or the TK console (GUI: Extension > Tk console). The main window, where molecules are displayed, is called VMD OpenGL Display. Visualization settings are manipulated by four main actions accessible via GUI: Graphics > Representation: Selection, Drawing Method (by which the representation style is chosen), Coloring Method, and Material (to set the effects of lighting, shading, and transparency).

The detailed protocol used to prepare Fig. 3 is described below.

5.1 System preparation and display settings

The topology PSF file (GUI: File > New Molecule... > Browse... > ace2_rbd_prot_glyc_memb_amarolab.psf > Load) and the trajectory file with a stride of 10 (GUI: File > Load data into Molecule... > Browse... > ace2_rbd_prot_glyc_memb_amarolab_1.dcd > Load) were loaded. During this process, the molecular system was hidden (GUI: IDo > D) to speed up the loading time. All MD frames were aligned to the Cα

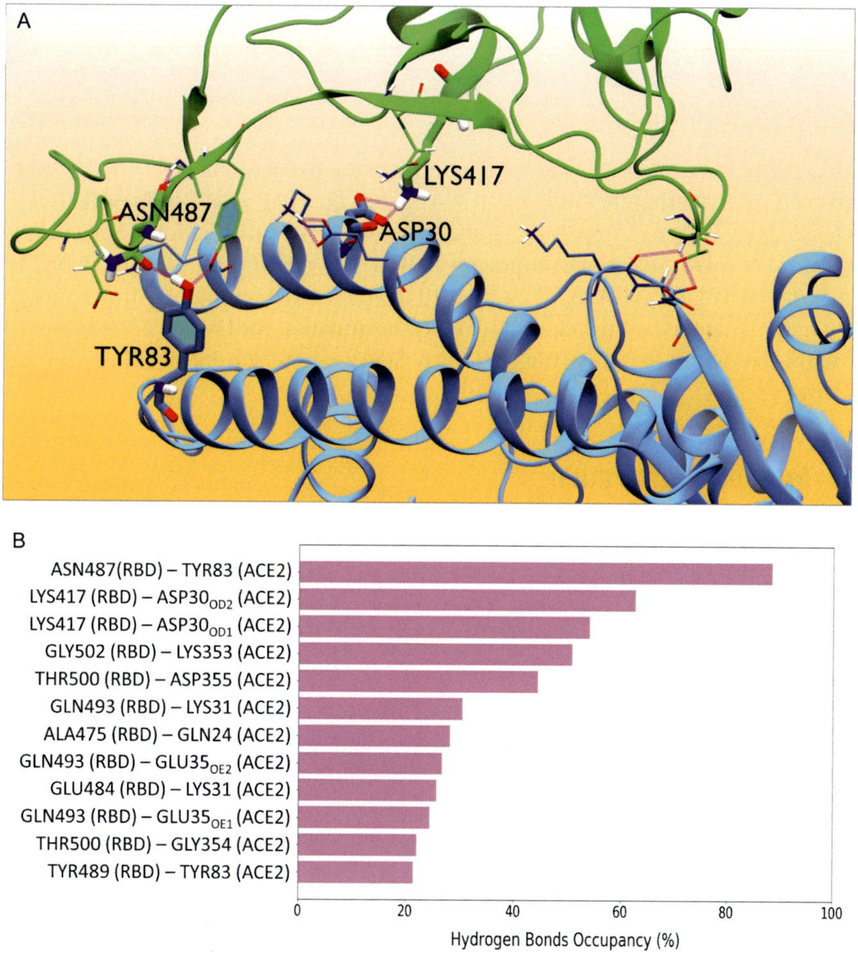

FIG. 3

Hydrogen bond network between the spike protein RBD and the ACE2 receptor. (A) The protein-protein interface obtained in the MD frame number 738 is displayed. RBD and ACE2 are represented as blue and green cartoons, respectively. Residues involved in H-bonds with occupancy between 20% and 60% are represented as licorice/bond radius 0.1, whereas residues involved in H-bonds with occupancy higher than 60% are represented as licorice/bond radius 0.3. H-bonds are shown as magenta dashed lines. (B) Bar plots of hydrogen bonds between the spike protein RBD and the ACE2 receptor with occupancy higher than 20%.

atoms of the first frame (GUI: Extensions > Analysis > RMSD Trajectory Tool > selection "protein and name CA"> ALIGN). We hid the axes (Display > Axes > off), set the background to white (GUI: Graphics > Colors > Display > Background > 8-white), the projection mode to orthographic (GUI: Display > Orthographic) and the render mode to GLSL (GUI: Display > Rendermode > GLSL) that allows the shading capability for higher quality molecular graphics/rendering.

5.2 H-bond analysis

HBonds Plugin, version 1.2, was used to monitor the hydrogen bonds between spike RBD and ACE2 during the simulation. We opened the Hydrogen Bonds panel (GUI: Extensions > Analysis > Hydrogen Bonds), set two different selections (Selection 1: segname ACE2A; Selection2: segname RBDB), modified the cut-off distance to 3.5 Å and the angle D-H-A of 70°, activated the options to compute hydrogen bonds only for polar atoms, set to calculate detailed information for *Unique hbond*, and set to write the output to a text file. We then launched the job by clicking on *Find hydrogen bonds*.

5.3 Protein visualization

The two proteins were represented as cartoons with a thickness of 0.10 and resolution of 30 (GUI: Graphics > Representation > Graphical Representations window > Drawing Method > New Cartoon). ACE2 was colored in blue (segname ACEA, GUI: Graphics > Representation > Coloring Method: ColorID24, red 0.12, green 0.56, blue 1.0) and spike RBD in green (segname RBDB, GUI: Graphics > Representation > Coloring Method: ColorID19, red 0.19, green 0.8, blue 0.19). Residues involved in H-bond interactions with occupancy higher than 20% were represented with the Licorice Drawing method (GUI: Graphics > Representation > Graphical Representations window > Drawing Method > Licorice): we set a bond radius of 0.3 (GUI: Graphics > Representation > Graphical Representations window > Drawing Method > Bond Radius) for residues involved in hydrogen bonds with occupancy higher than 60% for both ACE2 (segname ACEA and resid and not (name HE1 HE2 HB1 HB2 HB3 HA HG2 HD3 HG1)) and RBD (segname RBDB and resid 417 487 and not (name HE1 HE2 HD1 HD2 HB1 HB2 HB3 HA HG2 HD3 HG1)), and we set a bond radius of 0.1 for residues involved in hydrogen bonds with occupancy between 20 and 60% for both RBD (segname RBDB and resid 500 502 493 484 489 475 and not (name HE1 HE2 HD1 HD2 HB1 HB2 HB3 HA HG2 HD3 HG23 HG22 HG21 HB)) and ACE2 (segname ACEA and resid 355 353 31 35 24 354 and not (name HE1 HE2 HD1 HD2 HB1 HB2 HB3 HA HG2 HD3 HG1)). We selected the aromatic residues $TYR89_{ACE2}$ and $TYR489_{RBD}$ (Selection: (segname ACEA and resid 89) and (segname RBDB and resid 489)) and used the PaperChain Drawing method (GUI: Graphics > Representation > Graphical Representations window > Drawing Method) to represent the internal rings as *mirror*.

5.4 H-bond visualization

We selected all residues with a total occupancy higher than 20% (Selection: (segname ACEA and resid 355 31 35 30 83 353 24 354 and not name "C.*") or (segname RBDB and resid 417 487 500 493 484 489 502 475 and not name "C.*")) and visualized the H-bonds (GUI: Graphics > Representation > Graphical Representations window > Drawing Method > Hbond). We changed the distance to 3.5 Å and the angle to 70°, and line thickness to 13. We colored the lines in magenta (GUI: Graphics > Colors > Categories: Element, Name: X, Colors: 23 Magenta).

5.5 View and rendering settings

AOChalky was used as a material for all elements with the exception of the aromatic rings, for which a new material (GUI: Graphics > Material... > Create New) was set. We used the following parameters: Ambient 0.70, Diffuse 0.44, Specular 0.88, Shiness 0.81, Mirror 0.35, Opacity 0.51 Outline and OutlineWidth to max. We activated Lights 0, 1, and 2 (GUI: Display) and enabled the ambient occlusion lighting (GUI: Display > Display Settings > Amb. Occl. > On). The Cue Mode was set to Linear (GUI: Display > Display Settings > Cue Mode > Linear) to have the fog effect and the final image was saved using Tachyon (GUI: File > Render... > Render the current scene using: > Tachyon).

5.6 Movie

Using VMD Movie Generator (GUI: Extension > Visualization > Movie Maker), we also created a movie of the MD simulation starting from frame 0 to 920. We set *Trajectory* in the Movies Setting (GUI: Extension > Visualization > Movie Maker > Movies Setting), wrote the video name and indicated the output format as .jpg (GUI: VMD Movie Generator > Name of Movie) with a trajectory step size of 1 and we created the video by clicking on *Make Movie*. The video can be found on the following website: https://github.com/dipizio/spike_rbd. It shows how the residues reported in Fig. 3 move and create a dynamic network of H-bond interactions.

The H-bond between $ASN487_{RBD}$ and $TYR83_{ACE2}$ (thick sticks in Fig. 3A) is maintained over all the trajectory with an occupancy of 88.39% (Fig. 3B), the H-bond between $LYS417_{RBD}$ and $ASP30_{ACE2}$ is also very stable, with occupancy of 62.60% (Fig. 3B). As oppositely charged residues, LYS417 and ASP30 are involved in a salt bridge interaction that was identified also in Fig. 2B. Indeed, $LYS417_{RBD}$ has been suggested as responsible for the increased binding affinity of the SARS-CoV-2 compared to the SARS-CoV whose spike protein has a valine in this position (Lan et al., 2020).

6 Photorealistic representations of protein complexes

As discussed above, proteins are live entities, moving and interacting within cellular compartments. To provide an example of the tools and techniques that can be used to visualize PPIs in their biological environment, we prepared a representation of the interaction between SARS-CoV-2 and ACE2 embedded in their membranes and surrounded by other proteins (Fig. 4). We adopted a photorealistic style, taking care of correctly sizing the molecular components and using light settings to emphasize the

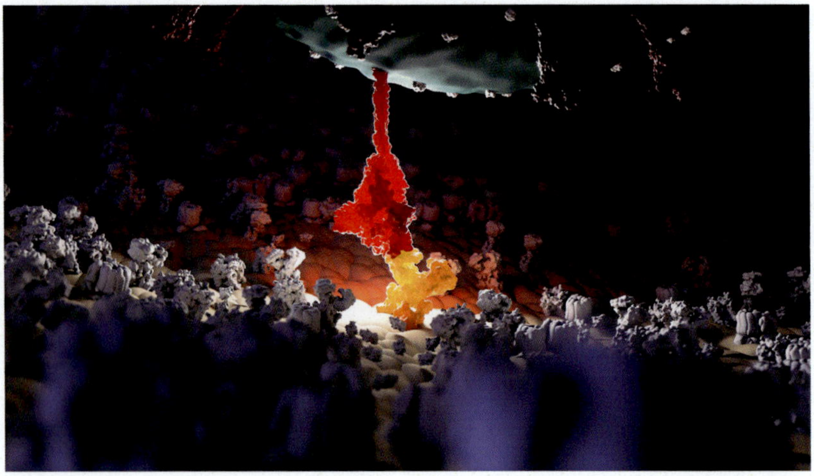

FIG. 4

Interaction between Sars-CoV-2 and ACE2 receptor. The interacting proteins, spike RBD (in red) and ACE2 (in yellow), are represented as inserted in the membrane of the viral capsid and the endosome, respectively, and are surrounded by membrane proteins (in gray).

binding event. This provides a visual representation that can facilitate the description of the binding process to both the scientific and the general public.

The first step to prepare the figure was to identify the ideal size of the figure components, such as the diameter of the viral capsid and the length of the spike protein. In the cryo-EM structure of the SARS-CoV (Neuman et al., 2006), the diameter of the viral particle ranges from 82 to 94 nm, excluding spike proteins. Based on this evidence, we used Rhinoceros (McNeel, USA), a 3D CAD software, to model a spherical viral particle with a diameter of 88 nm. The length of the spike protein was measured on the electron microscopy images reported in the same paper after properly resizing them based on the 88 nm of virion capsid measurement: the longitudinal size was established to be about 12.40 nm. The Image scaling was then processed in Rhinoceros software (Rhino3d, 2021) using *Scale in all dimension* function.

To model the SARS-CoV-2, we used the structure of SARS-CoV-2 spike glycoprotein containing the two RBD domains erect (PDB ID: 7A93) (Benton et al., 2020). The tail and the transmembrane domain of the S protein are not present in this structure and were retrieved from the three-dimensional model available in the Amaro Lab (Casalino, Gaieb, et al., 2020) and structurally aligned to the 7A93.pdb with MatchMaker implemented in UCSF Chimera (Pettersen et al., 2004) (Tools > Structure Comparison > MatchMaker, 7A93.pdb was used as the *Reference Structure* while the spike protein from the Amaro Lab as the *Structure to Match*). Although the spike's top and tail 3D structures are bound to some approximation, in the final representation, they appear as a continuity.

To represent the ACE2 receptor, we used the three-dimensional structure of the full-length human ACE2 in complex with the spike RBD (PDB ID: 6M17).

This structure was superimposed to 7A93.pdb with MatchMaker. We then used Chimera (Pettersen et al., 2004) to export the complex components we are interested in: we loaded the prepared ACE2 structure and the S protein PDB file (File> Open>7A93.pdb; File> Open> ace2.pdb); similarly, we opened the prepared structural file of the spike with tail, and we selected the visualization on the S protein (Select> Chain A, B, C; Select> Invert; Action> Atom/Bonds> Hide); then, the structures were saved as individual PDB files (for the spike: Select > Chain A, B, C > File > Salve PDB, taking care to check the *Save selected atoms only* checkbox; for ACE2: Select> Chain A, B, C, D > File > Salve PDB, taking care to check the *Save selected atoms only* checkbox).

To render the surfaces of the three protein structures, the structural files were imported into the Autodesk Maya software (Autodesk, USA), using the free mMaya plugin (2021) (https://clarafi.com/tools/mmaya/), and the option *PDB Import*. In the Atomic tab, we deselected the *Atomic* checkbox, and in the *Surface* tab we selected the *Mesh Vis* checkbox and set the preset of mesh quality to high. The higher the definition, the higher the number of polygons that make up the mesh and, therefore, the detail of the obtained surface. We then saved the generated objects (File> Save Selected, we choose "OBJ" as the export format for each of the 3D models that we were going to insert into the final scene rendering). This procedure was then applied to the additional protein structures we added in the figure: the membrane proteins E Protein (PDB ID: 5X29) and ORF3a (PDB ID: 6XDC) reported in the SARS-CoV-2 viral capsid, and the transient receptor potential mucolipin 1 (TRPML1, PDB ID: 5WJ5) and the lysosome-associated membrane protein 1 (LAMP-1, PDB ID: 5GV0) inserted in the endosomal surface. The choice of the proteins was functional only to the illustration.

The OBJs models were loaded in Rhinoceros (File> Import). We then created an 88 nm sphere to represent the viral capsid and a second upward-facing concaveness sphere to represent the endosome's inner surface (Solid> Sphere). The distribution of membrane proteins on the viral capsid's surface and on the endoplasmic membrane was randomized. The proteins were distributed in space through the drawing of orientation curves directly on the sphere representing the virus and on the hemisphere representing the endosome (Curve> Polyline> Interpolate curve on the surface).

Once we distributed all protein structures on the scene we assigned the rendering materials to the molecular components.

To realize the final rendering of the figure, we used Rhinoceros software with the *V-Ray for Rhino* plugin (ChaosGroup, Bulgaria). We assigned materials to all components of the figure (V-Ray> Asset Editor> Materials): we assigned a neutral white color for all proteins except for spike and ACE2, which were colored in red and yellow, respectively. A displacement map was used for the endosome membrane in order to recreate the appearance of the phospholipid layer (V-ray> Asset Editor> Displacement). A slight displacement map was used also for the surface of the SARS-CoV-2 capsid. To emphasize the focus on the spike-ACE2 interaction, one light was placed behind the two protein structures and directed toward the viewer's eye (V-Ray> V-Ray Lights> Rectangle Light). This effect allowed us to create a "luminous contour" on the two proteins' surface. A focus was established for the camera on the spike-ACE2 binding interface (V-Ray> Asset Editor> Settings> Camera> Depth of Field). We, then, launched the rendering of the final image (V-Ray> Asset Editor> V-Ray Rendering> Rendering).

7 Protein-protein interactions: hot spots and small molecule design

Protein-protein interfaces generally cover an area ranging from 1200 to 2000 Å^2, which is larger than small molecule binding sites of 100–600 Å^2. Many PPIs are not contiguous, with several patches present on the surface. However, not all residues involved in the interaction contribute equally to the binding. As it was demonstrated for the first time by Clarkson and Wells, it is possible to recognize within this large interaction surface the presence of key residues contributing the most to the binding free energy (Clackson & Wells, 1995). These "hot spot" residues are usually in a central position (core) of the interacting surfaces, separated from the bulk solvent and are surrounded by the first shell of residues (rim region) enclosed, in turn, by another shell of residues called support region (Chakrabarti & Janin, 2002). Computational methods are being developed to predict hot spots in protein complexes considering evolutionary and energetic parameters, also starting from sequence information (Marchetti, Capelli, Rizzato, Laio, & Colombo, 2019; Moreira et al., 2017; Preto & Moreira, 2020). Indeed, it has been demonstrated that the hot spot composition relies on defined residue properties (Yan, Wu, Jernigan, Dobbs, & Honavar, 2008). Residues composing the core are generally hydrophobic/aromatic (Moreira, Martins, Ramos, Fernandes, & Ramos, 2013); the evolutionarily conserved tryptophan, tyrosine and arginine are the most common residues in the hot spot regions (Hu, Li, Chen, & Zhang, 2016; Moreira, Fernandes, & Ramos, 2007). The rim region is represented by more polar residues; while the support region is composed of buried hydrophobic residues (Levy, 2010).

An example of PPIs where hot spot residues were clearly identified, is represented by the complex of the oncoprotein Mouse double minute 2 (MDM2) bound to the N-terminal transactivation domain of p53, shown in Fig. 5. The tumor suppressor p53 is also known as the "guardian of the genome" and is expressed by cells under stress conditions (Haupt, 2004). p53 results mutated in 50% of human tumors, in the remaining 50%, its activity is frequently impaired by overexpression of negative regulators, such as MDM2 (Hainaut & Hollstein, 2000). The binding with this oncoprotein causes p53 translocation outside the nucleus, blocks the transactivation, and favors its degradation (Wade, Wang, & Wahl, 2010). From a structural point of view, p53 binds the hydrophobic cleft of MDM2 with the N-terminal transactivation domain, representing an example of an intrinsically disordered protein that gets organized in a secondary structure upon interaction with a partner protein: p53, in fact, arranges in an amphipathic α-helix inserting its apolar residues in the hydrophobic cleft of MDM2 (Demir, Barros, Offutt, Rosenfeld, & Amaro, 2021; Kussie et al., 1996).

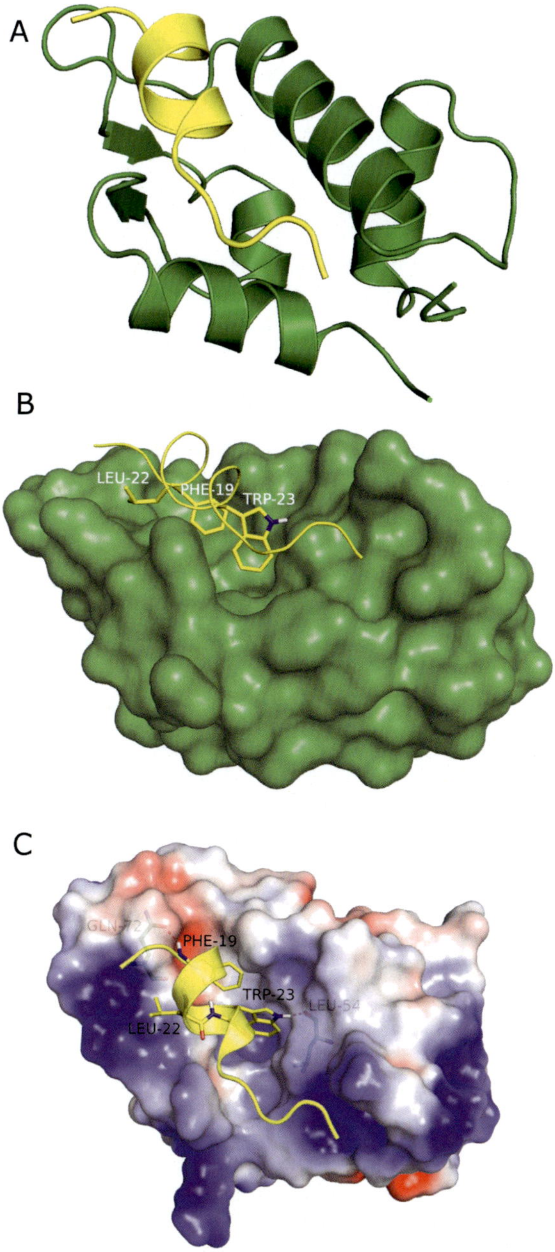

FIG. 5

p53-MDM2 complex. (A) p53 and MDM2 are represented as yellow and dark green cartoons, respectively. (B) To highlight the p53 occupancy of the hydrophobic cleft of MDM2, the complex has been rotated; p53 is depicted as yellow ribbons and hot spot residues are depicted as sticks and are labeled. MDM2 is represented as a dark green solid surface. (C) p53 is represented as yellow cartoons with hot spot residues as labeled sticks, while the MDM2 surface is colored according to the APBS calculated electrostatic potential. H-bonds are depicted as red dashed lines and interacting residues are shown in sticks.

Studies highlighted the role of p53 key residues (PHE19, LEU22 and TRP23) bearing most of the binding energy (Kussie et al., 1996). This evidence paved the way to the identification of small molecule binders of MDM2 that, mimicking the key residues, are able to hinder the PPI restoring p53 oncosuppressor activity. The first example of interfering small molecules is represented by nutlin (Vassilev et al., 2004), but, since then, a large number of compounds have been identified, reaching clinical phases (Beloglazkina, Zyk, Majouga, & Beloglazkina, 2020; Sun et al., 2014). Almost all ligands binding MDM2 have a definite shape that overlaps to p53 key residues, are quite big and hydrophobic. An analysis of their chemical space pinpoints these properties as common among most small molecular binders of PPIs (Sperandio, Reynès, Camproux, & Villoutreix, 2010).

The structure of the p53-MDM2 complex (PDB ID: 1YCR (Kussie et al., 1996)) is reported in different representation styles in Fig. 5. To prepare the panels we used PyMOL, v. 1.8.4.0. PyMOL is a Python-based molecular visualization tool, developed as open-source software by Warren Lyford DeLano in 2000 and released initially by DeLano Scientific LLC. A literature analysis carried out in Elsevier's Scopus using the query "pymol AND molecular AND graphics AND system", retrieved more than 20,000 items in 20 years (2001–21). In 2010, Schrödinger Inc., after the acquisition of PyMOL, has taken over the development and maintenance, including all current subscriptions. The requested license fee is used to support the Open source project in line with DeLano's philosophy, but free educational licenses are available for teachers and students, and most of the source code is available through the Schrödinger GitHub repository (https://github.com/schrodinger/pymol-open-source). As open-source software, users can develop or modify various functions to meet specific requirements. One of the most used plugins was developed to solve the Poisson-Boltzmann equation with the Adaptive Poisson Boltzmann Solver (APBS electrostatic Plugin, https://github.com/Electrostatics/apbs-pdb2pqr). Useful online tutorials, frequently updated, can be found at the web site http://www.pymolwiki.org.

When PyMOL is launched, the graphic window (viewer) appears together with the internal GUI at its right side, whereas the external GUI, at the top, appears as a separate window. In the recent versions of PyMOL, the external GUI and the viewer have been incorporated. A command-line interface is associated with both the external GUI and the viewer. Most of the visualization functions can be accessed via the menu bar of the internal GUI. In particular, it has five tabs: "action (A)" to change orientation, apply presets, rename, duplicate, remove, view sequence, add/delete hydrogen, perform computations, etc…; "show (S)", to show a unique style of the structure as lines, sticks, ribbon, cartoon, mesh, surface, etc…; "hide (H)" to hide any representation that has been previously selected; "label (L)" to display information related to selected atoms, residues, chain, or segment; "color (C)" to change coloring methods of single elements, chains, or ss, with a large choice of applicable color and color scheme. These functions can be activated for each entry or applied to all objects when activated directly from the menu bar.

Necessary steps to generate Fig. 5 are described below.

7.1 Structure visualization—Panel A

The p53-MDM2 structural file was opened through the external GUI (external GUI: File> Open> 1YCR.pdb), but it could have been alternatively retrieved directly from the PDB (command line: fetch 1YCR, type=pdb). The entry is listed in the menu bar of the internal GUI as a new object loaded into PyMOL. Our entry contains chain A (MDM2) and chain B (p53). To distinguish the protein subunits, we represented the structures as cartoons and hid the lines (internal GUI: S> as> cartoon; H> lines), we then separated the chains, and assigned them appropriate names (left click of the mouse on a residue of chain A; on the "sele" menu bar in the internal GUI: A> modify> complete> chains). The *sele* menu bar must be re-named because the program will overwrite successive selections (internal GUI: A> extract object; A> rename object/MDM2). At the non-extracted chain B, we assigned the name *p53* (internal GUI: "1YCR"/A> rename object> p53). Another possible approach to select residues from different chains is visualizing the one-letter amino acids sequence by pressing the button "S" (sequence) at the bottom of the Internal GUI, then clicking on one residue of the sequence and proceeding as already explained to complete chain, extract object and rename the new object. The same operations can be carried out by command line (fetch 1YCR, type=pdb| show cartoon| hide lines| get_chains| extract MDM2, chain A| set_name 1ycr, p53). We assigned the forest color to MDM2 and the yellow to p53 (internal GUI: "MDM2"> C> forest; "p53"> C> yellow). The cartoon representation was improved as ribbons with tubular edges (external GUI: Setting > Cartoon> Fancy Helices), and the transparency of cartoon was set to 20% (external GUI: Setting> Transparency> Cartoon> 20%). By default, the program produces shadows during the ray rendering, therefore, we eliminated the shadow option (external GUI > Setting > Rendering > Shadows > None) and then rendered the figure (external GUI: Display> Quality> Maximum quality; external GUI: ray). The representation in the viewer was then saved as an image file (external GUI: File> save images as > choose the format, select the directory, and type the name of the file> save). All operations carried out by GUI can be performed by command line: color forest, MDM2| color yellow, p53| set cartoon_fancy_helices, 1| set cartoon_transparency, 0.2| set ray_shadow, off| ray| save filename.png.

7.2 Surface and interface visualization—Panel B

To visualize the details of inter-molecular packing in the complex, we calculated the molecular surface of MDM2 and left p53 in the cartoon representation (internal GUI: "MDM2"> S> surface; H> cartoon). In the MDM2 surface, a cleft appears, and the p53 loop is clearly located inside this cleft. The hot spot residues of p53 (PHE19, LEU22, TRP23) were then selected in the protein sequence in the display area (movie control: S), re-named (internal GUI: "sele" menu bar> A> rename object> p53_hs), showed as sticks (internal GUI: "p53_hs"> S> as> stick), and labeled with their residue name (internal GUI: "p53_hs"> L> residues). To better visualize the occupancy of p53 in the MDM2 cleft, the non-polar hydrogens were hidden (internal GUI: "p53_hs"> H> hydrogens> non-polar) and the cartoon representation was changed to ribbon (internal GUI: "p53"> H> cartoon; S> ribbon). The backbone of the selected residues is not displayed in the ribbon representation (external GUI: Setting> Ribbon> Side Chain Helper). This setting could be used to visualize the disposition of the side chain, but it could be avoided for the H-bond interactions. (command line: show surface, MDM2|show lines, p53|sele p53_hs, resi 19+22 +23| hide lines, p53|show ribbon, p53|show sticks, p53_hs|set ribbon_side_chain_helper, on|label p53_hs).

7.3 Polar contacts analysis—Panel C

To visualize the polar contacts between p53 and MDM2, we used the command "find polar contacts" (internal GUI: p53 > A > find > polar contacts > to any atoms) and a group of dashed lines appeared in yellow to show the contacts between all the atoms in the viewer while a "p53_polar_co" menu bar appeared in the internal GUI. To better visualize the contacts, proteins should be visualized as lines, including the backbone (external GUI: Setting> Ribbon> Side Chain Helper; internal GUI: "p53"> H> ribbon; S> lines; internal GUI: "MDM2"> H> surface; S> lines). After examining all polar contacts in the graphic window, two H-bonds between p53 and MDM2 residues were identified: one between PHE19$_{p53}$ and GLN72$_{MDM2}$, and the other between TRP23$_{p53}$ and LEU54$_{MDM2}$. The residues of p53 and MDM2 were separately selected, renamed as "p53_pc" and "MDM2_pc," showed in sticks, colored by elements, and labeled with their residue name (internal GUI: "sele" bar> A> rename selection> p53_pc (or MDM2_pc); "p53_pc" or "MDM2_pc"> S> as> stick; C> by element> choose the first color mode to make carbon the same color as the chain; L> residues). The label font, size, and color can be easily changed by the external GUI > Setting > Edit all > label_font_id, label_size, or label_color. The *Edit all* command allows to modify and customize all PyMOL settings. The labels were moved in the viewer activating the *3-Botton Editing* in Mouse Mode in the internal GUI and using the left click in the mouse plus control press in the keyboard. The labels in the inner part of the surface are not shown. The H-bonds and the distance measurements can be easily visualized in the graphic viewer as dash yellow lines and label by the selection of the two residue atoms, between PHE19$_{p53}$ and GLN72$_{MDM2}$, and between TRP23$_{p53}$ and LEU54$_{MDM2}$ (external GUI: Wizard> Measurement; in "measurement menu" in the internal GUI> distances; select the atoms; internal GUI: distances). Once the residues were identified, the line-style representation of proteins was eliminated, MDM2 was visualized as surface and p53 as cartoon (internal GUI: "MDM2"> H> lines; S> Surface; internal GUI: "p53"> H> lines; S> cartoon). The transparency of the surface was set to 20% edges (external GUI: Setting> Transparency> Surface> 20%). In this way, the H-bonds and the residues can be clearly identified. Same operations from command line: hide surface, MDM2| show lines, MDM2| hide ribbon, p53| show lines, p53| set cartoon_side_chain_helper, off| dist polar_contacts, p53, MDM2, mode = 2| sele p53_pc, resi 19+23| hide label| hide dashes| dist H_bond, p53_pc, MDM2, mode = 2| sele MDM2_pc, resi 54+72| show surface, MDM2| set transparency, 0.2| hide lines| show sticks, MDM2_pc| show sticks, p53_pc| label H_p53| label MDM2_pc.

In panel C, the surface of MDM2 was rendered as an electrostatic potential molecular surface, using the APBS electrostatic Plugin (external GUI: Plugin> APBS Tool; tab main> use PyMOL generated PQR and PyMOL generated Hydrogens and termini > set grid > Run APBS). In the "visualization tab," the solvent-accessible surface was removed and molecular surface colored by potential in the range of −2/+2 (tab visualization> uncheck solvent accessible surface> check color by potential on sol. Acc. surf.> Low −2/High +2 > Update). The MDM2 was colored according to the electrostatic potential, from red in the region containing an excess of negative charges to blue when the surface is positively charged, and in white the regions with a fairly neutral potential. In the menu bar of the internal GUI, more entries are included (apbs-map, iso_pos_0_1, e_lvl_0_1) that allow to visualize and modify the energy range of the electrostatic surface, show the positive and negative contour fields at an established value, or visualize the legend.

8 Selectivity of protein interactions

As revealed by interactome maps, one protein can interact with diverse partners. The analysis of the biomolecular interactions established between the interacting proteins is determinant to identify the interface residues that play a role in driving the selectivity toward specific protein partners. Specifically, residues that upon point-mutation cause a decrease of the affinity of the protein to one partner and an increase of affinity to a different partner are defined selectivity-switch residues (Arafeh et al., 2019; Shirian, Sharabi, & Shifman, 2016), but often a cluster of residues is responsible for the selectivity of the protein to one specific partner, and we refer to them as correlated-selectivity residues (Israeli, Asli, Avital-Shacham, & Kosloff, 2019; Pazos, Helmer-Citterich, Ausiello, & Valencia, 1997).

G protein-coupled receptors (GPCRs), with 800 members, are the largest family in the human genome (Fredriksson, Lagerstrom, Lundin, & Schioth, 2003). They mediate the communication between the cell and the extracellular environment and are therefore involved in almost all physiological functions. GPCRs can mediate cellular signaling through G protein-dependent (Bourne, Sanders, & McCormick, 1990; Gilman, 1987) and G protein-independent pathways, involving, for example, arrestins (Chutkow et al., 2010; van Gastel et al., 2018). The design of "biased" drugs that cause GPCRs to prefer one signaling pathway over the others could lead to more effective and safer treatments for a wide range of diseases (Omieczynski et al., 2019; Violin, Crombie, Soergel, & Lark, 2014). However, GPCR biased agonism can involve also more than one downstream effector (Dwivedi, Baidya, & Shukla, 2018). Computational and experimental structural analyses are attempting to unravel this complex protein interaction system (Chaturvedi, Maharana, & Shukla, 2020; Lally, Bauer, Selent, & Sommer, 2017; Latorraca et al., 2018; Marti-Solano, Sanz, Pastor, & Selent, 2014; Preto et al., 2020). The advancements in cryo-EM technologies are strongly contributing to reveal structural details of the interaction of different GPCR subtypes with downstream partners (Garcia-Nafria & Tate, 2019, 2020; Safdari, Pandey, Shukla, & Dutta, 2018).

In Fig. 6, we provide an example of a GPCR structure, the Neurotensin receptor 1 (NTR1), that was experimentally solved in complex with two of its most frequent interaction partners, G_{i1} protein (PDB ID: 6OS9) (Kato et al., 2019) and β-arrestin 1 (PDB ID: 6UP7) (Huang et al., 2020). NTSR1 is a promiscuous GPCR, which couples to G_s, $G_{q/11}$, $G_{i/o}$ and $G_{12/13}$. Both active- and inactive-state structures of NTSR1 were determined by crystallography (Deluigi et al., 2021; Egloff et al., 2014; Krumm, White, Shah, & Grisshammer, 2015; White et al., 2012) and the structures of the NTSR1 in complex with G_i and β-arrestin by cryo-EM (Kato et al., 2019; Zhang et al., 2021).

NTR1 is a class A GPCR that responds to neurotensin (NT), a 13-residue peptide, and, in addition to other applications, is suggested as a valid target for treating meth abuse (Frankel et al., 2011; Griebel & Holsboer, 2012). NTR1 agonists β-arrestin biased were found to reduce stimulant effects without the side effects of unbiased NTSR1 agonists (Barak et al., 2016; Slosky et al., 2020).

To visualize NTR1 residues selectively involved in G_i and β-arrestin binding, we used the Schrödinger Maestro visualizer via the BioLuminate Suite (Schrödinger Release 2020-4: BioLuminate, Schrödinger, LLC, New York, NY, 2020).

Schrödinger's BioLuminate is a comprehensive user interface that is designed to specifically address the key questions associated with the molecular design of biologics. Building on a solid foundation of comprehensive protein modeling tools, BioLuminate provides access to additional advanced tools for protein (and antibody) modeling, protein engineering, and more advanced simulations. To prepare Fig. 6, we used the tools for the analysis of protein-protein interfaces (Protein Interaction Analysis) and visualization tools from the Bioluminate GUI (Maestro).

Through Maestro (File > Import structures) we opened the two structures: NTR1-G_i (PDB ID: 6OS9) and NTR1-β-arrestin 1 (PDB ID: 6UP7). In order to have the same view for both complexes, we superimposed the two structures (Tasks > Protein Structure Alignment) and saved our preferred view for the figure (View > Save Camera View). We then set the visualization options; Maestro offers a multitude of options for visualization customization.

8.1 Full-size view representation

The cartoon style was chosen and the color was set to constant using the *style* toolbar (Style > Ribbon > Cartoon/Single Color). The color of NTR1 was set to light gray, the β-arrestin to green and the G protein α, β and γ subunits to pink, orange and dark blue, respectively. Furthermore, the interior helix was colored according to the exterior (Edit > Preferences > Molecular Representation > Ribbons > Helix interior:

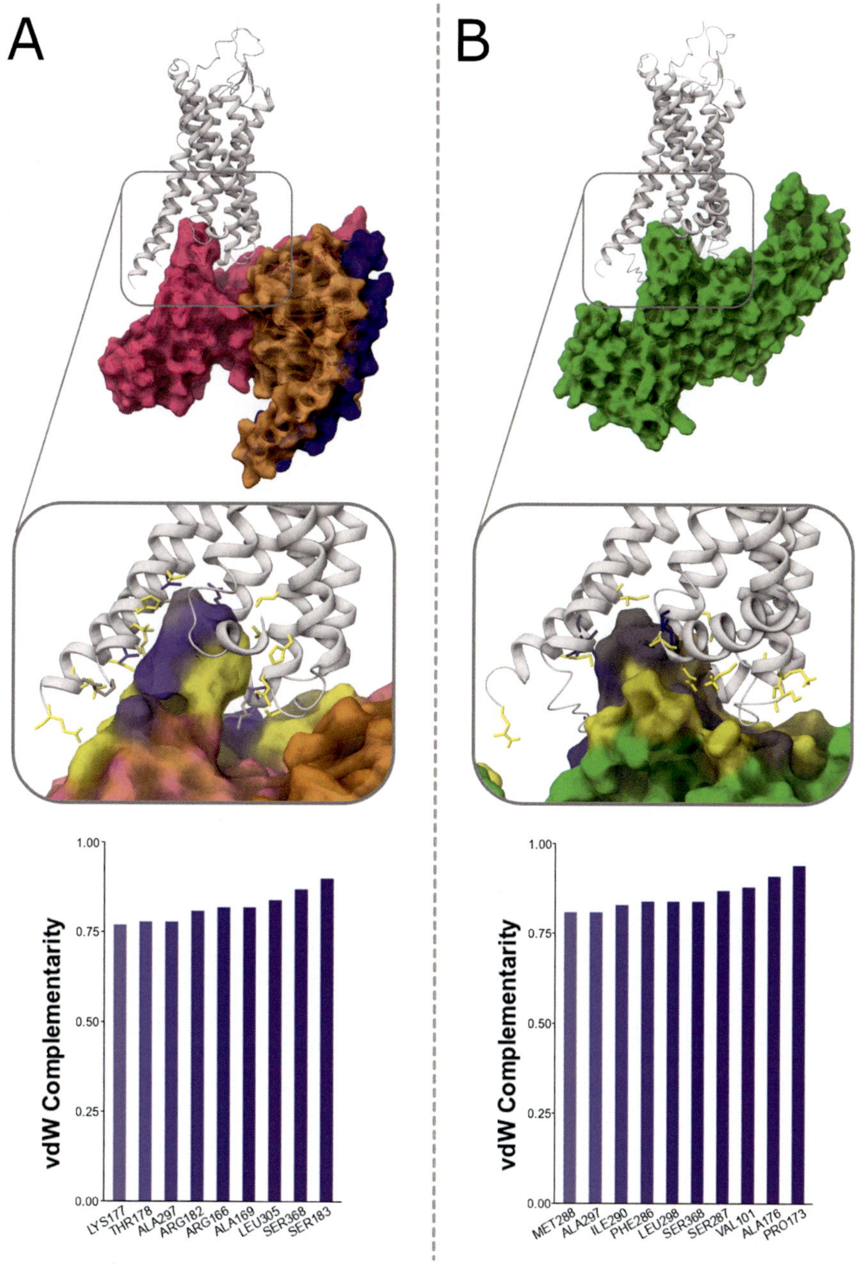

FIG. 6

NTR1 in complex with Gi protein (A) and β-arrestin (B). (A) In the full-size representation, NTR1 is represented as gray cartoons, and G_i protein α, β and γ subunits as pink, orange and dark blue surfaces, respectively. In the zoomed view, we represented the interacting residues of NTR1 as sticks and the interacting surface of G_i protein. (B) In the full-size representation, NTR1 is represented as gray cartoons, and the β-arrestin as orange surface, respectively. In the zoomed view, we report the interacting residues of NTR1 as sticks and the interacting surface of β-arrestin. In both panels, the coloring of the zoomed views follows the vdW complementarity, from yellow to blue. The histograms on the bottom show the NTR1 residues with high vdW complementarity with Gi protein (LYS177, THR178, ALA297, ARG182, ARG166, ALA169, LEU305, SER368, SER183) and β-arrestin (MET288, ALA297, ILE290, PHE286, LEU298, SER368, SER287, VAL101, ALA176, PRO173).

same as exterior), and cartoon width and thickness were set to 0.35 and 0.60, respectively (Edit> Preferences> Molecular Representation> Ribbons: Default style: Cartoon, Ribbon width: 0.35, Ribbon thickness: 0.60). Using the Quick Surface tool implemented in BioLuminate, we generated the molecular surfaces of G_i protein and β-arrestin (Tasks> Quick Surface). We hid the cartoon and colored the surfaces according to the atom colors (Surface> Manage> Display Options> Color Scheme> Other Property> Atom Color) and applied a transparency of 30% for the front and 10% for the backside (right-click the S icon on the entry > Manage> Display Options> Transparency: Front surface: 30, Back surface: 10).

8.2 Zoomed view representation

We zoomed-in on the protein-protein interfaces of the two complexes and we saved a new view (View> Save Camera View). Using the Protein Interaction Analysis tool (Tasks> Protein Interaction Analysis), a unique feature of the Schrödinger BioLuminate Suite, we calculated all contact-type interactions between NTR1 (chain R) and G_i protein (G protein α subunit, chain A) and between NTR1 (chain R) and β-arrestin (chain B). This way, we obtained the list of interacting residues, and for each residue the details of the interactions (H-bonds, salt bridges, π stacking, disulfides, vdW clash, surface complementarity, buried SASA, vdW complementarity). We focused our analysis on the vdW complementarity and colored accordingly the residues at the protein-protein interfaces with a gradient color scheme (Tasks> Protein Interaction Analysis Panel> Color by: van der Waals complementarity> Color Residues). Specifically, we set the color ramp ranging from yellow for minimal vdW complementarity to dark blue for the highest. The minimum cut-off value was set to 0.5 for both complexes while the maximum cut-off value was set to the highest calculated values, i.e., 0.97 for β-arrestin and 0.98 for the G_{i1} protein. The corresponding residues on the NTR1 were highlighted according to the vdW complementarity in the same color scheme as the surface (right-click the S icon on the entry> Manage> Display Options> Color scheme: other property: Atom Color). Additional accentuation was accomplished by changing the molecular representation of the residues to Ball and Stick, thereby hiding the bond orders and resetting the stick radius to 0.20 and the ball percentage to 4% (Edit> Preferences> Molecular Representation> Atoms and Bonds: Stick radius: 0.2, Ball percentage: 4).

As Schrödinger BioLuminate offers two directional lights which can be altered with respect to position, ambience, diffusion and specular reflexion, the ambience parameter was set to 70 and specular reflexion to 0 for both light sources (Edit> Preferences> Lighting: Ambient: 70, Specular: 0).

Through Fig. 6, NTR1 residues involved in the interaction with G_i protein and β-arrestin are highlighted according to the vdW complementarity. The zoomed view of the interaction depicts a different selectivity profile, the distribution of yellow (lower complementarity) and blues (higher complementarity) residues in NTR1 is different if it interacts with G_i protein or β-arrestin. NTR1 residues differently involved in these interactions are listed in the histograms at the bottom of the figure, and interestingly only one residue, $SER368_{NTR1}$, was found to have high vdW complementarity with both Gi protein and β-arrestin.

9 Summary and outlook

Life is the result of the chemical activity of proteins. In this chapter, we briefly reviewed the history of PPI studies, the advancements in experimental and computational techniques, highlighting the potential breakthroughs we can witness in the next future. The case studies we presented support the biomedical significance of structural studies. We provided methodological protocols for some visualization tools, but many more are available (3DProteinImaging, 2021; DiscoveryStudio, 2021; ICM-Pro, 2021; Samson, 2021; Unitymol, 2021). A comprehensive list can be found on the PDB website (PDB-moleculargraphics, 2021).

Nowadays, computer graphics evolution allows us to develop powerful tools to study and analyze macromolecules and their interactions. An interesting approach is represented by the application of augmented reality or virtual reality to the molecular visualization field. Virtual reality overcomes the limitations of perceiving the spatial arrangements of interacting molecules by displaying models in stereoscopic 3D (Deeks et al., 2020; Ratamero, Bellini, Dowson, & Romer, 2018; Yiu & Chen, 2021). A number of virtual reality software dedicated to proteins have been recently released, e.g., ProteinVR (Cassidy, Sefcik, Raghav, Chang, & Durrant, 2020), MoleculARweb, VRmol (Xu et al., 2021), *Autodesk Molecular Viewer* (Balo, Wang, & Ernst, 2017), YASARA (Land & Humble, 2018).

Protein illustrations are effective tools for scientific communication and are currently created with most diverse techniques, from virtual reality to hand painting. Prof. David Goodsell (Scripps Research Institute) has developed a signature style of scientific drawing that uses flat shading and simple color-schemes. Similar colors and shadows characterize the style of Illustrate, the non-photorealistic visualizer available also as webserver (Goodsell, Autin, & Olson, 2019). CellPAINT was recently realised as an interactive tool to create illustrations of the molecular structure of cells and viruses (Gardner, Autin, Barbaro, Olson, & Goodsell, 2018). All representations reported in this manuscript (Figs. 1–6) demonstrate that many structural details can only be unraveled by visual inspection. Protein visualization is not only the last step but integrative part of the structural analysis. A good use of available visualization tools, together with the integration of techniques currently being developed, will contribute to get insights on molecular interactions and foster scientific communication.

Acknowledgments

The authors thank Schrödinger GmbH, Mannheim, Germany for the license of BioLuminate and Chaos Group, Bulgaria, for the license of V-ray. ADP participates in the European COST action ERNEST (CA 18133).

References

Zhang, M., Gui, M., Wang, Z.-F., Gorgulla, C., Yu, J. J., Wu, H., et al. (2021). Cryo-EM structure of an activated GPCR-G protein complex in lipid nanodiscs. *Nature Structural & Molecular Biology*, *28*(3), 258–267.

3DProteinImaging. 2021. 3DProteinImaging, https://3dproteinimaging.com/.

Alonso-López, D., Campos-Laborie, F. J., Gutiérrez, M. A., Lambourne, L., Calderwood, M. A., Vidal, M., et al. (2019). APID database: Redefining protein-protein interaction experimental evidences and binary interactomes. *Database*, *2019*, 1–8.

Amaro, R. E., & Mulholland, A. J. (2020). A community letter regarding sharing biomolecular simulation data for COVID-19. *Journal of Chemical Information and Modeling*, *60*(6), 2653–2656.

Arafeh, R., Di Pizio, A., Elkahloun, A. G., Dym, O., Niv, M. Y., & Samuels, Y. (2019). RASA2 and NF1; two-negative regulators of Ras with complementary functions in melanoma. *Oncogene*, *38*(13), 2432–2434.

Arthanari, H., Takeuchi, K., Dubey, A., & Wagner, G. (2019). Emerging solution NMR methods to illuminate the structural and dynamic properties of proteins. *Current Opinion in Structural Biology*, *58*, 294–304.

Balo, A. R., Wang, M., & Ernst, O. P. (2017). Accessible virtual reality of biomolecular structural models using the Autodesk Molecule Viewer. *Nature Methods*, *14*(12), 1122–1123.

Barak, L. S., Bai, Y., Peterson, S., Evron, T., Urs, N. M., Peddibhotla, S., et al. (2016). ML314: A biased neurotensin receptor ligand for methamphetamine abuse. *ACS Chemical Biology*, *11*(7), 1880–1890.

Barradas-Bautista, D., Rosell, M., Pallara, C., & Fernandez-Recio, J. (2018). Structural prediction of protein-protein interactions by docking: Application to biomedical problems. *Advances in Protein Chemistry and Structural Biology*, *110*, 203–249.

Barros, E. P., Casalino, L., Gaieb, Z., Dommer, A. C., Wang, Y., Fallon, L., et al. (2021). The flexibility of ACE2 in the context of SARS-CoV-2 infection. *Biophysical Journal*, *120*(6), 1072–1084.

Bauvois, B. (2012). New facets of matrix metalloproteinases MMP-2 and MMP-9 as cell surface transducers: Outside-in signaling and relationship to tumor progression. *Biochimica et Biophysica Acta*, *1825*(1), 29–36.

Beloglazkina, A., Zyk, N., Majouga, A., & Beloglazkina, E. (2020). Recent small-molecule inhibitors of the p53–MDM2 protein–protein interaction. *Molecules*, *25*(5), 1211.

Benton, D. J., Wrobel, A. G., Xu, P., Roustan, C., Martin, S. R., Rosenthal, P. B., et al. (2020). Receptor binding and priming of the spike protein of SARS-CoV-2 for membrane fusion. *Nature*, *588*(7837), 327–330.

Berman, H. M., Westbrook, J., Feng, Z., Gilliland, G., Bhat, T. N., Weissig, H., et al. (2000). The protein data bank. *Nucleic Acids Research*, *28*(1), 235–242.

Bourne, H. R., Sanders, D. A., & McCormick, F. (1990). The GTPase superfamily: A conserved switch for diverse cell functions. *Nature*, *348*(6297), 125–132.

Braun, P., & Gingras, A. C. (2012). History of protein-protein interactions: From egg-white to complex networks. *Proteomics*, *12*, 1478–1498.

Burley, S. K., Bhikadiya, C., Bi, C., Bittrich, S., Chen, L., Crichlow, G. V., et al. (2021). RCSB Protein Data Bank: Powerful new tools for exploring 3D structures of biological macromolecules for basic and applied research and education in fundamental biology, biomedicine, biotechnology, bioengineering and energy sciences. *Nucleic Acids Research*, *49*(D1), D437–D451.

Cai, Y., Zhang, J., Xiao, T., Peng, H., Sterling, S. M., Walsh, R. M., Jr., et al. (2020). Distinct conformational states of SARS-CoV-2 spike protein. *Science*, *369*(6511), 1586–1592.
Callaway, E. (2020a). 'It will change everything': DeepMind's AI makes gigantic leap in solving protein structures. *Nature*, *588*(7837), 203–204.
Callaway, E. (2020b). Revolutionary cryo-EM is taking over structural biology. *Nature*, *578*(7794), 201.
Cao, L., Goreshnik, I., Coventry, B., Case, J. B., Miller, L., Kozodoy, L., et al. (2020). De novo design of picomolar SARS-CoV-2 miniprotein inhibitors. *Science*, *370*(6515), 426–431.
Casalino, L., Dommer, A., Gaieb, Z., Barros, E. P., Sztain, T., Ahn, S. H., et al. (2020). *AI-driven multiscale simulations illuminate mechanisms of SARS-CoV-2 spike dynamics. bioRxiv*.
Casalino, L., Gaieb, Z., Goldsmith, J. A., Hjorth, C. K., Dommer, A. C., Harbison, A. M., et al. (2020). Beyond shielding: The roles of glycans in the SARS-CoV-2 spike protein. *ACS Central Science*, *6*(10), 1722–1734.
Cassidy, K. C., Sefcik, J., Raghav, Y., Chang, A., & Durrant, J. D. (2020). ProteinVR: Web-based molecular visualization in virtual reality. *PLoS Computational Biology*, *16*(3), e1007747.
Chakrabarti, P., & Janin, J. (2002). Dissecting protein-protein recognition sites. *Proteins*, *47*(3), 334–343.
Chandler, D. (2005). Interfaces and the driving force of hydrophobic assembly. *Nature*, *437*(7059), 640–647.
Chaturvedi, M., Maharana, J., & Shukla, A. K. (2020). Terminating G-protein coupling: Structural snapshots of GPCR-beta-arrestin complexes. *Cell*, *180*(6), 1041–1043.
ChimeraX-website. 2021. https://www.rbvi.ucsf.edu/chimerax/.
Chutkow, W. A., Birkenfeld, A. L., Brown, J. D., Lee, H. Y., Frederick, D. W., Yoshioka, J., et al. (2010). Deletion of the alpha-arrestin protein Txnip in mice promotes adiposity and adipogenesis while preserving insulin sensitivity. *Diabetes*, *59*(6), 1424–1434.
Cianfrocco, M. A., & Kellogg, E. H. (2020). What could go wrong? A practical guide to single-particle cryo-EM: From biochemistry to atomic models. *Journal of Chemical Information and Modeling*, *60*, 2458–2469.
Clackson, T., & Wells, J. A. (1995). A hot spot of binding energy in a hormone-receptor interface. *Science*, *267*, 383–386.
Cressey, D., & Callaway, E. (2017). Cryo-electron microscopy wins chemistry nobel. *Nature*, *550*(7675), 167.
Das, S., & Chakrabarti, S. (2021). Classification and prediction of protein-protein interaction interface using machine learning algorithm. *Scientific Reports*, *11*(1), 1761.
Deeks, H. M., Walters, R. K., Hare, S. R., O'Connor, M. B., Mulholland, A. J., & Glowacki, D. R. (2020). Interactive molecular dynamics in virtual reality for accurate flexible protein-ligand docking. *PLoS One*, *15*(3), e0228461.
Deganutti, G., Prischi, F., & Reynolds, C. A. (2021). Supervised molecular dynamics for exploring the druggability of the SARS-CoV-2 spike protein. *Journal of Computer-Aided Molecular Design*, *35*, 195–207.
Deluigi, M., Klipp, A., Klenk, C., Merklinger, L., Eberle, S. A., Morstein, L., et al. (2021). Complexes of the neurotensin receptor 1 with small-molecule ligands reveal structural determinants of full, partial, and inverse agonism. *Science Advances*, *7*(5), eabe5504.
Demir, Ö., Barros, E. P., Offutt, T. L., Rosenfeld, M., & Amaro, R. E. (2021). An integrated view of p53 dynamics, function, and reactivation. *Current Opinion in Structural Biology*, *67*, 187–194.

DiscoveryStudio. 2021. BIOVIA discovery studio visualizer. https://discover.3ds.com/discovery-studio-visualizer-download.

Douangamath, A., Fearon, D., Gehrtz, P., Krojer, T., Lukacik, P., Owen, C. D., et al. (2020). Crystallographic and electrophilic fragment screening of the SARS-CoV-2 main protease. *Nature Communications*, *11*(1), 5047.

Duarte, J. M., Srebniak, A., Scharer, M. A., & Capitani, G. (2012). Protein interface classification by evolutionary analysis. *BMC Bioinformatics*, *13*, 334.

Dwivedi, H., Baidya, M., & Shukla, A. K. (2018). GPCR signaling: The interplay of galphai and beta-arrestin. *Current Biology*, *28*(7), R324–R327.

Egloff, P., Hillenbrand, M., Klenk, C., Batyuk, A., Heine, P., Balada, S., et al. (2014). Structure of signaling-competent neurotensin receptor 1 obtained by directed evolution in Escherichia coli. *Proceedings of the National Academy of Sciences of the United States of America*, *111*(6), E655–E662.

Ehrlich, P. (1891). Experimentelle Untersuchungen über Immunität. II. Ueber Abrin. *Deutsche Medizinische Wochenschrift*, *17*, 1218–1219.

Elez, K., Bonvin, A. M. J. J., & Vangone, A. (2020). Biological vs. Crystallographic protein interfaces: An overview of computational approaches for their classification. *Crystals*, *10*, 1–15.

Frankel, P. S., Hoonakker, A. J., Alburges, M. E., McDougall, J. W., McFadden, L. M., Fleckenstein, A. E., et al. (2011). Effect of methamphetamine self-administration on neurotensin systems of the basal ganglia. *The Journal of Pharmacology and Experimental Therapeutics*, *336*(3), 809–815.

Fredriksson, R., Lagerstrom, M. C., Lundin, L. G., & Schioth, H. B. (2003). The G-protein-coupled receptors in the human genome form five main families. Phylogenetic analysis, paralogon groups, and fingerprints. *Molecular Pharmacology*, *63*(6), 1256–1272.

Garcia-Nafria, J., & Tate, C. G. (2019). Cryo-EM structures of GPCRs coupled to Gs, Gi and Go. *Molecular and Cellular Endocrinology*, *488*, 1–13.

Garcia-Nafria, J., & Tate, C. G. (2020). Cryo-electron microscopy: Moving beyond X-ray crystal structures for drug receptors and drug development. *Annual Review of Pharmacology and Toxicology*, *60*, 51–71.

Gardner, A., Autin, L., Barbaro, B., Olson, A. J., & Goodsell, DS. (2018). CellPAINT: interactive illustration of dynamic mesoscale cellular environments. *EEE Computer Graphics and Applications*, *38*(6), 51–66. https://doi.org/10.1109/MCG.2018.2877076.

Ghose, A. K., Viswanadhan, V. N., & Wendoloski, J. J. (1998). Prediction of hydrophobic (lipophilic) properties of small organic molecules using fragmental methods: An analysis of ALOGP and CLOGP methods. *The Journal of Physical Chemistry A*, *102*(21), 3762–3772.

Gilman, A. G. (1987). G proteins: Transducers of receptor-generated signals. *Annual Review of Biochemistry*, *56*, 615–649.

Glashagen, G., de Vries, S., Uciechowska-Kaczmarzyk, U., Samsonov, S. A., Murail, S., Tuffery, P., et al. (2020). Coarse-grained and atomic resolution biomolecular docking with the ATTRACT approach. *Proteins*, *88*(8), 1018–1028.

Goddard, T. D., Huang, C. C., Meng, E. C., Pettersen, E. F., Couch, G. S., Morris, J. H., et al. (2018). UCSF ChimeraX: Meeting modern challenges in visualization and analysis. *Protein Science*, *27*(1), 14–25.

Goodsell, D. S., Autin, L., & Olson, A. J. (2019). Illustrate: Software for biomolecular illustration. *Structure*, *27*(11), 1716–1720 e1711.

Griebel, G., & Holsboer, F. (2012). Neuropeptide receptor ligands as drugs for psychiatric diseases: The end of the beginning? *Nature Reviews. Drug Discovery*, *11*(6), 462–478.

Hainaut, P., & Hollstein, M. (2000). p53 and human cancer: The first ten thousand mutations. *Advances in Cancer Research*, *77*, 81–137.

Harmalkar, A., & Gray, J. J. (2020). Advances to tackle backbone flexibility in protein docking. *Current Opinion in Structural Biology*, *67*, 178–186.

Haupt, Y. (2004). p53 regulation: A family affair. *Cell Cycle*, *3*, 884–885.

Hendsch, Z. S., & Tidor, B. (1994). Do salt bridges stabilize proteins? A continuum electrostatic analysis. *Protein Science*, *3*(2), 211–226.

Hermjakob, H., Montecchi-Palazzi, L., Lewington, C., Mudali, S., Kerrien, S., Orchard, S., et al. (2004). IntAct: An open source molecular interaction database. *Nucleic Acids Research*, *32*(Database issue), D452–D455.

Hoffmann, M., Kleine-Weber, H., Schroeder, S., Kruger, N., Herrler, T., Erichsen, S., et al. (2020). SARS-CoV-2 cell entry depends on ACE2 and TMPRSS2 and is blocked by a clinically proven protease inhibitor. *Cell*, *181*(2), 271–280 e278.

Hu, J., Li, J., Chen, N., & Zhang, X. (2016). Conservation of hot regions in protein–protein interaction in evolution. *Methods*, *110*, 73–80.

Huang, S. Y. (2014). Search strategies and evaluation in protein-protein docking: Principles, advances and challenges. *Drug Discovery Today*, *19*(8), 1081–1096.

Huang, W., Masureel, M., Qu, Q., Janetzko, J., Inoue, A., Kato, H. E., et al. (2020). Structure of the neurotensin receptor 1 in complex with beta-arrestin 1. *Nature*, *579*(7798), 303–308.

Huang, Y., Yang, C., Xu, X. F., Xu, W., & Liu, S. W. (2020). Structural and functional properties of SARS-CoV-2 spike protein: Potential antivirus drug development for COVID-19. *Acta Pharmacologica Sinica*, *41*(9), 1141–1149.

Humphrey, W., Dalke, A., & Schulten, K. (1996). VMD: Visual molecular dynamics. *Journal of Molecular Graphics*, *14*(1), 33–38.

Huttlin, E. L., Bruckner, R. J., Navarrete-Perea, J., Cannon, J. R., Baltier, K., Gebreab, F., et al. (2020). *Dual proteome-scale networks reveal cell-specific remodeling of the human interactome. bioRxiv*.

Iacobucci, I., Monaco, V., Cozzolino, F., & Monti, M. (2021). From classical to new generation approaches: An excursus of -omics methods for investigation of protein-protein interaction networks. *Journal of Proteomics*, *230*, 103990.

ICM-Pro. 2021. Molsoft ICM-Pro, http://www.molsoft.com/icm_pro.html.

Interactome3d-stats. (2021). https://interactome3d.irbbarcelona.org/statistics.php#stats_inter. Accessed on February 3rd, 2021.

Israeli, R., Asli, A., Avital-Shacham, M., & Kosloff, M. (2019). RGS6 and RGS7 discriminate between the highly similar galphai and galphao proteins using a two-tiered specificity strategy. *Journal of Molecular Biology*, *431*(17), 3302–3311.

Jimenez-Garcia, B., Pons, C., & Fernandez-Recio, J. (2013). pyDockWEB: A web server for rigid-body protein-protein docking using electrostatics and desolvation scoring. *Bioinformatics*, *29*(13), 1698–1699.

Jones, S., & Thornton, J. M. (1996). Principles of protein-protein interactions. *Proceedings of the National Academy of Sciences of the United States of America*, *93*(1), 13–20.

Jones, S., & Thornton, J. M. (1997). Analysis of protein-protein interaction sites using surface patches. *Journal of Molecular Biology*, *272*(1), 121–132.

Joshi, S., Joshi, M., & Degani, M. S. (2020). Tackling SARS-CoV-2: Proposed targets and repurposed drugs. *Future Medicinal Chemistry*, *12*(17), 1579–1601.

Kaczor, A. A., Bartuzi, D., Stepniewski, T. M., Matosiuk, D., & Selent, J. (2018). Protein-protein docking in drug design and discovery. *Methods in Molecular Biology*, *1762*, 285–305.

Kalathiya, U., Padariya, M., Mayordomo, M., Lisowska, M., Nicholson, J., Singh, A., et al. (2020). Highly conserved homotrimer cavity formed by the SARS-CoV-2 spike glycoprotein: A novel binding site. *Journal of Clinical Medicine*, *9*(5), 1473.

Kato, H. E., Zhang, Y., Hu, H., Suomivuori, C. M., Kadji, F. M. N., Aoki, J., et al. (2019). Conformational transitions of a neurotensin receptor 1-Gi1 complex. *Nature*, *572*(7767), 80–85.

Khan, F. M., Marquardt, S., Gupta, S. K., Knoll, S., Schmitz, U., Spitschak, A., et al. (2017). Unraveling a tumor type-specific regulatory core underlying E2F1-mediated epithelial-mesenchymal transition to predict receptor protein signatures. *Nature Communications*, *8*(1), 198.

Khan, F. M., Sadeghi, M., Gupta, S. K., & Wolkenhauer, O. (2018). A network-based integrative workflow to unravel mechanisms underlying disease progression. *Methods in Molecular Biology*, *1702*, 247–276.

Kozakov, D., Hall, D. R., Xia, B., Porter, K. A., Padhorny, D., Yueh, C., et al. (2017). The ClusPro web server for protein-protein docking. *Nature Protocols*, *12*(2), 255–278.

Krumm, B. E., White, J. F., Shah, P., & Grisshammer, R. (2015). Structural prerequisites for G-protein activation by the neurotensin receptor. *Nature Communications*, *6*, 7895.

Kumar, A., & Snyder, M. (2002). Protein complexes take the bait. *Nature*, *415*(6868), 123–124.

Kundrotas, P. J., & Alexov, E. (2006). Electrostatic properties of protein-protein complexes. *Biophysical Journal*, *91*(5), 1724–1736.

Kussie, P. H., Gorina, S., Marechal, V., Elenbaas, B., Moreau, J., Levine, A. J., et al. (1996). Structure of the MDM2 oncoprotein bound to the p53 tumor suppressor transactivation domain. *Science*, *274*, 948–953.

Lally, C. C., Bauer, B., Selent, J., & Sommer, M. E. (2017). C-edge loops of arrestin function as a membrane anchor. *Nature Communications*, *8*, 14258.

Lan, J., Ge, J., Yu, J., Shan, S., Zhou, H., Fan, S., et al. (2020). Structure of the SARS-CoV-2 spike receptor-binding domain bound to the ACE2 receptor. *Nature*, *581*(7807), 215–220.

Land H., Humble M.S., YASARA: a tool to obtain structural guidance in biocatalytic investigations. Methods in Molecular Biology. 2018;1685:43–67. doi: 10.1007/978-1-4939-7366-8_4. PMID: 29086303.

Latorraca, N. R., Wang, J. K., Bauer, B., Townshend, R. J. L., Hollingsworth, S. A., Olivieri, J. E., et al. (2018). Molecular mechanism of GPCR-mediated arrestin activation. *Nature*, *557*(7705), 452–456.

Lensink, M. F., Brysbaert, G., Nadzirin, N., Velankar, S., Chaleil, R. A. G., Gerguri, T., et al. (2019). Blind prediction of homo- and hetero-protein complexes: The CASP13-CAPRI experiment. *Proteins: Structure, Function, and Bioinformatics*, *87*, 1200–1221.

Levy, E. D. (2010). A simple definition of structural regions in proteins and its use in analyzing interface evolution. *Journal of Molecular Biology*, *403*(4), 660–670.

Li, W., Moore, M. J., Vasilieva, N., Sui, J., Wong, S. K., Berne, M. A., et al. (2003). Angiotensin-converting enzyme 2 is a functional receptor for the SARS coronavirus. *Nature*, *426*(6965), 450–454.

Licata, L., Briganti, L., Peluso, D., Perfetto, L., Iannuccelli, M., Galeota, E., et al. (2012). MINT, the molecular interaction database: 2012 update. *Nucleic Acids Research*, *40*, 857–861.

Lu, H., Zhou, Q., He, J., Jiang, Z., Peng, C., Tong, R., et al. (2020). Recent advances in the development of protein-protein interactions modulators: Mechanisms and clinical trials. *Signal Transduction and Targeted Therapy*, *5*(1), 213.

Marchetti, F., Capelli, R., Rizzato, F., Laio, A., & Colombo, G. (2019). The subtle trade-off between evolutionary and energetic constraints in protein-protein interactions. *Journal of Physical Chemistry Letters*, *10*(7), 1489–1497.

Marti-Solano, M., Sanz, F., Pastor, M., & Selent, J. (2014). A dynamic view of molecular switch behavior at serotonin receptors: Implications for functional selectivity. *PLoS One*, *9*(10), e109312.

Marze, N. A., Roy Burman, S. S., Sheffler, W., & Gray, J. J. (2018). Efficient flexible backbone protein-protein docking for challenging targets. *Bioinformatics*, *34*(20), 3461–3469.

Merz, K. M., Jr., Amaro, R., Cournia, Z., Rarey, M., Soares, T., Tropsha, A., et al. (2020). Editorial: Method and data sharing and reproducibility of scientific results. *Journal of Chemical Information and Modeling*, *60*(12), 5868–5869.

Moal, I. H., Barradas-Bautista, D., Jimenez-Garcia, B., Torchala, M., van der Velde, A., Vreven, T., et al. (2017). IRaPPA: Information retrieval based integration of biophysical models for protein assembly selection. *Bioinformatics*, *33*(12), 1806–1813.

Moreira, I. S., Fernandes, P. A., & Ramos, M. J. (2007). Hot spots—A review of the protein–protein interface determinant amino-acid residues. *Proteins: Structure, Function, and Bioinformatics*, *68*, 803–812.

Moreira, I. S., Koukos, P. I., Melo, R., Almeida, J. G., Preto, A. J., Schaarschmidt, J., et al. (2017). SpotOn: High accuracy identification of protein-protein interface hot-spots. *Scientific Reports*, *7*(1), 8007.

Moreira, I. S., Martins, J. M., Ramos, R. M., Fernandes, P. A., & Ramos, M. J. (2013). Understanding the importance of the aromatic amino-acid residues as hot-spots. *Biochimica et Biophysica Acta*, *1834*(1), 404–414.

Mosca, R., Pons, T., Céol, A., Valencia, A., & Aloy, P. (2013). Towards a detailed atlas of protein-protein interactions. *Current Opinion in Structural Biology*, *23*, 929–940.

Moult, J., Pedersen, J. T., Judson, R., & Fidelis, K. (1995). A large-scale experiment to assess protein structure prediction methods. *Proteins*, *23*(3), ii–v.

mMaya plugin, 2021. https://clarafi.com/tools/mmaya/.

Mugnai, M. L., Templeton, C., Elber, R., & Thirumalai, D. (2020). *Role of long-range allosteric communication in determining the stability and disassembly of SARS-COV-2 in complex with ACE2. bioRxiv*.

Nakane, T., Kotecha, A., Sente, A., McMullan, G., Masiulis, S., Brown, P., et al. (2020). Single-particle cryo-EM at atomic resolution. *Nature*, ***587***, 152–156.

Neuman, B. W., Adair, B. D., Yoshioka, C., Quispe, J. D., Orca, G., Kuhn, P., et al. (2006). Supramolecular architecture of severe acute respiratory syndrome coronavirus revealed by electron cryomicroscopy. *Journal of Virology*, *80*(16), 7918–7928.

Niedzialkowska, E., Gasiorowska, O., Handing, K. B., Majorek, K. A., Porebski, P. J., Shabalin, I. G., et al. (2016). Protein purification and crystallization artifacts: The tale usually not told. *Protein Science*, *25*(3), 720–733.

Northrop, J. H. (1922). The inactivation of trypsin. *The Journal of General Physiology*, *4*, 245–260.

O'Donoghue, S. I., Goodsell, D. S., Frangakis, A. S., Jossinet, F., Laskowski, R. A., Nilges, M., et al. (2010). Visualization of macromolecular structures. *Nature Methods*, *7*(3 Suppl), S42–S55.

Olson, A. J. (2018). Perspectives on structural molecular biology visualization: From past to present. *Journal of Molecular Biology*, *430*(21), 3997–4012.

Omieczynski, C., Nguyen, T. N., Sribar, D., Deng, L., Stepanov, D., Schaller, D., et al. (2019). *BiasDB: A comprehensive database for biased GPCR ligands*. bioRxiv.

Oughtred, R., Stark, C., Breitkreutz, B. J., Rust, J., Boucher, L., Chang, C., et al. (2019). The BioGRID interaction database: 2019 update. *Nucleic Acids Research*, *47*, D529–D541.

Padhorny, D., Kazennov, A., Zerbe, B. S., Porter, K. A., Xia, B., Mottarella, S. E., et al. (2016). Protein-protein docking by fast generalized Fourier transforms on 5D rotational manifolds. *Proceedings of the National Academy of Sciences of the United States of America*, *113*(30), E4286–E4293.

Pan, A. C., Jacobson, D., Yatsenko, K., Sritharan, D., Weinreich, T. M., & Shaw, D. E. (2019). Atomic-level characterization of protein-protein association. *Proceedings of the National Academy of Sciences of the United States of America*, *116*(10), 4244–4249.

Pazos, F., Helmer-Citterich, M., Ausiello, G., & Valencia, A. (1997). Correlated mutations contain information about protein-protein interaction. *Journal of Molecular Biology*, *271*(4), 511–523.

PDB-moleculargraphics. (2021). https://www.rcsb.org/pages/thirdparty/molecular_graphics.

PDB-stats. (2021). https://www.rcsb.org/stats. Accessed on February 3rd, 2021.

Perricone, U., Gulotta, M. R., Lombino, J., Parrino, B., Cascioferro, S., Diana, P., et al. (2018). An overview of recent molecular dynamics applications as medicinal chemistry tools for the undruggable site challenge. *MedChemComm*, *9*(6), 920–936.

Pettersen, E. F., Goddard, T. D., Huang, C. C., Couch, G. S., Greenblatt, D. M., Meng, E. C., et al. (2004). UCSF Chimera–A visualization system for exploratory research and analysis. *Journal of Computational Chemistry*, *25*(13), 1605–1612.

Pettersen, E. F., Goddard, T. D., Huang, C. C., Meng, E. C., Couch, G. S., Croll, T. I., et al. (2021). UCSF ChimeraX: Structure visualization for researchers, educators, and developers. *Protein Science*, *30*(1), 70–82.

Phillips, J. C., Hardy, D. J., Maia, J. D. C., Stone, J. E., Ribeiro, J. V., Bernardi, R. C., et al. (2020). Scalable molecular dynamics on CPU and GPU architectures with NAMD. *The Journal of Chemical Physics*, *153*(4), 044130.

Pierce, B. G., Wiehe, K., Hwang, H., Kim, B. H., Vreven, T., & Weng, Z. (2014). ZDOCK server: Interactive docking prediction of protein-protein complexes and symmetric multimers. *Bioinformatics*, *30*, 1771–1773.

Preto, A. J., Barreto, C. A. V., Baptista, S. J., Almeida, J. G., Lemos, A., Melo, A., et al. (2020). Understanding the binding specificity of G-Protein coupled receptors toward G-proteins and arrestins: Application to the dopamine receptor family. *Journal of Chemical Information and Modeling*, *60*(8), 3969–3984.

Preto, A. J., & Moreira, I. S. (2020). SPOTONE: Hot spots on protein complexes with extremely randomized trees via sequence-only features. *International Journal of Molecular Sciences*, *21*(19), 7281.

Raghuvamsi, P., Tulsian, N., Samsudin, F., Qian, X., Purushotorman, K., Yue, G., et al. (2021). SARS-CoV-2 S protein:ACE2 interaction reveals novel allosteric targets. *Elife*, *10*, e63646.

Rakers, C., Bermudez, M., Keller, B. G., Mortier, J., & Wolber, G. (2015). Computational close up on protein-protein interactions: How to unravel the invisible using molecular dynamics simulations? *Wiley Interdisciplinary Reviews: Computational Molecular Science*, *5*(5), 345–359.

Ratamero, E. M., Bellini, D., Dowson, C. G., & Romer, R. A. (2018). Touching proteins with virtual bare hands: Visualizing protein-drug complexes and their dynamics in self-made virtual reality using gaming hardware. *Journal of Computer-Aided Molecular Design*, *32*(6), 703–709.

Rhino3d. 2021. https://www.rhino3d.com/, McNeel, USA.

Safdari, H. A., Pandey, S., Shukla, A. K., & Dutta, S. (2018). Illuminating GPCR signaling by cryo-EM. *Trends in Cell Biology*, *28*(8), 591–594.

Samson. 2021. Samson. https://www.samson-connect.net/.

Sanchez, C., Lachaize, C., Janody, F., Bellon, B., Roder, L., Euzenat, J., et al. (1999). Grasping at molecular interactions and genetic networks in Drosophila melanogaster using FlyNets, an Internet database. *Nucleic Acids Research*, *27*(1), 89–94.

SciFinder, 2021. https://scifinder.cas.org, consulted on February 7th, 2021.

Service, R. F. (2020). 'The game has changed.' AI triumphs at solving protein structures. *Science*, *370*, 1144–1145.

Sheinerman, F. B., & Honig, B. (2002). On the role of electrostatic interactions in the design of protein–protein interfaces. *Journal of Molecular Biology*, *318*(1), 161–177.

Sheinerman, F. B., Norel, R., & Honig, B. (2000). Electrostatic aspects of protein-protein interactions. *Current Opinion in Structural Biology*, *10*(2), 153–159.

Shirian, J., Sharabi, O., & Shifman, J. M. (2016). Cold spots in protein binding. *Trends in Biochemical Sciences*, *41*(9), 739–745.

Siebenmorgen, T., Engelhard, M., & Zacharias, M. (2020). Prediction of protein-protein complexes using replica exchange with repulsive scaling. *Journal of Computational Chemistry*, *41*(15), 1436–1447.

Siebenmorgen, T., & Zacharias, M. (2019a). Computational prediction of protein–protein binding affinities. *WIREs Computational Molecular Science*, *10*(3), e1448.

Siebenmorgen, T., & Zacharias, M. (2019b). Evaluation of predicted protein-protein complexes by binding free energy simulations. *Journal of Chemical Theory and Computation*, *15*(3), 2071–2086.

Slosky, L. M., Bai, Y., Toth, K., Ray, C., Rochelle, L. K., Badea, A., et al. (2020). beta-arrestin-biased allosteric modulator of NTSR1 selectively attenuates addictive behaviors. *Cell*, *181*(6), 1364–1379 e1314.

Sperandio, O., Reynès, C. H., Camproux, A.-C. C., & Villoutreix, B. O. (2010). Rationalizing the chemical space of protein-protein interaction inhibitors. *Drug Discovery Today*, *15*, 220–229.

Sun, D., Li, Z., Rew, Y., Gribble, M., Bartberger, M. D., Beck, H. P., et al. (2014). Discovery of AMG 232, a potent, selective, and orally bioavailable MDM2-p53 inhibitor in clinical development. *Journal of Medicinal Chemistry*, *57*, 1454–1472.

Szklarczyk, D., Gable, A. L., Lyon, D., Junge, A., Wyder, S., Huerta-Cepas, J., et al. (2019). STRING v11: Protein-protein association networks with increased coverage, supporting functional discovery in genome-wide experimental datasets. *Nucleic Acids Research*, *47*, D607–D613.

Thul, P. J., Akesson, L., Wiking, M., Mahdessian, D., Geladaki, A., Ait Blal, H., et al. (2017). A subcellular map of the human proteome. *Science*, *356*(6340), eaal3321.

Tiwari, V., Beer, J. C., Sankaranarayanan, N. V., Swanson-Mungerson, M., & Desai, U. R. (2020). Discovering small-molecule therapeutics against SARS-CoV-2. *Drug Discovery Today*, *25*(8), 1535–1544.

Torchala, M., Moal, I. H., Chaleil, R. A. G., Fernandez-Recio, J., & Bates, P. A. (2013). SwarmDock: A server for flexible protein-protein docking. *Bioinformatics*, *29*, 807–809.

Turonova, B., Sikora, M., Schurmann, C., Hagen, W. J. H., Welsch, S., Blanc, F. E. C., et al. (2020). In situ structural analysis of SARS-CoV-2 spike reveals flexibility mediated by three hinges. *Science*, *370*(6513), 203–208.

Unitymol. 2021. Unitymol. http://www.baaden.ibpc.fr/umol/.

van Gastel, J., Hendrickx, J. O., Leysen, H., Santos-Otte, P., Luttrell, L. M., Martin, B., et al. (2018). beta-Arrestin based receptor signaling paradigms: Potential therapeutic targets for complex age-related disorders. *Frontiers in Pharmacology*, *9*, 1369.

Van Zundert, G. C. P., Rodrigues, J. P. G. L. M., Trellet, M., Schmitz, C., Kastritis, P. L., Karaca, E., et al. (2016). The HADDOCK2.2 web server: User-friendly integrative modeling of biomolecular complexes. *Journal of Molecular Biology*, *428*, 720–725. The Authors.

Vassilev, L. T., Vu, B. T., Graves, B., Carvajal, D., Podlaski, F., Filipovic, Z., et al. (2004). In vivo activation of the p53 pathway by small-molecule antagonists of MDM2. *Science*, *303*(5659), 844–848.

Violin, J. D., Crombie, A. L., Soergel, D. G., & Lark, M. W. (2014). Biased ligands at G-protein-coupled receptors: Promise and progress. *Trends in Pharmacological Sciences*, *35*(7), 308–316.

Wade, M., Wang, Y. V., & Wahl, G. M. (2010). The p53 orchestra: Mdm2 and Mdmx set the tone. *Trends in Cell Biology*, *20*(5), 299–309.

Walls, A. C., Park, Y. J., Tortorici, M. A., Wall, A., McGuire, A. T., & Veesler, D. (2020). Structure, function, and antigenicity of the SARS-CoV-2 spike glycoprotein. *Cell*, *181*(2), 281–292 e286.

Wang, X., Terashi, G., Christoffer, C. W., Zhu, M., & Kihara, D. (2020). Protein docking model evaluation by 3D deep convolutional neural networks. *Bioinformatics*, *36*(7), 2113–2118.

White, J. F., Noinaj, N., Shibata, Y., Love, J., Kloss, B., Xu, F., et al. (2012). Structure of the agonist-bound neurotensin receptor. *Nature*, *490*(7421), 508–513.

WHO-coronavirus2019. (2021). *World Health Organization, Coronavirus disease (COVID-19) pandemic*. https://www.who.int/emergencies/diseases/novel-coronavirus-2019. Accessed on February 3rd, 2021.

Wodak, S. J., & Janin, J. (1978). Computer analysis of protein-protein interaction. *Journal of Molecular Biology*, *124*(2), 323–342.

Wodak, S. J., Velankar, S., & Sternberg, M. J. E. (2020). Modeling protein interactions and complexes in CAPRI: Seventh CAPRI evaluation meeting, April 3-5 EMBL-EBI, Hinxton, UK. *Proteins*, *88*(8), 913–915.

Wrapp, D., Wang, N., Corbett, K. S., Goldsmith, J. A., Hsieh, C. L., Abiona, O., et al. (2020). Cryo-EM structure of the 2019-nCoV spike in the prefusion conformation. *Science*, *367*(6483), 1260–1263.

Wrobel, A. G., Benton, D. J., Xu, P., Roustan, C., Martin, S. R., Rosenthal, P. B., et al. (2020). SARS-CoV-2 and bat RaTG13 spike glycoprotein structures inform on virus evolution and furin-cleavage effects. *Nature Structural & Molecular Biology*, *27*(8), 763–767.

Xu, K., Liu, N., Xu, J., Guo, C., Zhao, L., Wang, H.-W., et al. (2021). VRmol: An integrative cloud-based virtual reality system to explore macromolecular structure. *Bioinformatics*, *37*(7), 1029–1031.

Yan, C., Wu, F., Jernigan, R. L., Dobbs, D., & Honavar, V. (2008). Characterization of protein-protein interfaces. *The Protein Journal*, *27*(1), 59–70.

Yan, R., Zhang, Y., Li, Y., Xia, L., Guo, Y., & Zhou, Q. (2020). Structural basis for the recognition of SARS-CoV-2 by full-length human ACE2. *Science*, *367*(6485), 1444–1448.

Yang, J., Petitjean, S. J. L., Koehler, M., Zhang, Q., Dumitru, A. C., Chen, W., et al. (2020). Molecular interaction and inhibition of SARS-CoV-2 binding to the ACE2 receptor. *Nature Communications*, *11*(1), 4541.

Yiu, C. B., & Chen, Y. W. (2021). Molecular data visualization with augmented reality (AR) on mobile devices. *Methods in Molecular Biology*, *2199*, 347–356.

Yueh, C., Hall, D. R., Xia, B., Padhorny, D., Kozakov, D., & Sandor, V. (2017). ClusPro-DC: Dimer classification by the cluspro server for protein-protein docking. *Journal of Molecular Biology*, *429*, 372–381.

CHAPTER

14

Interactions between noncoding RNAs as epigenetic regulatory mechanisms in cardiovascular diseases

Bruno Moukette[a], Nipuni P. Barupala[a], Tatsuya Aonuma[a], Marisa Sepulveda[a], Satoshi Kawaguchi[a], and Il-man Kim[a,b,c,*]

[a]*Department of Anatomy, Cell Biology and Physiology, Indiana University School of Medicine, Indianapolis, IN, United States*

[b]*Krannert Institute of Cardiology, Indiana University School of Medicine, Indianapolis, IN, United States*

[c]*Wells Center for Pediatric Research, Indiana University School of Medicine, Indianapolis, IN, United States*

**Corresponding author: e-mail address: ilkim@iu.edu*

Chapter outline

Methods in Cell Biology, Volume 166, ISSN 0091-679X, https://doi.org/10.1016/bs.mcb.2021.06.002

Abstract

Cardiovascular diseases (CVDs) represent the foremost cause of mortality in the United States and worldwide. It is estimated that CVDs account for approximately 17.8 million deaths each year. Despite the advances made in understanding cellular mechanisms and gene mutations governing the pathophysiology of CVDs, they remain a significant cause of mortality and morbidity. A major segment of mammalian genomes encodes for genes that are not further translated into proteins. The roles of the majority of such noncoding ribonucleic acids (RNAs) have been puzzling for a long time. However, it is becoming increasingly clear that noncoding RNAs (ncRNAs) are dynamically expressed in different cell types and have a comprehensive selection of regulatory roles at almost every step involved in DNAs, RNAs and proteins. Indeed, ncRNAs regulate gene expression through epigenetic interactions, through direct binding to target sequences, or by acting as competing endogenous RNAs. The profusion of ncRNAs in the cardiovascular system suggests that they may modulate complex regulatory networks that govern cardiac physiology and pathology. In this review, we summarize various functions of ncRNAs and highlight the recent literature on interactions between ncRNAs with an emphasis on cardiovascular disease regulation. Furthermore, as the broad-spectrum of ncRNAs potentially establishes new avenues for therapeutic development targeting CVDs, we discuss the innovative prospects of ncRNAs as therapeutic targets for CVDs.

1 Introduction

Cardiovascular disease (CVD) is known as the primary cause of death in the United States and worldwide. It is estimated that CVDs account for approximately 17.8 million deaths each year and about 655,000 deaths annually in the United States (Virani et al., 2020). Although this number has remained somewhat stable, there has been the growing prevalence of heart failure in the US population, and most developed countries and developing countries face an epidemic not yet to materialize fully. Despite the advances made in understanding cellular mechanisms and gene mutations governing the pathophysiology of CVDs, they remain a significant cause of mortality and morbidity.

Recent innovations in science and technology have enabled more detailed studies of the whole genome and its inferences in physiology and pathology (Park et al., 2016). Advanced techniques such as high throughput RNA-sequencing analyses of the genomic functional elements have allowed scientists to demonstrate that most of the genome is transcribed into RNAs. Early reports on the human genome estimated that it consisted of at least two million of protein-coding genes. However, the establishment of the Human Genome Project in 1990 re-estimated this number to 20,500 coded genes, representing only about 2% of the genome. Therefore, most (98%) of the human genome consisted of RNAs without coding potential, denoted as noncoding RNAs (ncRNAs) (Park et al., 2016).

These ncRNAs were considered as "transcriptional noises" or "genetic junks" based on the assumption that they elicited no essential cellular functions. However, emerging pieces of evidence indicate that ncRNAs, translated from noncoding regions of the genome, are involved in gene transcription and translation (Bayoumi et al., 2017). This notion has revolutionized our understanding of gene regulatory processes. There is now a mounting recognition of critical roles of ncRNAs in the development and maintenance of the cardiovascular system. Over the past decade, multiple studies have shown that ncRNAs are necessary for the normal development and physiology of various organs, including the heart. Several reports have characterized the roles of ncRNAs in different aspects of cardiogenesis, such as chamber morphogenesis, conduction, and contraction. Moreover, congenital abnormalities of the heart are known to be associated with the dysregulation of some ncRNAs (Bayoumi et al., 2017).

In general, ncRNAs are involved in many biological processes, including physiological regulation, developmental processes, and diseases. NcRNAs could be categorized into two major groups based on their regulatory roles. The first group, referred to as housekeeping ncRNAs, are profusely and ubiquitously expressed in cells, and predominantly regulate common cellular functions. On the other hand, regulatory ncRNAs are generally known to be key regulatory RNA molecules, functioning as regulators of gene expression at epigenetic, transcriptional, and post-transcriptional levels (Fig. 1). Regulatory ncRNAs have gained a growing interest in the cardiovascular community because of their critical roles in processes involved in CVDs, including neovascularization, atherosclerosis, hypertension, and aneurysm formation (van der Kwast et al., 2020).

In previous reports, we have shown the cardioprotective roles of microRNAs (miRNAs or miRs), including miR-150 (Tang et al., 2015), miR-199a-3p and miR-214 (Park et al., 2016), miR-532-5p (Bayoumi et al., 2017), and miR-125b (Bayoumi, Park, et al., 2018). We have also summarized the roles of miRs, long noncoding RNAs (lncRNAs), and circular noncoding RNAs (circRNAs) in human diseases in published reviews (Aonuma, Bayoumi, Tang, & Kim, 2018; Archer et al., 2015; Bayoumi et al., 2016; Bayoumi, Aonuma, Teoh, Tang, & Kim, 2018). In this article, we summarize more recent literature to describe the roles of ncRNAs in the

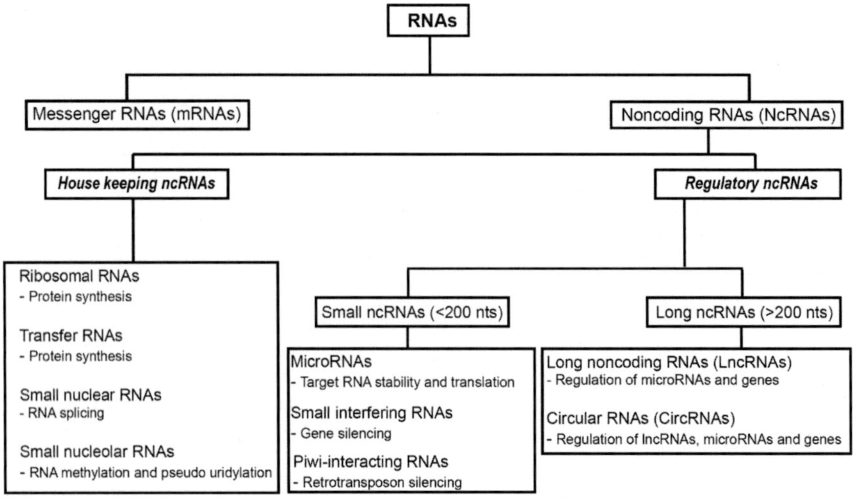

FIG. 1

Summary of the different groups of ncRNAs and their functions. Regulatory RNAs comprise most noncoding RNAs (ncRNAs) such as miRNAs, long noncoding RNAs (lncRNAs), and circular RNAs (circRNAs). These ncRNAs are categorized for practical reasons based on their size (*i.e.*, more or less than 200 nucleotides [nts] in length) into short- and long ncRNAs.

pathology of CVDs, and their potential as novel diagnostic and therapeutic agents. We also discuss the current knowledge of the implications of ncRNA interactions on several pathophysiological pathways associated with atherosclerosis, cardiac arrhythmia, cardiac hypertrophy, myocardial infarction, cardiac fibrosis, and pulmonary hypertension. We then emphasize the therapeutic potential of ncRNAs and suggest directions for future work.

2 Molecular functions of the different classes of regulatory ncRNAs

Regulatory ncRNAs consist of various RNA categories (Fig. 1), which could be classified based on their average size. These include small noncoding RNAs (sncRNAs), which are comprised of transcripts fewer than 200 nucleotides (nts) and lncRNAs with more than 200 nts. The primary classes of small ncRNAs are miRNAs, small interfering RNAs (siRNAs) and piwi-interacting RNAs (piRNAs). LncRNAs include circRNAs and linear lncRNAs (Gomes et al., 2020).

2.1 Small noncoding RNAs

2.1.1 MicroRNAs (miRNAs or miRs)

MiRNAs are the most known class of small ncRNAs with an average of 22 nucleotides in length (O'Brien, Hayder, Zayed, & Peng, 2018). The biogenesis of miRNA commences with the transcription of a primary (pri-) miRNA gene product, which includes a sequence of one to six precursor (pre-) miRNA repeats (Fig. 2). A sequence of nuclear processes by the enzyme complex of drosha ribonuclease III (DROSHA) and DiGeorge syndrome chromosomal region 8 (DGCR8) then synthesizes the pre-miRNA, a double-stranded RNA molecule comprising the mature miRNA connected to a complementary (star-) strand by a hairpin loop. The pre-miRNA is then translocated to the cytoplasm by the nucleocytoplasmic

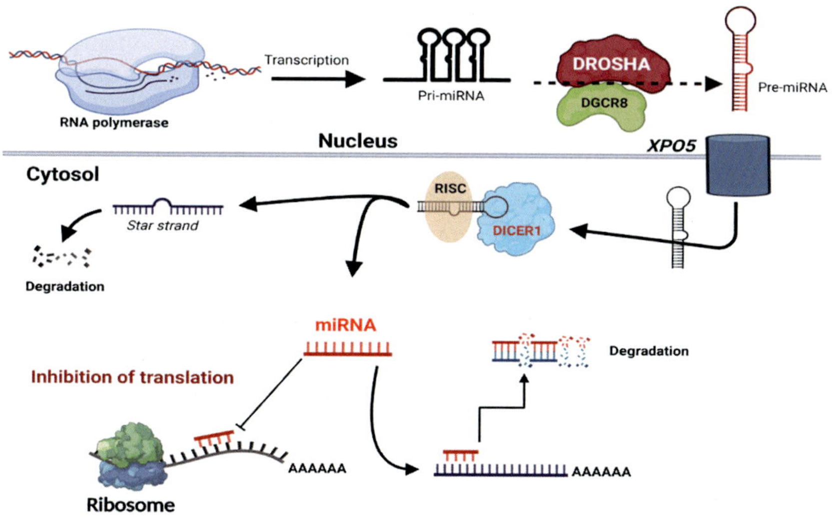

FIG. 2

Summary of microRNA (miRNA) biogenesis. MiRNA is at the outset synthesized as a primary (pri-) miRNA gene product comprising 1–6 repeat sequences of the precursor (pre-) miRNA, which includes the mature miRNA sequence linked to the complementary (star) strand by a hairpin loop. The pre-miRNA is cleaved from the pri-miRNA transcript by the DROSHA/DGCR8 enzyme complex and transported out of the nucleus by the XPO5. The hairpin loop is cleaved from the pre-miRNA by DICER1, and the miRNA is loaded into the RNA-induced silencing complex (RISC), where the star strand is sometimes deleted, granting the mature miRNA's capacity to bind to target mRNAs. The mature miRNA can inhibit protein synthesis by blocking translation by the ribosome or flagging the mRNA for degradation. Abbreviations: DGCR8, DiGeorge syndrome chromosomal region 8; DICER1, endoribonuclease dicer; DROSHA, drosha ribonuclease III; XPO5, exportin-5.

protein exportin-5 (XPO5). This step is followed by cleavage of the hairpin loop with endoribonuclease dicer (DICER1), enabling the miRNA to be inserted into the RNA-induced silencing complex (RISC) where the star-strand is sometimes deleted, and the mature miRNA is released and ready for binding to target mRNA transcripts (Gibbons, Udawela, & Dean, 2018). Generally, miRNAs bind with the 3′-untranslated region (UTR) of target mRNAs to regulate their expression in the cytoplasm and nucleus *via* distinct mechanisms. MiRNAs mediate gene silencing at the post-transcriptional level (Ha & Kim, 2014). However, interactions between miRNAs and other DNA regions, including 5′-UTRs, coding sequences, and gene promoters, have also been reported (Chipman & Pasquinelli, 2019). Moreover, miRNAs have been reported to activate gene expression under certain conditions (Vasudevan, 2012). Recent studies have also indicated that miRNAs are transported between distinct subcellular compartments to regulate the rate of translation and even transcription (Makarova et al., 2016).

2.1.2 PIWI-interacting RNAs (piRNAs)

The PIWI-interacting RNAs (piRNAs) are a distinct class of single-stranded regulatory ncRNAs with 23–36nts. They are known to interact with the piwi family of proteins, resulting in functional complexes to inhibit transposons (Wu et al., 2020). piRNAs are represented as clusters all over the genome. Each group can include around a dozen to a few thousand piRNAs. Similarly to miRNAs, piRNAs are derived from the noncoding regions of the genome or within protein-coding regions. They primarily target RNAs originating from transposons (namely retrotransposons) to cleave the transcripts, and preclude them from the reverse-transcription and an unexpected integration into the DNA (Vella et al., 2016). piRNAs are considered as "the guardians" of the genome as they prevent excessive mutations (Wu et al., 2020). Early studies in germ cells demonstrated that piRNAs are essential for spermatogenesis and embryonic development (Houwing et al., 2007). However, recent pieces of evidence indicate that piRNAs are not only limited to the germline but can be active also in somatic cells (Sharma et al., 2001), including cardiomyocytes (Sun & Han, 2019).

Additionally, it has been recently shown that piRNAs regulate AKT signaling pathway (Vella et al., 2016), which is critical in heart pathophysiology (Lin et al., 2015) and oxidative stress-mediated cell degeneration (Zheng, Zheng, et al., 2020). Vella et al. also suggested that piRNAs may possess a functional role in cardiac cell proliferation and regeneration, similar as microRNAs (Vella et al., 2016). A more recent study reported that piRNAs are differentially expressed in a mouse model of non-alcoholic fatty liver disease (NAFLD) and may regulate the development of NAFLD (Ma et al., 2020).

2.2 Long noncoding RNAs

2.2.1 Long noncoding RNAs (LncRNAs)

LncRNAs represent a large class of RNA transcripts, which are greater than 200nts in length. They possess more spatial and temporal specificity, and less interspecies conservation when compared with mRNAs (Ponting, Oliver, & Reik, 2009).

LncRNAs had been previously believed to be the by-products of the transcription process. However, it has been widely accepted that lncRNAs are involved in the process of cell differentiation and growth, and the pathogenesis of many diseases such as CVDs (Hu et al., 2018). Based on the relative location of lncRNAs to coding genes in the genome, they are grouped into several classes, which include sense lncRNAs, antisense lncRNAs, bidirectional lncRNAs, intron lncRNAs, intergenic lncRNAs, and enhancer lncRNAs (Wang & Chang, 2011).

Hitherto, lncRNAs have been reported to play crucial roles in the regulation of gene expression at the epigenetic, transcriptional, and post-transcriptional levels (Deans & Maggert, 2015). Indeed, they regulate gene expression through chromatin modification and remodeling, histone modification, and nucleosome localization changes (Peterson & Workman, 2000). The mechanism underlying chromosome remodeling includes the regulation of SWItch/Sucrose Non-Fermentable (SWI/SNF) complex, which is a chromosome reconstitution complex consisted of numerous subunits to process the transformation of nucleosome localization (Peterson & Workman, 2000). LncRNAs interact with SWI/SNF complex to modify the structure of the chromosome and therefore regulate gene expression (Chen et al., 2018). Also, lncRNAs are capable of controlling the chromatin remodeling by shaping the DNA methylation with S-adenosylhomocysteine hydrolase, by promoting DNA demethylation with growth arrest and DNA-damage-inducible alpha (GADD45A), and by acetylation of histone with Sirtuin 6 (SIRT6) (Chi, Wang, Wang, Yu, & Yang, 2019). Histone is a fundamental core unit of the nucleosome, which has a vital role in forming the body and function of chromosomes. Various lncRNAs have been shown to regulate histone methylation by reacting with polycomb repressive complex 2 (PRC2), and modifying the structure and function of chromosomes (Peterson & Workman, 2000). Moreover, lncRNAs act as competing endogenous RNAs (ceRNAs) that can indirectly regulate the activity of miRNAs *via* competitively binding with their target mRNAs (Fig. 3). This competitive binding of miRNAs is also known as miRNA sponges (Thomson & Dinger, 2016). Through this mechanism, lncRNAs can act as miRNA sponges to modulate gene expression and the development of CVDs (Hobuß, Bär, & Thum, 2019).

2.2.2 Circular noncoding RNAs (CircRNAs)

CircRNAs are single-stranded RNAs that were first reported in virus. They are characterized by a continuously closed RNA strand through covalent joining on the 5′ and 3′ ends (Hsu & Coca-Prados, 1979). They, therefore, possess several biological properties that are not yet to be fully uncovered; these include protein scaffolding, gene modulation, and miRNA sponging (Lu, 2020). Because of their closely-looped structures, which have no 5′ and 3′ ends, circRNAs are known to be relatively immune to exonuclease cleavage that improves their molecular efficiency. Similar to other ncRNAs and because of their singular structure, circRNAs do not code for proteins. Accumulating pieces of evidence have recently highlighted the ability of some circRNAs to act as miRNA sponges and thus efficiently hinder their actions. Indeed, circRNAs have the complementarity to bind themselves onto

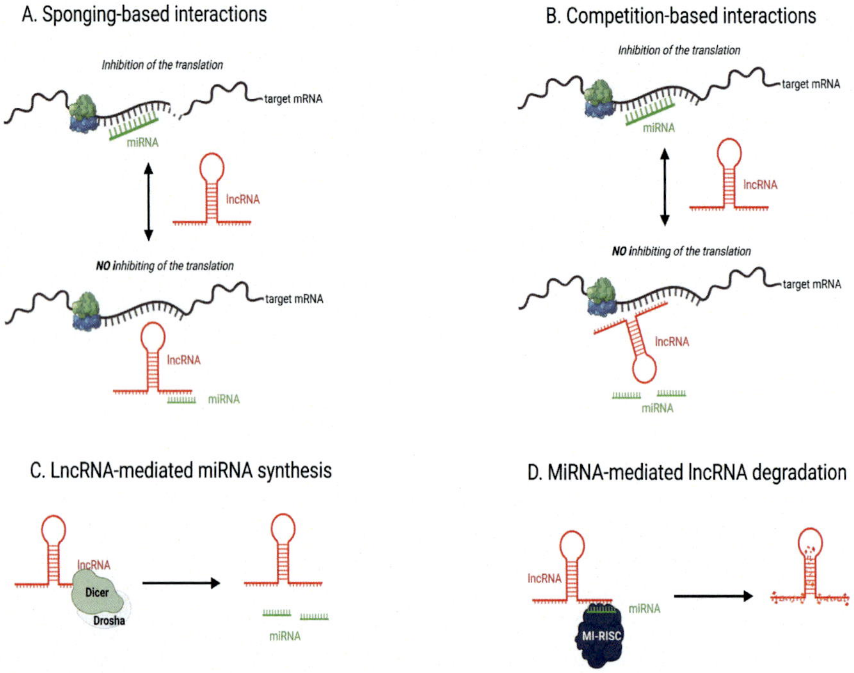

FIG. 3

Molecular impact of lncRNA–miRNA interaction on mRNA expression. There are four primary interaction mechanisms between lncRNAs and miRNAs to regulate gene expression. (A) LncRNAs can act as molecular sponges for miRNAs. LncRNAs bind miRNAs to inhibit their interaction with their target mRNAs, thus precluding silencing of target genes. (B) LncRNAs and miRNAs initiate a competition to bind the same target mRNAs. Thus, based on the stimulus, gene expression can be differentially expressed. (C) MiRNAs can be formed from lncRNAs by endoribonucleases such as Dicer and Drosha. (D) MiRNAs suppress lncRNA activity by targeting lncRNAs for degradation using the same mechanism by which they target mRNAs. Mi-RISC, miRNA-induced silencing complex.

miRNAs by attaching to seed regions on the miRNAs and competitively inhibiting their activity (Lu, 2020). One of the most cited examples illustrating the sponging ability of circRNAs is the cerebellar degeneration associated protein 1 antisense transcript (CDR1as). It has been reported that CDR1as comprises 70 conserved binding sites for miRNA-7, far important than any other known linear miRNA sponges. Its sponging ability was demonstrated in a report by Memczak et al. (2013). The researchers used the sequestering of CDR1as molecules against the raised expression of miR-7 in zebrafish brains, and found CDR1as's inhibitory properties against its targeted miRNA and its subsequent effects on zebrafish midbrain development.

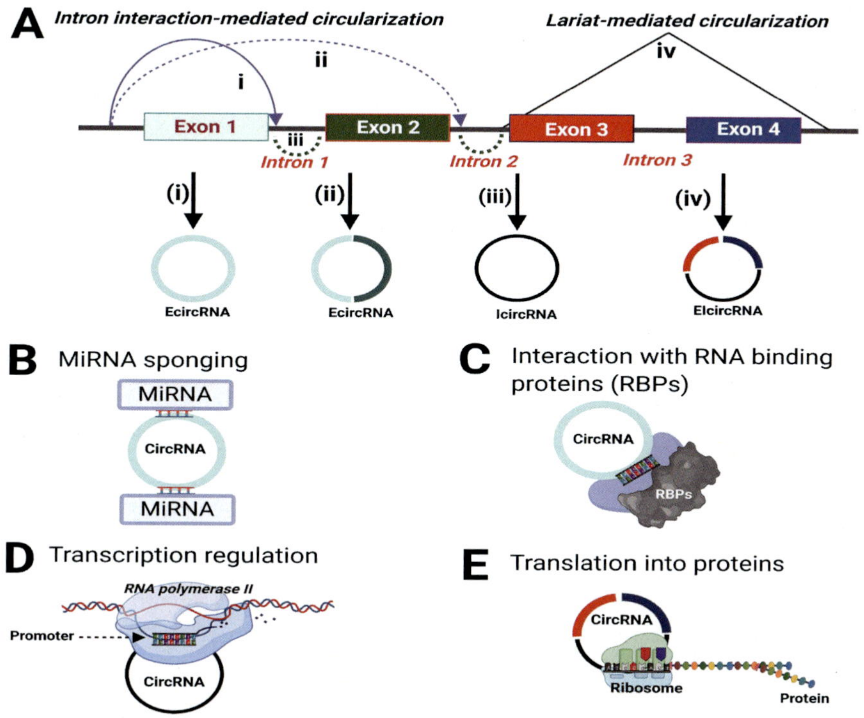

FIG. 4

Illustration of circRNA biogenesis and cellular function. (A) Exonic circular RNAs (EcircRNAs) originate from back-splicing of the 5′splice (donor) site to a 3′splice (acceptor) site *via* 3′,5′-phosphodiester bond. (i) Exon can omit splicing, leading to intron-free transcripts. (ii) The first intron is detached and results in the 5′ splice site of Exon 2 to be closer to 3′ splice site of Exon 1, leading to an ecircRNA with 2 exons. (iii) Intronic circular RNAs (IcircRNAs) originate from intron lariats that skip the standard intron debranching and degeneration. (iv) Exonic-intronic circular RNAs (EIcircRNAs) are generated by lariat-mediated circularization of exons and introns. (B) CircRNAs can act as miRNA sponges which possess several miRNA binding sites to indirectly regulate gene expression. (C) RNA binding proteins (RBPs) can interact with circRNAs, leading to the regulation of multiple cellular processes such as cell cycle progression. (D) CircRNAs act as transcriptional regulators through direct interactions with RNA polymerase II in the promotor region. (E) Some circRNAs have an open reading frame that can be translated by ribosome into proteins.

Recently, circRNAs have become the spotlight of research in heart disease, as they have been shown to elicit significant roles in CVDs. Our previous review provided a comprehensive summary of their biological function and mechanism, as well as their potential as biomarkers for heart diseases (Bayoumi et al., 2016; Bayoumi, Aonuma, et al., 2018). Here, we accordingly focus on the impact of their interactions with miRNAs and genes in CVDs as summarized in Fig. 4.

3 Interactions between the different types of noncoding RNAs in cardiovascular diseases

3.1 Atherosclerosis

Atherosclerosis, a leading cause of CVD, is a chronic affection of various types of arteries, including elastic and muscular-elastic types (Shemiakova et al., 2020). The pathophysiology of atherogenesis is characterized by multifocal structural modifications in the wall of large and medium-sized arteries, and the consequent development of atherosclerotic plaques. Many molecular processes are mechanistically implicated in the pathogenesis and development of atherosclerosis. These include angiogenesis, inflammatory responses, adipogenesis, lipid metabolism, cell proliferation, apoptosis and oxidative stress (Yang, Wu, et al., 2020). These processes are important for the development of atherosclerosis and eventually trigger thrombotic plaque complications, which have been shown to be regulated by a wide array of ncRNAs (Ou, Li, Zhao, Cui, & Tu, 2020).

The pathophysiology of atherosclerosis is characterized by the deposition of lipid in the intima and subintima of large and middle arteries (Raggi et al., 2018). Additionally, the migration of smooth muscle cells (SMCs) to the intima, their proliferation, and the initiation of the inflammatory pathways, play important roles in atherosclerosis (Orekhov, Andreeva, Mikhailova, & Gordon, 1998). These factors contribute to the thickening of the intima and the development of atherosclerotic lesions or fibro-lipid plaque lesions. Another factor involved in the progression of atherosclerosis is the regulation of vascular smooth muscle cells (VSMCs), a critical cellular component of the vascular wall. During development and maturation, VSMCs regulate vasoconstriction and relaxation, and react to the stimulation of various hemodynamic and environmental signals to stabilize the blood pressure and control vascular homeostasis in the body (Zhu, Liu, Zhao, & Yan, 2020). VSMCs are characterized by strong plasticity. In a healthy state, VSMCs have a limited capacity of proliferation, migration, and a controlled secretion of the extracellular matrix, known as contractile-differentiated VSMCs. However, in pathological states, VSMCs display significant proliferative and migratory activities at a very early stage of maturation when physiological conditions change. VSMCs also synthesize a large quantity of extracellular matrix, which is termed secretory-proliferative VSMCs. The proliferation and differentiation of VSMCs are important regulatory processes in the maturation and development of the vascular system. Therefore, alterations of vascular intima are followed by increased proliferation and migration of VSMCs, and the dysregulated synthesis of the cellular matrix, which significantly contribute to the pathology of atherosclerosis (Zhu, Liu, et al., 2020).

Recent evidence has suggested that a diversity of ncRNAs are involved in the regulation of atherosclerosis. For example, Zhu et al. reported that lncRNA-SNHG14, also known as UBE3A-ATS, is associated with dysfunction of VSMCs in atherosclerosis. Indeed, their results showed that lncRNA-SNHG14 is significantly decreased in atherosclerotic plaque tissues of ApoE$^{-/-}$ mice. Also, the authors observed that overexpression of lncRNA-SNHG14 led to reduced VSMC proliferation while promoting apoptosis. Also, they reported a potential regulatory interaction between lncRNA-SNHG14 and miR-19a-3p, in which lncRNA-SNHG14 acts as a ceRNA of miR-19a-3p and inhibits its expression in VSMCs. Additionally, using luciferase reporter analysis, they showed that miR-19a-3p binds to the 3′UTR of RORα to regulate its expression. This was further confirmed by qRT-PCR analysis, which showed a significant increase of RORα in the aortas treated with miR-19a-3p and SNHG14 compared with those treated with miR-19a-3p alone (Zhu, Liu, et al., 2020). Overall, this study suggests that the lncRNA-SNHG14/microRNA-19a-3p/RORα axis may regulate the dysfunction of atherosclerotic VSMCs.

Another study showed that downregulation of lncRNA HOXA-AS3 reduces the progression of atherosclerosis through the sponging of miR-455-5p. Indeed, using human vascular endothelial cells (HUVECs) treated with oxidized low-density lipoprotein (ox-LDL) as a model of atherosclerosis *in vitro*, authors showed that lncRNA HOXA-AS3 was upregulated in oxLDL-treated HUVECs. Also, the suppression of lncRNA HOXA-AS3 increased the growth of HUVECs following their exposure to ox-LDL by preventing its ox-LDL-induced G1 cell cycle arrest. Mechanistically, they used RNA pull-down and luciferase reporter assays to verify that HOXA-AS3 bound to miR-455-5p (Chi et al., 2020). In conclusion, this study demonstrated that suppression of lncRNA-HOXA-AS3 downregulates the progression of atherosclerosis through the regulation of miR-455-5p/p27 Kip1 axis.

Another type of ncRNAs involved in atherosclerosis includes circRNAs, which have been reported to impact the progression of atherosclerosis and regulate the activity of other ncRNAs such as miRNAs and lncRNAs as well as protein-coding genes. For instance, Ding et al. demonstrated the functions and molecular mechanisms of circ_0010283 in ox-LDL-induced VSMC function and atherosclerosis. The authors demonstrated that the expression of circ_0010283 was increased in ox-LDL-induced VSMCs as compared with those in VSMCs without ox-LDL, whereas the expression of miR-370-3p, a potential functional target of circ_0010283 was downregulated. In the same token, they observed an increase of HMGB1, which is regulated by miR-370-3p. Also, their results showed that knockdown of circ_0010283 reduced VSMC viability and migration. This was further demonstrated by a decrease of the viability-associated proteins (cyclin D1 and proliferating cell nuclear antigen), as well as migration-associated proteins [matrix metalloproteinase 2 (MMP2) and MMP9] in oxLDL-induced VSMCs compared with untreated VSMCs. Moreover, using RNA immunoprecipitation (RIP) and luciferase reporter assays, they demonstrated the interaction between circ_0010283 and miR-370-3p, and revealed HMGB1 as a target of miR-370-3p. Based on siRNA- and miRNA mimic-based strategies, they also showed that the inhibition of miR-370-3p reversed the circ_0010283′s inhibitory effects on VMSC viability and migration. Their functional assay also revealed that miR-370-3p's suppressive effects on cell viability and migration were rescued after the upregulation of HMGB1. Overall, this study showed that circ_0010283 regulates the progression of atherosclerosis through the miR-370-3p/HMGB1 axis (Ding, Ding, Tian, & Lei, 2020).

Another study reported the importance of one circRNA in diabetic atherosclerosis and its mechanisms of actions (Wei et al., 2020). The authors first demonstrated upregulation of circVEGFC in high glucose (HG)-induced HUVECs using circRNA microarray analysis. Functionally, they demonstrated that knockdown of circVEGFC reduced apoptosis and recovered HUVEC proliferation induced by HG stimulation. Mechanistically, they showed that circVEGFC acts as a ceRNA of miR-338-3p, repressing its expression. Also, they revealed that miR-338-3p regulated its functional target hypoxia-inducible factor 1 alpha (HIF-1α), a critical transcription factor by binding to its 3′-UTR, establishing the circVEGFC/miR-338-3p/HIF-1α axis. Taken together, this study shows the role of circVEGFC/miR-338-3p/HIF-1α/VEGFA axis in the HG-induced EC apoptosis and provides a potential treatment strategy for diabetic atherosclerosis.

Accumulating evidence has also highlighted that miRNAs play critical roles in various pro-atherosclerotic processes, including inflammation, cell proliferation, and apoptosis by binding to the 3′-UTR of target mRNAs. Therefore, the dysfunction of miRNA metabolism may represent one of the molecular events underlying the development and progression of atherosclerosis. This notion has been supported by several studies. For example, Wang et al. investigated the role of miR-16 in atherosclerosis and explored its anti-inflammatory-mediated therapeutic potential on atherosclerosis in a $ApoE^{-/-}$ mouse model fed with a high-fat diet. The results from this study showed that miR-16 was downregulated in the plasma and peripheral blood mononuclear cells of arteriosclerotic patients as compared to normal subjects. Its expression level was negatively associated with IL-6 and the severity of atherosclerosis. Conversely, miR-16 expression was positively correlated to the IL-10 level. After injecting $ApoE^{-/-}$ mice with miR-16 agomiR, the authors observed a decrease in the formation of atherosclerotic plaque and the accumulation of proinflammatory factors (IL-6, TNF-α, MCP-1, and IL-1β) in the plasma and tissues. However, miR-16 agomiR injection promoted the secretion of anti-inflammatory factors (IL-10 and TGF-β). Using luciferase reporter and functional assays, they revealed that overexpression of miR-16 downregulated PDCD4, its functional target, as well as activated p38 and ERK1/2, but inactivated the JNK pathway (Wang, Li, Cai, et al., 2020). In conclusion, these findings suggest that miR-16 may play a key role in the regulation of inflammation during atherosclerosis.

de Yébenes et al. also reported the importance of miRs in atherosclerosis by using gain- and loss-of-function strategies in mice. The authors showed that miR-217 promotes endothelial dysfunction and atherosclerotic-mediated vascular dysfunction (de Yébenes et al., 2020). In this study, they developed an inducible endothelium-specific knock-in mouse model to demonstrate the role of miR-217 in vascular function and atherosclerosis. Their results showed that miR-217 expression is upregulated during physiological aging and that increasing endothelial miR-217 was associated with downregulation of the metabolic sensor *Sirt1*. The inducible endothelium-specific knock-in mice also showed the impairment of left ventricular diastolic function, which was characterized by decreased left ventricular relaxation. In opposition, inhibition of endogenous vascular miR-217 in ApoE$^{-/-}$ mice improved vascular contractility and suppressed atherosclerosis. Mechanistically, they reported that miR-217 inhibited endothelial NO synthase (eNOS) pathways in endothelial cells, and this was associated with reduced NO production, augmented blood pressure, and exacerbated atherosclerosis in proatherogenic ApoE$^{-/-}$ mice. Also, transcriptome analysis confirmed that miR-217 downregulates eNOS activators, such as VEGF and apelin receptor pathways, leading to decreased eNOS expression. Altogether, these data suggest that miR-217 may play an important role in atherosclerosis by interacting with eNOS pathways. Similar mechanisms of actions have been observed with a diversity of ncRNAs shown by recent studies published in 2020 (Table 1).

3.2 Cardiac arrhythmia

Cardiac arrhythmia (CA) originates from the dysfunction in electrical excitation processes that regulate coordinated and effective cardiac contraction. Defects of cardiac rhythm are linked to considerable morbidity and economic costs. One of the most common forms of CAs, atrial fibrillation (AF) affects at least 2.3 million people in the US and is correlated with increased risks of stroke and mortality. Ventricular arrhythmias are associated with about 75–80% of cases of sudden cardiac death, which are anticipated to result in 184,000–450,000 deaths in the US per year (Khurshid et al., 2018). Compelling published reports have demonstrated that AF has a substantial genetic basis. Even though factors such as sex, aging, and comorbidities contribute to AF risk, several studies have shown that a family history of AF grants an increased risk to the disease (Xia et al., 2005).

Indeed, several studies have reported AF candidate genes and risk loci, such as ion channels, genes implicated in fibrosis and the extracellular matrix (ECM) remodeling, genes involved in cardiogenesis, and genes implicated in the cell-cell coupling or nuclear structure. In 2003, Chen et al. discovered the first mutation in a gene associated with familial AF, which is a gain-of-function mutation in KCNQ1 coding for the α-subunit of the slowly repolarizing potassium current, IKS (Chen et al., 2003). Secondarily to these initial discoveries, various genes associated with AF development have since been identified. Diverse potassium channel mutations have now been associated with AF (Yang, Yang, Roden, & Darbar, 2010). In addition to mutations in the channels themselves, mutations in channel accessory proteins have been associated with AF. A large amount of the mutations in these proteins could be linked to an increased channel activity, which in turn, might decrease action potential duration and refractoriness in the atria. Mutations associated with reduced channel activity can also prolong the atrial action potential and lead to AF.

Table 1 Roles of noncoding RNAs in atherosclerosis.

NcRNA type	Experimental model	Mechanisms of actions	Pathology	Role	References
LncRNA Ang362	Human	Lipid metabolism	Atherosclerosis	Upregulated during atherosclerosis	Wang, Gong, Liu and Feng (2020)
LncRNA CASC2	oxLDL-treated VSMCs	MiR-532-3p/PAPD5	Atherosclerosis	Apoptosis	Lin, Yang, et al. (2020)
MiR-16	$ApoE^{-/-}$ mice	JNK pathway	Atherosclerosis	Inhibition of inflammatory pathways	Wang, Li, Cai, et al. (2020)
CircVEGFC	HG-induced vascular endothelial cells	MiR-338-3p/HIF-1α/VEGFA axis	Atherosclerosis	Apoptosis	Wei et al. (2020)
MiR-122	Rats	SIRT6/ELABELA/ACE2 axis	Atherosclerosis	Promotion of angiotensin II-mediated apoptosis	Song et al. (2020)
LncRNA Sox2ot	$Fbn1^{C1039G/+}$ mice	MiR-330-5p/MYH11 axis	Atherosclerosis	Downregulation of vascular dysfunctions	Xiao, Li, Ji, Shi and Pan (2020)
MiR-181b	Monocytes	PIAS1-KLF4 axis	Atherosclerosis	Repression of coronary heart diseases	Wang, Li, Sun, et al. (2020)
MiR-9	HUVEC	Sphingosine 1-phosphate receptor 1	Atherosclerosis	Promotion of angiogenesis	Yao, Xie and Zeng (2020)
LncRNA H19	$ApoE^{-/-}$ mice	p53-mediated VSMC apoptosis	Atherosclerosis	Prevention of atherosclerosis	Safaei, Tahmasebi-Birgani, Bijanzadeh and Seyedian (2020) and Sun, Jiang, Sheng and Cui (2020)
LncRNA PEBP1P2	Human aortic smooth muscle cell line and rat carotid artery injury model	CDK9	Atherosclerosis	Suppression of vascular smooth muscle cell (VSMC) proliferation	He et al. (2020)

Circ_0010283	oxLDL-induced VSMCs	MiR-370-3p/HMGB1 axis	Atherosclerosis	Promotion of VSMC proliferation	Ding et al. (2020)
MiR-448	ApoE$^{-/-}$ mice and aortic smooth muscle cells (ASMCs)	PRDM16 and TGF-β signaling pathway	Atherosclerosis	Repression of ASMC proliferation	Liu, Song, et al. (2020)
MiR-217	MiR-217$^{KI/+}$; VE-Cad CreERT2$^{TG/+}$ mice and ApoE$^{-/-}$ mice	VEGF and eNOS signaling pathways	Atherosclerosis	Promotes cardiovascular dysfunction	de Yébenes et al. (2020)
LncRNA-SNHG14	ApoE$^{-/-}$ mice and H_2O_2-treated cardiomyocytes	MiR-19a-3p/RORα axis and miR-770-5p	Atherosclerosis and myocardial infarction	Promotes atherosclerosis and proliferation of primary cardiomyocytes	Yang, Lu, et al. (2020) and Zhu, Liu, et al. (2020)
MiR-29a-3p	ApoE$^{-/-}$ male mice and human VSMCs	TNF receptor-1 (TNFRSF1A)	Atherosclerosis	Reduction of oxLDL-induced VSMC activation	You, Chen, Xu and Li (2020)
LncRNA HOXA-AS3	Mice with high-fat diet and HUVECs	MiR-455-5p/p27 Kip1 axis	Atherosclerosis	Suppression of the progression of atherosclerosis	Chi et al. (2020)
LncRNA THRIL	Endothelial progenitor cells	AKT pathway and FUS	Atherosclerosis	Inhibition of proliferation and mediation of autophagy in endothelial progenitor cells	Xiao, Lu and Yang (2020)
MiR-200a	oxLDL-induced lipid storage in macrophages	Nrf2 signaling pathway	Atherosclerosis	Decreases oxLDL levels in the artery	Mu, Zhang and Lei (2020)

NcRNAs' primary function is the regulation of gene expression post-transcriptionally *via* various mechanisms, which generally include the binding to complementary target sites within mRNAs resulting in the inhibition of translation or degradation of target transcripts. Indeed, ncRNAs have been shown to regulate genes involved in electrical and structural remodeling directly linked to the development of AF (Table 2). For instance, Wang et al. showed that LINC00472 was downregulated in the plasma of AF patients. Also, methylation levels of all five CpG islands in its promoter were significantly elevated in AF tissues as compared to normal tissues. This observation guided to the hypothesis that DNA methylation may play an essential regulatory role in the expression of LINC00472 and may be linked to the progression of AF. Conversely, the expression of miR-24 was found to be increased in the plasma of AF patients. By means of *in silico* analysis and luciferase reporter assays, they also showed that LINC00472 is a competing endogenous RNA of miR-24 and might negatively regulate the expression of JP2 by directly binding to its 3′UTR, contributing to the development of AF (Wang, Shen, et al., 2019). Overall, this study showed that the dysregulation of LINC00472 in AF patients increases the expression of miR-24, a competing endogenous RNA of LINC00472, and thus leads to the suppression of JP2 and RyR2, which may contribute to the pathogenesis of AF.

In the same token, Yao et al. demonstrated the function of lncRNA myocardial infarction-associated transcript (MIAT) in AF (Yao, Zhou, You, Hu, & Xie, 2020). Using a rat model of AF, the authors showed that the expression of MIAT was significantly increased. This increase was associated with downregulation of miR-133a-3p in atrium tissues of rats with AF induction. This result was further confirmed by data from human samples, showing that MIAT was highly expressed, and miR-133a-3p was decreased in peripheral blood leukocytes of AF patients. From a functional perspective, MIAT downregulation significantly lessened AF, improved atrial effective refractory period (AERP), and decreased the duration of AF as well as cardiomyocyte apoptosis. Interestingly, these effects of MIAT downregulation were reversed by antimiR-133a-3p injection. A luciferase reporter assay also revealed that miR-133a-3p was a direct functional target of MIAT. More importantly, MIAT knockdown efficiently decreased AF-induced atrial fibrosis by reducing collagen in the right atria and repressing the expression of fibrosis-related genes, including *collagen I*, *collagen III*, *connective tissue growth factor (Ctgf)*, and *transforming growth factor-β1* (*Tgf-β1*) in rats with AF. In conclusion, this study demonstrated that MIAT interaction with miR-133a-3p may contribute to the development of AF, and this axis may represent a potential therapeutic target for the management of AF.

Recent studies of circRNAs in AF patients have also provided important insights on the pathogenesis of AF (Table 2). As an example, Ruan et al. used microarray analyses to screen the differentially expressed circRNAs in peripheral blood monocytes between four AF patients and four healthy subjects. The results revealed that 120 circRNAs were differentially expressed in AF patients (with fold-change >2 and $P < 0.05$) in comparison to the healthy control group, of which 65 circRNAs were

Table 2 Roles of noncoding RNAs in atrial fibrillation.

NcRNA type	Experimental model	Mechanisms of actions	Pathology	Role	References
LncRNA LINC00472	Human, HCMs and H9C2 cells	MiR-24/JP2/RyR2 signaling pathway	Atrial fibrillation	Downregulated in AF	Wang, Shen, et al. (2019)
LncRNA MIAT	Rat model of AF	MiR-133a-3p	Atrial fibrillation	Upregulated in AF	Yao, Zhou, et al. (2020)
MiR-106b, -26a-5p, -484, and -20a-5p	Human	Atrial remodeling	Atrial fibrillation	Upregulated in AF	Vaze et al. (2020)
CircRNA_0031, CircRNA_1837, CircRNA_5901, and CircRNA_7571	Human and peripheral blood monocytes	Phosphatidylcholine acyl chain remodeling and phospholipid biosynthesis	Atrial fibrillation	Upregulated in AF	Ruan et al. (2020)
CircRNA_5801, CircRNA_7386, and CircRNA_7577	Human and peripheral blood monocytes	Phosphatidylcholine acyl chain remodeling and phospholipid biosynthesis	Atrial fibrillation	Downregulated in AF	Ruan et al. (2020)
LncRNA DSCAM-AS1	Mouse or Rat CMs	MiR-188-5p/GRK2 axis	Atrial fibrillation	Upregulated in AF	Chen and Cai (2020)
MiR-206	Mouse hearts and CMs	GJA1/Connexin43 axis	Atrial fibrillation	Induces abnormal heart rate and PR interval	Jin et al. (2019)
MiR-32-5p, miR-107, and miR-144-3p	Human	Biomarker	Atrial fibrillation	Downregulated in AF	Kiliszek et al. (2020)
MiR-338-5, miR-10a-5p, and miR-200b-5	Human	Biomarker	Atrial fibrillation	Downregulated in AF	Larupa Santos, Rodríguez, Olesen, Bentzen and Schmitt (2020)
MiR-429	Langendorf perfusion of isolated rat hearts/ Ischemia reperfusion	Regulation of GJA1 and KCNJ2	Reperfusion and arrhythmia	Downregulated in arrhythmia	Tang, Gao, et al. (2020)
MiR-17	Langendorf perfusion of isolated rat hearts/ Ischemia reperfusion	Regulation of GJA1 and KCNJ2	Reperfusion and arrhythmia	Upregulated in arrhythmia	Tang, Gao, et al. (2020)
LncRNA PANCR	Human atrial tissues	Regulation of PITX2c	Atrial fibrillation	Upregulated in AF	Gore-Panter et al. (2016)
MiR-499	Human atrial tissues and HL-1 cells	Regulation of CACNB2	Atrial fibrillation	Upregulated in AF	Ling et al. (2017)
LncRNA TCONS_00202959	Acetylcholine/CaCl2-stimulated rats	Regulation of cardiac autonomic nervous function	Atrial fibrillation	Downregulated in AF	Zhao, Zhu, Lei, Zhang and Li (2018)

upregulated and 55 circRNAs were downregulated (Ruan et al., 2020). These findings suggest that these differentially expressed circRNAs may play essential roles in the pathogenesis of AF. Four upregulated circRNAs (circRNA_0031, circRNA_1837, circRNA_5901, and circRNA_7571), and 3 out of 4 downregulated circRNAs (circRNA_5801, circRNA_7386, and circRNA_7577) were validated by qRT-PCR analyses, showing an expression pattern that was consistent with microarray results. To further assess the roles of differentially expressed circRNAs in AF, *in silico* analysis of genes associated with the differentially expressed circRNAs was conducted. GO analysis of differentially expressed circRNAs suggests that the most extensively enriched biological process was primarily involved in phosphatidylethanolamine acyl chain remodeling, phosphatidylcholine acyl chain remodeling, and phospholipid biosynthesis. The most extensively enriched cell composition was mainly engaged in unique particles, cell surfaces, and extracellular areas. The most considerably enriched molecular function was primarily involved in 2-acyl glycerol-3-phosphate acyltransferase activity, transmembrane signal receptor activity, and 1-acyl glycerol-3-phosphate acyltransferase activity. Also, miR-328 was the largest node that interacts with the dysregulated circRNAs in the circRNA-miRNA network. Overall, this report suggests that interactions between identified circRNAs and miR-328 might play essential roles in AF, and a better understanding of these interactions may provide therapeutic benefits.

MiRNAs also play important roles in the development of AF (Table 2). As an example, Jin et al. investigated the role of miR-206 in cardiac arrhythmias by using a mouse model. In this study, the authors demonstrated that miR-206 was upregulated in AF. MiR-206 overexpression also induced abnormal heart rate and PR interval, and shortened lifespan of AF mice. Using a luciferase reporter assay, they showed that *Connexin43* (*Cx43*), an essential cardiac gap junction gene, was a direct functional target of miR-206. Also, overexpression of miR-206 in the mouse heart downregulated *Cx43* expression, mainly in the atria and ventricle (Jin et al., 2019). This study suggests that miR-206 is an upstream regulator of *Cx43*, and its overexpression downregulates *Cx43*, leading to the worsening of AF. Several miRNAs have been also found to participate in the regulation of the fibrotic remodeling happening during AF. MiR-21 suppresses sprouty RTK signaling antagonist 1 (SPRY1), a repressor of the extracellular signal-regulated kinase (ERK) pathway. In AF, ERK pathway stimulates fibrosis indirectly through miR-21-induced SPRY1 suppression (Cardin et al., 2012).

In addition, miR-21 regulates cardiac fibrosis *via* the transcription factor signal transducer and activator of transcription 3 (STAT3) to reduce the expression of cell adhesion molecule 1 (CADM1) (Cao, Shi, & Ge, 2017). It is also known that WW domain-containing E3 ubiquitin-protein ligase 1 (WWP1) expression levels are suppressed to stimulate TGF-β1/Smad2 signaling pathway, which promotes cardiac fibroblast proliferation in AF patients (Tao, Zhang, Yang, & Shi, 2018). Lastly, miR-29 targets several extracellular matrix genes, comprising *collagens*, *fibrillins*, and *elastins*. This miRNA is inhibited, and its expression is inversely correlated with extracellular matrix protein levels and the development of AF (Dawson et al., 2013).

3.3 Cardiac hypertrophy

Cardiac hypertrophy is defined as an adaptive response to stress-induced damages linked to pathological pressure overload or neurohormonal stimulation (Frey & Olson, 2003). The first step in this response is compensatory and aims to decrease wall stress through the thickening of the ventricular wall, which in turn maintains normal cardiac function (Collins, Binder, Chen, & Wang, 2020). Nevertheless, continuous hypertrophy often leads to maladaptive cardiac remodeling, which can result in heart failure. Cardiac hypertrophy and dysfunction are followed by the activation of diverse signaling pathways, leading to increased cardiomyocyte size, protein synthesis, organelle alterations, and expression of fetal genes (Nakamura & Sadoshima, 2018). However, the molecular mechanisms underlying the pathogenesis of cardiac hypertrophy are still not fully understood. Various studies have reported that ncRNAs are important players in cardiac hypertrophy in different experimental models.

The role of miRNAs in cardiac hypertrophy was reported for the first time in mouse cardiac tissues in response to pressure overload and activated calcineurin upregulation, which led to pathological cardiac remodeling. Several dysregulated miRNAs were then identified in failing human hearts (Kumarswamy & Thum, 2013). MiRNA-195 was one of the identified miRNAs, and its expression was found to be increased during cardiac hypertrophy. The upregulation of this miRNA contributed to cardiac pathological hypertrophy, which results in heart failure in mice. These findings showed important roles of miRNAs in hypertrophic growth associated with remodeling of the heart following the activation of pathological signaling. Another report from the same group demonstrated that the cardiac-specific conserved miRNA, miR-208 is essential for hypertrophy, fibrosis, and βMHC upregulation in cardiomyocytes during stress conditions (van Rooij et al., 2006, 2007). A study led by Zhang et al. also demonstrated that miRNA-29 has an important regulatory effect on cardiac hypertrophy, as its overexpression appeared to inhibit angiotensin II-induced cardiac hypertrophy. Collagen I and III secretions, and TGFβ and pSMAD2/3 levels were also suppressed, indicating that cardiac hypertrophy was actually repressed (Zhang, Yun, et al., 2020).

Notably, a recent study reported the role a lncRNA ZEB2 antisense RNA 1 (ZEB2-AS1) in regulating the hypertrophic process of cardiomyocytes (Cheng, Liu, & Li, 2020). Using a mouse model through transverse aortic constriction procedures and an *in vitro* model of cardiac hypertrophy by stimulating mouse cardiomyocytes with phenylephrine, the authors showed that ZEB2-AS1 was upregulated in cardiac hypertrophy. Its overexpression was positively correlated to ANP and BNP and negatively with PTEN expression both *in vivo* and *in vitro*. Also, ZEB2-AS1 knockdown reduced cell surface area, and downregulated ANP and BNP expression levels in phenylephrine-treated cardiomyocytes. PTEN overexpression also downregulated ANP and BNP expression levels. Moreover, PTEN reversed the regulatory activity of ZEB2-AS1 on hypertrophic cardiomyocytes. Overall, this study shows that lncRNA ZEB2-AS1 may regulate the progression of cardiac hypertrophy by altering PTEN expression.

3.4 Myocardial infarction

Acute coronary syndrome (ACS) is the most hazardous and lethal form of coronary heart disease. One such condition is a heart attack (myocardial infarction: MI). In the United States, there were around 633,000 patients discharged with ACS as the first diagnosis in 2014. The 30-day death rate of ACS patients was 5.3% for men and 9.6% in women. ACS can be grouped into two main categories, which are ST-Segment Elevation ACS (STE-ACS) and Non-ST-Segment Elevation ACS (NSTE-ACS) based on the existence of ST-segment elevation in the electrocardiogram. Although the mechanism underlying the pathology of ACS is not fully uncovered, a mounting number of evidences suggest that ncRNAs may play crucial roles in the development and progression of ACS (Wang, Guo, Hu, et al., 2020). In a recent study, five microRNAs were found to be upregulated in patients with ACS. Among those, four (miR-122, miR-150, miR-195, and miR-16) were found to be involved in pathophysiological mechanisms of ACS (Elbaz et al., 2020). Another investigation demonstrated that miR-21 is upregulated in plasma samples from ACS patients as compared to normal subjects. Mechanistically, miR-21 represses its direct functional target, sprout homolog 1 (SPRY1) to activate the pathogenesis of ACS (He et al., 2019). Another relevant study also showed that miR-101 is upregulated in ACS patients. This study indicated that the upregulation of miR-101 in HUVECs increased apoptosis and reduced migration. Mechanistic studies showed that miR-101 binds to the 3′-UTR of CHD5 and inhibits its expression. Indeed, genetic manipulation studies revealed that CHD5 knockdown activated apoptosis and repressed the migration of HUVECs (Cao et al., 2019).

The lncRNA THRIL is another ncRNA that has been shown to be upregulated in ACS, suggesting its potential as a diagnostic and therapeutic target (Qi, Shen, & Zhou, 2020). The circRNA-miRNA-mRNA regulatory network has been also shown to play a key role in the pathology of ACS. A recent microarray analysis of plasma samples from ACS patients identified 266 differentially expressed circRNAs, among which 121 were upregulated and 145 were downregulated. Furthermore, *in silico* study revealed that they might regulate key miRNAs implicated in the development of ACS such as miR-133a-3p (Lin, Yang, et al., 2020). In conclusion, ncRNAs represent important targets for the development of biomarkers to detect ACS early as well as therapeutic agents to treat ACS.

MI causes a complex remodeling process that involves fibrosis, eventually initiating ischemic heart failure (Prabhu & Frangogiannis, 2016). NcRNAs have been shown to play important roles in this process. NcRNAs represent important targets in therapeutic approaches for diseases such as MI. Indeed, there are growing evidences supporting the significant impact of ncRNAs' interactions in MI. As an example, a recent study demonstrated the molecular mechanism underlying the role of circHipk3 during cardiac regeneration (Si et al., 2020). The results showed that circHipk3 is overexpressed in the fetal or neonatal hearts of mice. The transcription factor GATA4 binds to the circHipk3 promoter and regulates circHipk3 expression, leading to decreased proliferation of cardiomyocytes post-MI. At the same time, a reverse effect was observed after its overexpression. Also, circHipk3 upregulation stimulated coronary vessel endothelial cell proliferation, migration, and subsequent angiogenesis. Moreover, circHipk3 upregulation suppressed cardiac dysfunction and reduced the fibrotic area after MI. Mechanistically, circHipk3 stimulated CM proliferation by upregulating Notch1 intracellular domain (N1ICD) acetylation, thereby amplifying N1ICD stability and inhibiting its degradation. Additionally, circHipk3 was shown to act as a sponge for miR-133a to initiate CTGF upregulation, which activated endothelial cells. Overall, these results suggest that circHipk3 might play a key role in the prevention of heart failure post-MI.

Another study reported the role of circANXA2 in the development of myocardial ischemia (Zong & Wang, 2020). CircANXA2 regulated CM apoptosis by inhibiting miR-133 expression. The results showed that the expression of CircANXA2 is increased in CMs during ischemic conditions. This led to reduced miR-133, which was shown to be a target miRNA of circANXA2. Also, circANXA2 increased the stress level and CM apoptosis as reflected by an upregulation of LDH, MDA, SOD, and GSH-PX activity in H/R-treated H9c2 cells, as well as increased levels of pro-apoptotic genes *Bax* and *Cytochrome c*. In H9c2 cells, upregulating miR-133 can reverse the inhibition of proliferation induced by circANXA2 overexpression.

Moreover, Trembinski et al. recently identified an aging-regulated lncRNA (ENSMUST00000140003) named Sarrah, an antiapoptotic lncRNA that is downregulated after heart failure. Sarrah inhibition led to impaired contractile force development in the mouse heart. The results demonstrated that Sarrah binds to the promotor of NRF2 and regulates its expression to mitigate apoptosis. Overexpression of NRF2 partially prevented maladaptive cardiac remodeling mediated by Sarrah knockdown. This finding suggests the regulatory role of Sarrah on genes implicated in post-MI remodeling (Trembinski et al., 2020). Moreover, Zheng et al. used peripheral blood mononuclear cells (PBMCs) and plasma samples from four patients to identify novel differentially expressed lncRNAs after MI. The results from PBMCs showed the upregulation of 2677 upregulated lncRNAs, while 458 lncRNAs were downregulated. In the subjects' plasma samples, 41 were upregulated, and 51 were downregulated, among which four lncRNAs, uc002ddj.1, NR_047662 (downregulated), and ENST00000581794.1 ENST00000509938.1 (upregulated) were validated by qRT-PCR and found to be critical for the development of MI (Zheng, Liu, et al., 2020). Further recent studies supporting the implication of ncRNAs in myocardial infarction and heart failure are summarized in Table 3.

3.5 Cardiac fibrosis

Cardiac fibrosis is a recurrent phenotype present in a plethora of cardiac diseases. Cardiac fibrosis is characterized by the pathological accumulation of collagens and other ECM proteins in the heart. The development of effective anti-fibrotic therapies targeting cardiac fibrosis have been hindered by several factors such as the limited regenerative potential of the heart. Also, the molecular mechanism of cardiac

Table 3 Roles of noncoding RNAs in myocardial infarction and heart failure.

NcRNA type	Experimental model	Mechanisms of actions	Pathology	Role	References
MiR-489	Ischemia–reperfusion injury on cardiomyocytes	Regulation of IGF1	Myocardial infarction	Increases cardiac injury	Tang, Zhong, et al. (2020)
MiR-129-1-3p	Pirarubicin-induced toxicity on H9C2 and HL-1 cardiomyocytes	GRIN2D-mediated Ca^{2+} pathway	Heart injury	Cardioprotection	Li, Qin, Tan, et al. (2020)
MiR-125b-1-3p	Pirarubicin-induced toxicity on HL-1 cardiomyocytes	JunD signaling pathway	Heart injury	Cardioprotection	Li, Qin, Li, et al. (2020)
MiR-22-5p	Rat model of pirarubicin-induced toxicity	RAP1/ERK signaling pathway	Cardiac injury	Cardioprotection	Qin et al. (2020)
CircRNA _0010729	Oxygen/glucose deprivation-induced injury on human cardiomyocytes	MiR-145-5p/mTOR and MEK/ERK pathway	Myocardial ischemia	Cardioprotection	Jin and Chen (2019)
LncRNA SNHG14	H_2O_2-treated cardiomyocytes	MiR-770-5p	Myocardial infarction	Proliferation of primary cardiomyocytes	Yang, Lu, et al. (2020) and Zhu, Liu, et al. (2020)
LncRNA H19	Human plasma	Lipid metabolism	Myocardial infarction	Biomarker	Safaei et al. (2020)
LncRNA SNHG8	Hypoxia-Induced cardiomyocyte Injury	NF-κB pathway	Myocardial infarction	Promotion of myocardial infarction	Zhang and Bian (2020)
LncRNA TCONS_00006679	Human	Lipid metabolism	Dyslipidemia	Upregulated during dyslipidemia	Wang, Guo, et al. (2020)
LncRNA TCONS_00011823	Human	Lipid metabolism	Dyslipidemia	Downregulated during dyslipidemia	Wang, Guo, et al. (2020)
MiR-150-5p	Rat model of septic shock	Bcl pathway	Sepsis-induced myocardial depression	Regulation of cardiac depression	Zhu, Zhang, Wen, and Liu (2020)
MiR-27b	Obese rats	Proliferator-activated receptor gamma (PPARγ) pathway	Coronary Heart diseases	Regulation of lipid metabolism during hypoxia	Wang, Lu, Zhu, et al. (2020)
LncRNA ZEB2-AS1	Transverse aortic constriction (TAC) mice and primary cardiomyocytes	Regulation of PTEN	Cardiac hypertrophy	Upregulated after cardiac injury	Cheng et al. (2020)
Enhancer RNAs (IRENE-SS and IRENE-div)	hiPSCs	Regulation of *Nkx2-5*	Heart failure	IRENE-SS activates, but IRENE-div represses transcription of Nkx2-5	Salamon et al. (2020)
MiR-29b-3p	Rat cardiac fibroblasts (CFs)	Regulation of FOS	Myocardial infarction	Anti-fibrotic potential	Xue, Fan, Yang, Jiao and Li (2020)
CircHIPK3	Primary CFs	MiR-152-3p/TGF-β2 axis	Heart failure	Regulation of cardiac fibroblast proliferation	Liu, Wang, et al. (2020)
LncRNA FOXD3-AS1	AC16 cardiomyocytes	Regulation of miR-150-5p	Congenital heart disease	Upregulated after cardiac injury	Zheng, Peng, Zhang, Ai and Hu (2020)

LncRNA NEAT1	Rats and cell ischemic models	MiR-214/PTEN/PI-3K/Akt pathway	Cerebral ischemia-reperfusion	Upregulated during ischemia	Shen, Ma, Shao, Jin and Bian (2020)
MiR-30e-3p	Rat primary cardiomyocytes	Regulation of *Egr-1*	Heart failure	Cardioprotective	Su et al. (2020)
CircRNA CDR1	Pig	Regulation of miRNA-671-5p	Myocardial infarction	Upregulated during ischemia	Mester-Tonczar et al. (2020)
LncRNA NEAT1	Ischemic mouse models and HL-1 cells	Regulation of pri-miRNA processing	Heart failure	Cardioprotective	Gidlöf et al. (2020)
LncRNA UCA1	Human mesenchymal stem cell (hMSC)	MiR-873-5p/XIAP axis	Ischemic heart disease	Cardioprotective	Sun, Zhu, et al. (2020)
LncRNAs (MCM3AP-AS1, LINC01089, ITPK1-AS1, and HCG27)	Human data	MiR-125a/FYN, MiR-19a/PXN, and Let-7i/RHOA/GRB2/STAT1 axis	Stroke	Upregulated	Zhang, Liu, Han, Wang, and Han (2020)
LncRNA TTTY15	AC16 cells	Let-7b/MAPK6 axis	Myocardial infarction	Upregulated	Xie, Ji, Han, Zhang and Li (2020)
MiR-181c-5p	H9C2 cardiomyocyte and post-ischemic myocardium of rats	Regulation of protein tyrosine phosphatase nonreceptor type 4 (PTPN4)	Myocardial infarction	Promotes inflammatory response during ischemia	Wang, Ge, Zhang, et al. (2020)
LncRNA Ang362	Human	Lipid metabolism	Coronary heart disease	Upregulated biomarker	Wang, Gong, Liu and Feng (2020)
MiR-96 and miR-183	Human and adult mouse cardiac endothelial cells (MCECs)	Regulation of anillin	Myocardial infarction	Regulation of post-infarction neovascularization	Castellan et al. (2020)
LncRNA MHRT	Human	Nrf2 pathway	Myocardial infarction	Single nucleotide polymorphism increases myocardial infarction	Zhang, Dou and Chen (2020)
MiR-451	Rat model of MI	Regulation of HMGB1	Myocardial infarction	Cardioprotective	Cao, Da, Li, Peng and Hu (2020)
MiR-150-5p	Ischemic rats and neonatal cardiomyocytes	Regulation of TXNIP	Myocardial infarction	Cardioprotective	Ou, Teng, et al. (2020)
LncRNA ZFAS1	Rat H9c2 cardiomyocyte cells	MiR-590-3p/NF-κB signaling pathway	Myocardial infarction	Upregulated	Huang, Yang, Yu and Shi (2020)
CircHIPK3	Mouse model of MI and neonatal mouse cardiomyocytes	MiR-29a/VEGFA axis	Myocardial infarction	Stimulation of cardiac angiogenesis	Wang, Zhao, Shen, et al. (2020)
LncRNAs (ENST00000556899.1 and ENST00000575985.1)	Human	Inflammatory biomarkers	Myocardial infarction	Upregulated	Zheng, Liu, et al. (2020)
MiR-377	Human and human neonatal cardiomyocytes	Leukocyte immunoglobulin-like receptor B2 (LILRB2) signaling	Myocardial infarction	Cardioprotective	Xie, Hu, Li and Li (2020)

fibrosis is complex, and because not fully understood, it is challenging to develop strategies aiming at reducing cardiac fibrosis (Park, Nguyen, Pezhouman, & Ardehali, 2019). Recent shreds of evidence have provided exciting data on the therapeutic potential of ncRNAs in cardiac fibrosis therapeutics. Indeed, a variety of studies have identified ncRNAs capable of the regulation of fibrosis (Park et al., 2019).

As an example, Hao et al. recently reported that lncRNA-AK137033 regulates cardiac fibrosis (Hao et al., 2019). Indeed, the authors identified 389 differentially expressed lncRNAs in cardiac fibroblasts (CFs) in the remote area of LV tissues isolated from mice after MI. Further analysis showed that AK137033 was enriched in the nuclei of fibroblasts and upregulated after MI and TGF-β stimulation. Also, knockdown of AK137033 impaired fibroblast differentiation into myofibroblast, and reduced their proliferation and the secretion of ECM *in vitro*. AK137033 KO mice also showed the impairment of cardiac function. Using *in vitro* genetic manipulation, they showed that AK137033 upregulation led to reduced expression of *Sfrp*, and overexpression of *Sfrp* reversed the effects of AK137033 on CFs both *in vitro* and *in vivo*. A dual-luciferase assay showed that AK137033 binds to *Sfrp2* at the 3′-end. EMSA and RNA-IP data revealed that the RNA binding protein HuR regulates the AK137033-Sfrp complex. In conclusion, this study showed that AK137033 is a key regulator of cardiac fibrosis *via* the AK137033/Sfrp2/HuR axis. Another lncRNA, the RNA component of mitochondrial RNA processing endoribonuclease (RMRP) has been reported to be upregulated by cardiac fibrosis. In an *in vitro* assay, the stimulation of CFs by angiotensin II upregulated RMRP and its inhibition decreased cell proliferation and migration. Mechanistically, miR-613 was identified as a direct functional target of RMRP using luciferase reporter assays. RMRP also downregulates miR-613 to promote CF activation (Zhang et al., 2019). Moreover, a lncRNA, Hoxa-as3 has been shown to promote fibrogenesis. Lin et al. reported that Hoxa-as3 inhibition attenuated miR-450b-5p expression. Moreover, miR-450b-5p inhibition promoted fibrogenesis by regulating runt-related transcription factor 1 (Runx1), while increased expression of miR-450b-5p lessened fibrogenesis. Mechanistically, the results showed that Hoxa-as3 regulates fibroblast activation and fibrogenesis by acting as a competing endogenous RNA for miR-450b-5p. Hoxa-as3 downregulated miR-450b-5p, leading to increased levels of Runx1 and induced fibrosis, while Runx1 inhibition attenuated the pro-fibrotic effect of Hoxa-as3. Additionally, Hoxa-as3 was modulated by TGF-β1/Smad4 pathway as its transcriptional target (Lin, Zhang, et al., 2020).

CircRNAs also play important regulatory roles in cardiac fibrosis. Liu et al. reported the implication of circHIPK3 expression in CFs during an ischemic event in the heart (Liu, Wang, et al., 2020). It was found that CircHIPK3 is upregulated in CFs after hypoxia. CircHIPK3 overexpression resulted in CF proliferation and migration. CircHIPK3 was shown to act as a competing endogenous RNA for miR-152-3p. The repression of miR-152-3p in turn led to upregulation of TGF-β2 expression associated with increased fibrotic phenotype. Overall, the results showed that circHIPK3 acts as a miR-152-3p sponge to upregulate fibrosis post-MI.

Another interesting finding revealed that miR-675 is downregulated in cardiac fibrosis (Wang, Jiang, et al., 2019). Indeed, in an *in vitro* model, the authors showed that TGF-β1 stimulation decreased the expression of miR-675 *via* upregulation of Smads, which binds to the promoter region of this microRNA. Overexpression of miR-675 in CFs decreased their proliferation and migration by binding to its direct functional target TGF-β receptor 1 (TGFβR1). In conclusion, this research suggests that the Smad/miR-675/TGFβR1 axis could be regulated to prevent cardiac fibrosis. Interestingly, Chiasson et al. reported the role of miR-1954 in angiotensin II-mediated cardiac remodeling and fibrosis using a mouse model. The results showed that miR-1954 is downregulated in the heart after angiotensin II stimulation. Inversely, miR-1954 upregulation significantly decreased cardiac fibrosis in angiotensin II-infused mouse hearts. A luciferase reporter assay enabled to identify THBS1 as a direct target of miR-1954, suggesting that the miR-1954/THBS1 axis represents a potential target for cardiac fibrosis (Chiasson, Takano, Guleria, & Gupta, 2019).

3.6 Pulmonary hypertension

Artery vessels consist of three tissue layers: intima, media, and adventitial, mainly formed with pulmonary artery endothelial cells (PAECs), smooth muscle cells (PASMCs), and fibroblasts, respectively (Pak, Aldashev, Welsh, & Peacock, 2007). The development of pulmonary arterial hypertension (PAH) is the combinatorial effect of the dysfunctional miRNA/lncRNA network in the epigenetic aspect. NcRNA interactions navigate different signaling pathways related to the regulatory network of PAH, and thus ncRNA dysregulation triggers the development of PAH (Poller et al., 2018). Currently, a limited number of therapeutics are available for PAH, which include phosphodiesterase type 5 inhibitors and vasodilators. However, these do not revert the pathophysiology to normal.

In PAH, lncRNAs act as key drivers to gatekeep cellular and molecular trafficking for endothelial dysfunction and PASMC dysregulation (Zahid, Raza, Chen, Raj, & Gou, 2020). There are numerous lncRNAs associated with PASMC dysregulation. For instance, MALAT1 activates multiple signaling pathways by various mechanisms. By sponging miR-124-3p, MALAT1 promoted the expression of KLF5 transcription factor to increase hyperactive cell cycle progression (Wang, Xu, et al., 2019). Also, in PAECs MALAT1 regulated endothelial-to-mesenchymal transition (EndMT) by activating Wnt signaling to upregulate β-catenin (Razak et al., 2019) and by sponging miR-145 (Neumann et al., 2018). MiRNA regulation also contributes to the pathogenesis of PAH. Known miRNAs that contribute to PASMC dysregulation are miR-17-92 cluster, miR-21, miR-124, miR-206, and miR-204 (Wang, Xu, et al., 2019). Also, the miRs contributing to endothelial cell proliferation and apoptosis resistance are miR-17-92 cluster, miR-21, and miR-27a (Zhou, Chen, & Raj, 2015). In both circumstances, ncRNAs can be also used as biomarkers for disease diagnosis and prognosis.

4 Perspectives and therapeutic applications of noncoding RNAs in CVDs

Most of the drugs used for CVDs are statins, which reduce endogenous cholesterols. Clopidogrel is a blood thinner that works as an antiplatelet drug for coronary artery disease (Hargraves, Palmer, Orav, & Wright, 1996). Also, widely used β-blockers, antagonists for β-adrenergic receptors, can improve cardiac function in patients with heart failure. There are many more Food and Drug Administration (FDA)-approved drugs available at the moment. However, the effectiveness of drugs for CVDs has a great variability depending upon the population it was used with (Leslie, McCowan, & Pell, 2019). Current clinical therapy for CVDs is limited due to various complications such as stent thrombosis and systemic toxicity. Therefore, molecular based-cell approaches and nanotechnology approaches have provided the tools to explore therapeutics at the cellular level and to offer unique treatment options for CVDs (Behera, Pramanik, & Nayak, 2015). Cardiac remodeling is a common concept defined as changes in gene expression, as well as molecular, cellular, and interstitial changes manifested after cardiac injury, which in turn clinically change size, shape, and function of the heart (Cohn, Ferrari, & Sharpe, 2000). NcRNAs have been proposed as novel therapies in CVDs. For example, cardiac-specific *in vivo* approaches of ncRNAs exert the potential to ameliorate cardiac dysfunction caused after post-cardiac remodeling and diminish pathological progression of injured hearts (Piccoli et al., 2017).

4.1 Approaches and tools

Viral and nonviral approaches are now considered for RNA-based diagnostic and therapeutic strategies, especially for CVDs. However, only very few reach the clinical trials for therapeutics; among those, very few also reach successes. Antisense oligonucleotide (ASO) drugs directing toward the liver through GalNAc conjugation has so far reached the clinical stage with FDA approval. Also, FDA approved Patisiran known as a RNAi therapeutic agent for transthyretin amyloidosis, which is shown to improve multiple clinical manifestations in patients with cardiac dysfunction (Adams et al., 2018) by decreasing left ventricular wall thickness (Solomon et al., 2019). Although cardiac-specific delivery of ASO drug still remains as a hurdle to overcome in the future, RNase H-dependent mechanism or RNAi path using small interference RNA for gene silencing is the most widely used ncRNA therapeutics.

Beyond the potential applications of ncRNA biomarkers as diagnostic tools, ncRNAs are also useful as therapeutics to prevent the disease progression. Numerous ncRNAs showed the potential as therapeutic targets based on their differential expression and abundance. Generally, two different approaches are used to alter ncRNA expression in the disease progression. The first approach is by upregulation of ncRNAs for therapeutic use. This gain-of-function approach is commonly achieved with the adeno-associated virus (AAV) or some other viruses. Viruses are injected into host to exert a long-lasting *in vivo* ncRNA expression (Knabel et al., 2015). It is possible to orient the drug to a specific location of an organ as these viruses exert tropism toward each organ depending on the serotype. Also, cell-specific promoters can directly target to exert specificity (Lovric et al., 2012). The gain-of-function strategy can also be achieved through mimics for miRNAs. Synthetic, small chemically structured double-stranded oligonucleotides can incorporate into RISC complex to mimic miRNAs endogenously. The second approach is to downregulate ncRNAs through ASOs. These short synthetic single-stranded oligomers are complementary to the target ncRNA. The real challenge is to modify the pharmacodynamics and pharmacokinetics of these ASOs to enhance their efficacy. The most widely used method in ASO and siRNA therapeutics is chemically modifying the backbone of phosphodiester linkage to increase nuclease stability, enhance pharmacokinetics, and improve the binding affinity to plasma proteins, hence reducing the urinary clearance (Levin, 1999). The chemical modification is converting phosphodiester into phosphorothiolate (PS) linkage by replacing non-bridging O atom with S atom. Sugar modification on the C2 position remarkably increases the binding affinity of ASO to complementary RNA and nuclease stability. Such modifications include 2′-O-methylation and 2′-O-F or 2′-O-methoxyethylation (Lima, Cerqueira, Figueiredo, Oliveira, & Azevedo, 2018). Also, 2′-O,4′-*C*-methylene bridge forms locked nucleic acids (LNA) to make locked configuration with duplex, thus improving the potency of ASO (Beermann, Piccoli, Viereck, & Thum, 2016).

Notably, miRNA therapeutics reaching for clinical use is thin on the ground. However, recent findings from Dr. Thomas Thum group (Foinquinos et al., 2020) demonstrated favorable pharmacokinetics and pharmacodynamics, proving safety and tolerance for the antimiR-132 treatment for heart failure. Synthetic LNA-based antisense oligonucleotide inhibitor (antimiR-132) is evident to have a high therapeutic efficacy when testing with murine and large animal models such as pigs. Consistent with the data, antimiR-132 is pursued in phase 1b human clinical trials for heart failure (Clinical research identifier: NCT04045405) under the name of its lead compound CDR132L. Interestingly, miR-132 was initially shown to elicit cardiac hypertrophy by regulating cardiomyocyte growth (Ucar et al., 2012).

LncRNAs are challenging to deliver along with existing drug vehicles such as AAV and lipid nanoparticles because they require modified gene-editing techniques. Also, the major drawback for most lncRNAs is not well-conserved across species, which limits the use as the translation from rodents to humans. There are only a few research studies so far performed on large animals. LncRNA CHROME is one example of atherosclerotic vascular diseases in nonhuman primates (Huang, Kafert-Kasting, & Thum, 2020).

Currently, it is standard practice in oncology to tailor therapies for individual groups to maximize the benefits associated with minimized secondary effects (Krzyszczyk et al., 2018). The opportunities opened along with such a precision approach to cardiovascular care can be enormous, but also challenging. Novel approaches, such as RNA therapies, span from human genetics to systems biology (MacRae, Roden, & Loscalzo, 2016). The development of novel ncRNA therapeutics has certain limits when it comes to clinical studies as no matter how genome of used models are closer to human; there may consist of differences in immune system and metabolism, which demote the predictability of use (Wright & Okusa, 1990). Difficulties yet to overcome with ncRNAs for therapeutics are delivering the active drug to the heart, improving the binding affinity to the target RNA, enhancing stability in circulation, minimizing off-target effects, improving pharmacokinetics and pharmacodynamics, and maximizing efficacy (Gomes et al., 2017). With the development of *ex vivo* approaches such as engineered heart tissues, living myocardial slice technology derived from human cells provides tools for drug screening and studying electrophysiology as it preserved electrical and mechanical connection (Ou et al., 2019). This would fasten the pharmaceutical drug development. With these new technological advances, ncRNA therapies have the potential to make remarkable progress in developing next-generation therapeutics for CVDs.

5 Conclusion

In summary, the progression of CVDs is preceded or/and followed by dynamic dysregulation of the ncRNA regulatory network. NcRNAs play key roles in a diversity of biological processes that contribute to maintaining the cardiovascular function. NcRNAs represent promising targets for early diagnosis and therapeutic development for CVDs, as numerous studies have provided novel insights into their impact on different disease states. Scientists have shown that genetic and pharmacological manipulation of ncRNAs could be beneficial in suppressing key genes required for CVD pathology. Yet, several preclinical studies remain untranslated to a relevant clinical stage, and the molecular mechanisms of numerous ncRNAs in CVD regulation are still not well understood. Nevertheless, as technological development continues to provide more insight on the function of ncRNAs and their regulatory network, it is very likely in the near future that ncRNAs become essential therapeutic agents for the management of CVDs.

Acknowledgments

We thank the editor for inviting us to write this review. Due to space restrictions, the authors cannot cite many important pieces of literatures in this field. The authors apologize to all other colleagues whose work contributed significantly. Figs. 2–4 were created using BioRender. BM and NPB drafted the manuscript and prepared the illustrations. TA, MS and SK helped to write the manuscript. I-mK supervised the manuscript preparation and critically revised it.

Sources of funding

This work was supported by the American Heart Association (AHA) Postdoctoral Fellowship 20POST34990024 to BM, 18POST34030054 to TA, and AHA Transformational Project Award 18TPA34170104 to I-mK, as well as the National Institutes of Health (NIH) R01HL124251 and R01HL146481 to I-mK.

Competing interest

The authors declare no conflict of interest.

References

Adams, D., Gonzalez-Duarte, A., O'Riordan, W. D., Yang, C. C., Ueda, M., Kristen, A. V., et al. (2018). Patisiran, an RNAi therapeutic, for hereditary transthyretin amyloidosis. *The New England Journal of Medicine*, *379*(1), 11–21. https://doi.org/10.1056/NEJMoa1716153.

Aonuma, T., Bayoumi, A. S., Tang, Y., & Kim, I.-M. (2018). A circular RNA regulator quaking: A novel gold mine to be unfolded in doxorubicin-mediated cardiotoxicity. *Non-Coding RNA Investigation*, *2*, 19. https://doi.org/10.21037/ncri.2018.04.02.

Archer, K., Broskova, Z., Bayoumi, A. S., Teoh, J.-p., Davila, A., Tang, Y., et al. (2015). Long non-coding RNAs as master regulators in cardiovascular diseases. *International Journal of Molecular Sciences*, *16*(10), 23651–23667. https://doi.org/10.3390/ijms161023651.

Bayoumi, A. S., Aonuma, T., Teoh, J.-P., Tang, Y.-L., & Kim, I.-M. (2018). Circular noncoding RNAs as potential therapies and circulating biomarkers for cardiovascular diseases. *Acta Pharmacologica Sinica*, *39*(7), 1100–1109. https://doi.org/10.1038/aps.2017.196.

Bayoumi, A. S., Park, K.-M., Wang, Y., Teoh, J.-P., Aonuma, T., Tang, Y., et al. (2018). A carvedilol-responsive microRNA, miR-125b-5p protects the heart from acute myocardial infarction by repressing pro-apoptotic bak1 and klf13 in cardiomyocytes. *Journal of Molecular and Cellular Cardiology*, *114*, 72–82. https://doi.org/10.1016/j.yjmcc.2017.11.003.

Bayoumi, A. S., Sayed, A., Broskova, Z., Teoh, J.-P., Wilson, J., Su, H., et al. (2016). Crosstalk between long noncoding RNAs and MicroRNAs in health and disease. *International Journal of Molecular Sciences*, *17*(3), 356. https://doi.org/10.3390/ijms17030356.

Bayoumi, A. S., Teoh, J.-P., Aonuma, T., Yuan, Z., Ruan, X., Tang, Y., et al. (2017). MicroRNA-532 protects the heart in acute myocardial infarction, and represses prss23, a positive regulator of endothelial-to-mesenchymal transition. *Cardiovascular Research*, *113*(13), 1603–1614. https://doi.org/10.1093/cvr/cvx132.

Beermann, J., Piccoli, M. T., Viereck, J., & Thum, T. (2016). Non-coding RNAs in development and disease: Background, mechanisms, and therapeutic approaches. *Physiological Reviews*, *96*(4), 1297–1325. https://doi.org/10.1152/physrev.00041.2015.

Behera, S. S., Pramanik, K., & Nayak, M. K. (2015). Recent advancement in the treatment of cardiovascular diseases: Conventional therapy to nanotechnology. *Current Pharmaceutical Design*, *21*(30), 4479–4497. https://doi.org/10.2174/1381612821666150817104635.

Cao, J., Da, Y., Li, H., Peng, Y., & Hu, X. (2020). Upregulation of microRNA-451 attenuates myocardial I/R injury by suppressing HMGB1. *PLoS One*, *15*(7), e0235614. https://doi.org/10.1371/journal.pone.0235614.

Cao, S., Li, L., Geng, X., Ma, Y., Huang, X., & Kang, X. (2019). The upregulation of miR-101 promotes vascular endothelial cell apoptosis and suppresses cell migration in acute coronary syndrome by targeting CDH5. *International Journal of Clinical and Experimental Pathology*, *12*(9), 3320–3328.

Cao, W., Shi, P., & Ge, J. J. (2017). miR-21 enhances cardiac fibrotic remodeling and fibroblast proliferation via CADM1/STAT3 pathway. *BMC Cardiovascular Disorders*, *17*(1), 88. https://doi.org/10.1186/s12872-017-0520-7.

Cardin, S., Guasch, E., Luo, X., Naud, P., Le Quang, K., Shi, Y., et al. (2012). Role for MicroRNA-21 in atrial profibrillatory fibrotic remodeling associated with experimental postinfarction heart failure. *Circulation. Arrhythmia and Electrophysiology*, *5*(5), 1027–1035. https://doi.org/10.1161/circep.112.973214.

Castellan, R. F., Vitiello, M., Vidmar, M., Johnstone, S., Iacobazzi, D., Mellis, D., et al. (2020). miR-96 and miR-183 differentially regulate neonatal and adult postinfarct neovascularization. *JCI Insight*, *5*(14), e134888. https://doi.org/10.1172/jci.insight.134888.

Chen, H., & Cai, K. (2020). DSCAM-AS1 mediates pro-hypertrophy role of GRK2 in cardiac hypertrophy aggravation via absorbing miR-188-5p. *Vitro Cellular & Developmental Biology—Animal*, *56*(4), 286–295. https://doi.org/10.1007/s11626-020-00441-w.

Chen, Z., Gao, Y., Yao, L., Liu, Y., Huang, L., Yan, Z., et al. (2018). LncFZD6 initiates Wnt/β-catenin and liver TIC self-renewal through BRG1-mediated FZD6 transcriptional activation. *Oncogene*, *37*(23), 3098–3112. https://doi.org/10.1038/s41388-018-0203-6.

Chen, Y. H., Xu, S. J., Bendahhou, S., Wang, X. L., Wang, Y., Xu, W. Y., et al. (2003). KCNQ1 gain-of-function mutation in familial atrial fibrillation. *Science*, *299*(5604), 251–254. https://doi.org/10.1126/science.1077771.

Cheng, Z., Liu, L., & Li, Q. (2020). lncRNA ZEB2-AS1 stimulates cardiac hypertrophy by downregulating PTEN. *Experimental and Therapeutic Medicine*, *20*(5), 92. https://doi.org/10.3892/etm.2020.9220.

Chi, Y., Wang, D., Wang, J., Yu, W., & Yang, J. (2019). Long non-coding RNA in the pathogenesis of cancers. *Cell*, *8*(9), 1015. https://doi.org/10.3390/cells8091015.

Chi, K., Zhang, J., Sun, H., Liu, Y., Li, Y., Yuan, T., et al. (2020). Knockdown of lncRNA HOXA-AS3 suppresses the progression of atherosclerosis via sponging miR-455-5p. *Drug Design, Development and Therapy*, *14*, 3651–3662. https://doi.org/10.2147/DDDT.S249830.

Chiasson, V., Takano, A. P. C., Guleria, R. S., & Gupta, S. (2019). Deficiency of microRNA miR-1954 promotes cardiac remodeling and fibrosis. *Journal of the American Heart Association*, *8*(21), e012880. https://doi.org/10.1161/JAHA.119.012880.

Chipman, L. B., & Pasquinelli, A. E. (2019). miRNA targeting: Growing beyond the seed. *Trends in Genetics*, *35*(3), 215–222. https://doi.org/10.1016/j.tig.2018.12.005.

Cohn, J. N., Ferrari, R., & Sharpe, N. (2000). Cardiac remodeling—Concepts and clinical implications: a consensus paper from an international forum on cardiac remodeling. Behalf of an International Forum on Cardiac Remodeling. *Journal of the American College of Cardiology*, *35*(3), 569–582. https://doi.org/10.1016/s0735-1097(99)00630-0.

Collins, L., Binder, P., Chen, H., & Wang, X. (2020). Regulation of long non-coding RNAs and microRNAs in heart disease: Insight into mechanisms and therapeutic approaches. *Frontiers in Physiology*, *11*, 798. https://doi.org/10.3389/fphys.2020.00798.

Dawson, K., Wakili, R., Ordög, B., Clauss, S., Chen, Y., Iwasaki, Y., et al. (2013). MicroRNA29: A mechanistic contributor and potential biomarker in atrial fibrillation. *Circulation*, *127*(14), 1466–1475. 1475e1461-1428 https://doi.org/10.1161/circulationaha.112.001207.

de Yébenes, V. G., Briones, A. M., Martos-Folgado, I., Mur, S. M., Oller, J., Bilal, F., et al. (2020). Aging-associated miR-217 aggravates atherosclerosis and promotes cardiovascular dysfunction. *Arteriosclerosis, Thrombosis, and Vascular Biology*, *40*(10), 2408–2424. https://doi.org/10.1161/ATVBAHA.120.314333.

Deans, C., & Maggert, K. A. (2015). What do you mean, "epigenetic"? *Genetics*, *199*(4), 887–896. https://doi.org/10.1534/genetics.114.173492.

Ding, P., Ding, Y., Tian, Y., & Lei, X. (2020). Circular RNA circ_0010283 regulates the viability and migration of oxidized low-density lipoprotein-induced vascular smooth muscle cells via an miR-370-3p/HMGB1 axis in atherosclerosis. *International Journal of Molecular Medicine*, *46*(4), 1399–1408. https://doi.org/10.3892/ijmm.2020.4703.

Elbaz, M., Faccini, J., Laperche, C., Grousset, E., Roncalli, J., Ruidavets, J.-B., et al. (2020). Identification of a miRNA based-signature associated with acute coronary syndrome: Evidence from the FLORINF study. *Journal of Clinical Medicine*, *9*(6), 1674. https://doi.org/10.3390/jcm9061674.

Foinquinos, A., Batkai, S., Genschel, C., Viereck, J., Rump, S., Gyongyosi, M., et al. (2020). Preclinical development of a miR-132 inhibitor for heart failure treatment. *Nature Communications*, *11*(1), 633. https://doi.org/10.1038/s41467-020-14349-2.

Frey, N., & Olson, E. N. (2003). Cardiac hypertrophy: The good, the bad, and the ugly. *Annual Review of Physiology*, *65*, 45–79. https://doi.org/10.1146/annurev.physiol.65.092101.142243.

Gibbons, A., Udawela, M., & Dean, B. (2018). Non-coding RNA as novel players in the pathophysiology of schizophrenia. *Non-Coding RNA*, *4*(2), 11. https://doi.org/10.3390/ncrna4020011.

Gidlöf, O., Bader, K., Celik, S., Grossi, M., Nakagawa, S., Hirose, T., et al. (2020). Inhibition of the long non-coding RNA NEAT1 protects cardiomyocytes from hypoxia in vitro via decreased pri-miRNA processing. *Cell Death & Disease*, *11*(8), 677. https://doi.org/10.1038/s41419-020-02854-7.

Gomes, C. P. C., Schroen, B., Kuster, G. M., Robinson, E. L., Ford, K., Squire, I. B., et al. (2020). Regulatory RNAs in heart failure. *Circulation*, *141*(4), 313–328. https://doi.org/10.1161/CIRCULATIONAHA.119.042474.

Gomes, C. P. C., Spencer, H., Ford, K. L., Michel, L. Y. M., Baker, A. H., Emanueli, C., et al. (2017). The function and therapeutic potential of long non-coding RNAs in cardiovascular development and disease. *Molecular Therapy—Nucleic Acids*, *8*, 494–507. https://doi.org/10.1016/j.omtn.2017.07.014.

Gore-Panter, S. R., Hsu, J., Barnard, J., Moravec, C. S., Van Wagoner, D. R., Chung, M. K., et al. (2016). PANCR, the PITX2 adjacent noncoding RNA, is expressed in human left atria and regulates PITX2c expression. *Circulation. Arrhythmia and Electrophysiology*, *9*(1), e003197. https://doi.org/10.1161/CIRCEP.115.003197.

Ha, M., & Kim, V. N. (2014). Regulation of microRNA biogenesis. *Nature Reviews. Molecular Cell Biology*, *15*(8), 509–524. https://doi.org/10.1038/nrm3838.

Hao, K., Lei, W., Wu, H., Wu, J., Yang, Z., Yan, S., et al. (2019). LncRNA-safe contributes to cardiac fibrosis through safe-Sfrp2-HuR complex in mouse myocardial infarction. *Theranostics*, *9*(24), 7282–7297. https://doi.org/10.7150/thno.33920.

Hargraves, J. L., Palmer, R. H., Orav, E. J., & Wright, E. A. (1996). Practice characteristics and performance of primary care practitioners. *Medical Care*, *34*(9 Suppl), SS67–76. https://doi.org/10.1097/00005650-199609002-00007.

He, X., Lian, Z., Yang, Y., Wang, Z., Fu, X., Liu, Y., et al. (2020). Long non-coding RNA PEBP1P2 suppresses proliferative VSMCs phenotypic switching and proliferation in atherosclerosis. *Molecular Therapy—Nucleic Acids*, *22*, 84–98. https://doi.org/10.1016/j.omtn.2020.08.013.

He, W., Zhu, L., Huang, Y., Zhang, Y., Shen, W., Fang, L., et al. (2019). The relationship of microRNA-21 and plaque stability in acute coronary syndrome. *Medicine*, *98*(47), e18049. https://doi.org/10.1097/MD.0000000000018049.

Hobuß, L., Bär, C., & Thum, T. (2019). Long non-coding RNAs: At the heart of cardiac dysfunction? *Frontiers in Physiology*, *10*, 30. https://doi.org/10.3389/fphys.2019.00030.

Houwing, S., Kamminga, L. M., Berezikov, E., Cronembold, D., Girard, A., van den Elst, H., et al. (2007). A role for Piwi and piRNAs in germ cell maintenance and transposon silencing in Zebrafish. *Cell*, *129*(1), 69–82. https://doi.org/10.1016/j.cell.2007.03.026.

Hsu, M. T., & Coca-Prados, M. (1979). Electron microscopic evidence for the circular form of RNA in the cytoplasm of eukaryotic cells. *Nature*, *280*(5720), 339–340. https://doi.org/10.1038/280339a0.

Hu, G., Niu, F., Humburg, B. A., Liao, K., Bendi, S., Callen, S., et al. (2018). Molecular mechanisms of long noncoding RNAs and their role in disease pathogenesis. *Oncotarget*, *9*(26), 18648–18663. https://doi.org/10.18632/oncotarget.24307.

Huang, C. K., Kafert-Kasting, S., & Thum, T. (2020). Preclinical and clinical development of noncoding RNA therapeutics for cardiovascular disease. *Circulation Research*, *126*(5), 663–678. https://doi.org/10.1161/CIRCRESAHA.119.315856.

Huang, P., Yang, D., Yu, L., & Shi, Y. (2020). Downregulation of lncRNA ZFAS1 protects H9c2 cardiomyocytes from ischemia/reperfusion-induced apoptosis via the miR-590-3p/NF-κB signaling pathway. *Molecular Medicine Reports*, *22*(3), 2300–2306. https://doi.org/10.3892/mmr.2020.11340.

Jin, Q., & Chen, Y. (2019). Silencing circular RNA circ_0010729 protects human cardiomyocytes from oxygen-glucose deprivation-induced injury by up-regulating microRNA-145-5p. *Molecular and Cellular Biochemistry*, *462*(1–2), 185–194. https://doi.org/10.1007/s11010-019-03621-9.

Jin, Y., Zhou, T.-Y., Cao, J.-N., Feng, Q.-T., Fu, Y.-J., Xu, X., et al. (2019). MicroRNA-206 downregulates connexin43 in cardiomyocytes to induce cardiac arrhythmias in a transgenic mouse model. *Heart, Lung & Circulation*, *28*(11), 1755–1761.

Khurshid, S., Choi, S. H., Weng, L.-C., Wang, E. Y., Trinquart, L., Benjamin, E. J., et al. (2018). Frequency of cardiac rhythm abnormalities in a half million adults. *Circulation. Arrhythmia and Electrophysiology*, *11*(7), e006273. https://doi.org/10.1161/CIRCEP.118.006273.

Kiliszek, M., Maciak, K., Maciejak, A., Krzyżanowski, K., Wierzbowski, R., Gora, M., et al. (2020). Serum microRNA in patients undergoing atrial fibrillation ablation. *Scientific Reports*, *10*(1), 4424. https://doi.org/10.1038/s41598-020-61322-6.

Knabel, M. K., Ramachandran, K., Karhadkar, S., Hwang, H. W., Creamer, T. J., Chivukula, R. R., et al. (2015). Systemic delivery of scAAV8-encoded MiR-29a ameliorates hepatic fibrosis in carbon tetrachloride-treated mice. *PLoS One*, *10*(4), e0124411. https://doi.org/10.1371/journal.pone.0124411.

Krzyszczyk, P., Acevedo, A., Davidoff, E. J., Timmins, L. M., Marrero-Berrios, I., Patel, M., et al. (2018). The growing role of precision and personalized medicine for cancer treatment. *Technology*, *6*(3–4), 79–100. https://doi.org/10.1142/S2339547818300020.

Kumarswamy, R., & Thum, T. (2013). Non-coding RNAs in cardiac remodeling and heart failure. *Circulation Research*, *113*(6), 676–689. https://doi.org/10.1161/CIRCRESAHA.113.300226.

Larupa Santos, J., Rodríguez, I., Olesen, S., Bentzen, B. H., & Schmitt, N. (2020). Investigating gene-microRNA networks in atrial fibrillation patients with mitral valve regurgitation. *PLoS One*, *15*(5), e0232719. https://doi.org/10.1371/journal.pone.0232719.

Leslie, K. H., McCowan, C., & Pell, J. P. (2019). Adherence to cardiovascular medication: A review of systematic reviews. *Journal of Public Health (Oxford, England)*, *41*(1), e84–e94. https://doi.org/10.1093/pubmed/fdy088.

Levin, A. A. (1999). A review of the issues in the pharmacokinetics and toxicology of phosphorothioate antisense oligonucleotides. *Biochimica et Biophysica Acta*, *1489*(1), 69–84. https://doi.org/10.1016/s0167-4781(99)00140-2.

Li, Q., Qin, M., Li, T., Gu, Z., Tan, Q., Huang, P., et al. (2020). Rutin protects against pirarubicin-induced cardiotoxicity by adjusting microRNA-125b-1-3p-mediated JunD signaling pathway. *Molecular and Cellular Biochemistry*, *466*(1–2), 139–148. https://doi.org/10.1007/s11010-020-03696-9.

Li, Q., Qin, M., Tan, Q., Li, T., Gu, Z., Huang, P., et al. (2020). MicroRNA-129-1-3p protects cardiomyocytes from pirarubicin-induced apoptosis by down-regulating the GRIN2D-mediated Ca(2+) signalling pathway. *Journal of Cellular and Molecular Medicine*, *24*(3), 2260–2271. https://doi.org/10.1111/jcmm.14908.

Lima, J. F., Cerqueira, L., Figueiredo, C., Oliveira, C., & Azevedo, N. F. (2018). Anti-miRNA oligonucleotides: A comprehensive guide for design. *RNA Biology*, *15*(3), 338–352. https://doi.org/10.1080/15476286.2018.1445959.

Lin, F., Yang, Y., Guo, Q., Xie, M., Sun, S., Wang, X., et al. (2020). Analysis of the molecular mechanism of acute coronary syndrome based on circRNA-miRNA network regulation. *Evidence-Based Complementary and Alternative Medicine: eCAM*, *2020*, 1584052. https://doi.org/10.1155/2020/1584052.

Lin, S., Zhang, R., Xu, L., Ma, R., Xu, L., Zhu, L., et al. (2020). LncRNA Hoxaas3 promotes lung fibroblast activation and fibrosis by targeting miR-450b-5p to regulate Runx1. *Cell Death & Disease*, *11*(8), 706. https://doi.org/10.1038/s41419-020-02889-w.

Lin, Z., Zhou, P., von Gise, A., Gu, F., Ma, Q., Chen, J., et al. (2015). Pi3kcb links Hippo-YAP and PI3K-AKT signaling pathways to promote cardiomyocyte proliferation and survival. *Circulation Research*, *116*(1), 35–45. https://doi.org/10.1161/circresaha.115.304457.

Ling, T. Y., Wang, X. L., Chai, Q., Lu, T., Stulak, J. M., Joyce, L. D., et al. (2017). Regulation of cardiac CACNB2 by microRNA-499: Potential role in atrial fibrillation. *BBA Clinical*, *7*, 78–84. https://doi.org/10.1016/j.bbacli.2017.02.002.

Liu, D., Song, J., Ji, X., Liu, Z., Li, T., & Hu, B. (2020). PRDM16 upregulation induced by microRNA-448 inhibition alleviates atherosclerosis via the TGF-β signaling pathway inactivation. *Frontiers in Physiology*, *11*, 846. https://doi.org/10.3389/fphys.2020.00846.

Liu, W., Wang, Y., Qiu, Z., Zhao, R., Liu, Z., Chen, W., et al. (2020). CircHIPK3 regulates cardiac fibroblast proliferation, migration and phenotypic switching through the miR-152-3p/TGF-β2 axis under hypoxia. *PeerJ*, *8*, e9796. https://doi.org/10.7717/peerj.9796.

Lovric, J., Mano, M., Zentilin, L., Eulalio, A., Zacchigna, S., & Giacca, M. (2012). Terminal differentiation of cardiac and skeletal myocytes induces permissivity to AAV transduction by relieving inhibition imposed by DNA damage response proteins. *Molecular Therapy*, *20*(11), 2087–2097. https://doi.org/10.1038/mt.2012.144.

Lu, M. (2020). Circular RNA: Functions, applications and prospects. *ExRNA*, *2*(1), 1. https://doi.org/10.1186/s41544-019-0046-5.

Ma, X., Huang, Y., Ding, Y., Shi, L., Zhong, X., Kang, M., et al. (2020). Analysis of piRNA expression spectra in a non-alcoholic fatty liver disease mouse model induced by a methionine- and choline-deficient diet. *Experimental and Therapeutic Medicine*, *19*(6), 3829–3839. https://doi.org/10.3892/etm.2020.8653.

MacRae, C. A., Roden, D. M., & Loscalzo, J. (2016). The future of cardiovascular therapeutics. *Circulation*, *133*(25), 2610–2617. https://doi.org/10.1161/CIRCULATIONAHA.116.023555.

Makarova, J. A., Shkurnikov, M. U., Wicklein, D., Lange, T., Samatov, T. R., Turchinovich, A. A., et al. (2016). Intracellular and extracellular microRNA: An update on localization and biological role. *Progress in Histochemistry and Cytochemistry*, *51*(3–4), 33–49. https://doi.org/10.1016/j.proghi.2016.06.001.

Memczak, S., Jens, M., Elefsinioti, A., Torti, F., Krueger, J., Rybak, A., et al. (2013). Circular RNAs are a large class of animal RNAs with regulatory potency. *Nature*, *495*(7441), 333–338. https://doi.org/10.1038/nature11928.

Mester-Tonczar, J., Winkler, J., Einzinger, P., Hasimbegovic, E., Kastner, N., Lukovic, D., et al. (2020). Association between circular RNA CDR1as and post-infarction cardiac function in pig ischemic heart failure: Influence of the anti-fibrotic natural compounds bufalin and lycorine. *Biomolecules*, *10*(8), 1180. https://doi.org/10.3390/biom10081180.

Mu, Z., Zhang, H., & Lei, P. (2020). Piceatannol inhibits pyroptosis and suppresses oxLDL-induced lipid storage in macrophages by regulating miR-200a/Nrf2/GSDMD axis. *Bioscience Reports*, *40*(9), BSR20201366. https://doi.org/10.1042/BSR20201366.

Nakamura, M., & Sadoshima, J. (2018). Mechanisms of physiological and pathological cardiac hypertrophy. *Nature Reviews Cardiology*, *15*(7), 387–407. https://doi.org/10.1038/s41569-018-0007-y.

Neumann, P., Jae, N., Knau, A., Glaser, S. F., Fouani, Y., Rossbach, O., et al. (2018). The lncRNA GATA6-AS epigenetically regulates endothelial gene expression via interaction with LOXL2. *Nature Communications*, *9*(1), 237. https://doi.org/10.1038/s41467-017-02431-1.

O'Brien, J., Hayder, H., Zayed, Y., & Peng, C. (2018). Overview of microRNA biogenesis, mechanisms of actions, and circulation. *Frontiers in Endocrinology*, *9*, 402. https://doi.org/10.3389/fendo.2018.00402.

Orekhov, A. N., Andreeva, E. R., Mikhailova, I. A., & Gordon, D. (1998). Cell proliferation in normal and atherosclerotic human aorta: Proliferative splash in lipid-rich lesions. *Atherosclerosis*, *139*(1), 41–48. https://doi.org/10.1016/s0021-9150(98)00044-6.

Ou, Q., Jacobson, Z., Abouleisa, R. R. E., Tang, X. L., Hindi, S. M., Kumar, A., et al. (2019). Physiological biomimetic culture system for pig and human heart slices. *Circulation Research*, *125*(6), 628–642. https://doi.org/10.1161/CIRCRESAHA.119.314996.

Ou, M., Li, X., Zhao, S., Cui, S., & Tu, J. (2020). Long non-coding RNA CDKN2B-AS1 contributes to atherosclerotic plaque formation by forming RNA-DNA triplex in the CDKN2B promoter. *eBioMedicine*, *55*, 102694. https://doi.org/10.1016/j.ebiom.2020.102694.

Ou, H., Teng, H., Qin, Y., Luo, X., Yang, P., Zhang, W., et al. (2020). Extracellular vesicles derived from microRNA-150-5p-overexpressing mesenchymal stem cells protect rat hearts against ischemia/reperfusion. *Aging*, *12*(13), 12669–12683. https://doi.org/10.18632/aging.102792.

Pak, O., Aldashev, A., Welsh, D., & Peacock, A. (2007). The effects of hypoxia on the cells of the pulmonary vasculature. *The European Respiratory Journal*, *30*(2), 364–372. https://doi.org/10.1183/09031936.00128706.

Park, S., Nguyen, N. B., Pezhouman, A., & Ardehali, R. (2019). Cardiac fibrosis: Potential therapeutic targets. *Translational Research: The Journal of Laboratory and Clinical Medicine*, *209*, 121–137. https://doi.org/10.1016/j.trsl.2019.03.001.

Park, K.-M., Teoh, J.-P., Wang, Y., Broskova, Z., Bayoumi, A. S., Tang, Y., et al. (2016). Carvedilol-responsive microRNAs, miR-199a-3p and -214 protect cardiomyocytes from simulated ischemia-reperfusion injury. *American Journal of Physiology. Heart and Circulatory Physiology*, *311*(2), H371–H383. https://doi.org/10.1152/ajpheart.00807.2015.

Peterson, C. L., & Workman, J. L. (2000). Promoter targeting and chromatin remodeling by the SWI/SNF complex. *Current Opinion in Genetics & Development*, *10*(2), 187–192. https://doi.org/10.1016/s0959-437x(00)00068-x.

Piccoli, M. T., Gupta, S. K., Viereck, J., Foinquinos, A., Samolovac, S., Kramer, F. L., et al. (2017). Inhibition of the cardiac fibroblast-enriched lncRNA Meg3 prevents cardiac fibrosis and diastolic dysfunction. *Circulation Research*, *121*(5), 575–583. https://doi.org/10.1161/CIRCRESAHA.117.310624.

Poller, W., Dimmeler, S., Heymans, S., Zeller, T., Haas, J., Karakas, M., et al. (2018). Non-coding RNAs in cardiovascular diseases: Diagnostic and therapeutic perspectives. *European Heart Journal*, *39*(29), 2704–2716. https://doi.org/10.1093/eurheartj/ehx165.

Ponting, C. P., Oliver, P. L., & Reik, W. (2009). Evolution and functions of long noncoding RNAs. *Cell*, *136*(4), 629–641. https://doi.org/10.1016/j.cell.2009.02.006.

Prabhu, S. D., & Frangogiannis, N. G. (2016). The biological basis for cardiac repair after myocardial infarction: From inflammation to fibrosis. *Circulation Research*, *119*(1), 91–112. https://doi.org/10.1161/CIRCRESAHA.116.303577.

Qi, H., Shen, J., & Zhou, W. (2020). Up-regulation of long non-coding RNA THRIL in coronary heart disease: Prediction for disease risk, correlation with inflammation, coronary artery stenosis, and major adverse cardiovascular events. *Journal of Clinical Laboratory Analysis*, *34*(5), e23196. https://doi.org/10.1002/jcla.23196.

Qin, M., Li, Q., Wang, Y., Li, T., Gu, Z., Huang, P., et al. (2020). Rutin treats myocardial damage caused by pirarubicin via regulating miR-22-5p-regulated RAP1/ERK signaling pathway. *Journal of Biochemical and Molecular Toxicology*, *35*, e22615. https://doi.org/10.1002/jbt.22615.

Raggi, P., Genest, J., Giles, J. T., Rayner, K. J., Dwivedi, G., Beanlands, R. S., et al. (2018). Role of inflammation in the pathogenesis of atherosclerosis and therapeutic interventions. *Atherosclerosis*, *276*, 98–108. https://doi.org/10.1016/j.atherosclerosis.2018.07.014.

Razak, N. A., Abu, N., Ho, W. Y., Zamberi, N. R., Tan, S. W., Alitheen, N. B., et al. (2019). Cytotoxicity of eupatorin in MCF-7 and MDA-MB-231 human breast cancer cells via cell cycle arrest, anti-angiogenesis and induction of apoptosis. *Scientific Reports*, *9*(1), 1514. https://doi.org/10.1038/s41598-018-37796-w.

Ruan, Z.-B., Wang, F., Bao, T.-T., Yu, Q.-P., Chen, G.-C., & Zhu, L. (2020). Genome-wide analysis of circular RNA expression profiles in patients with atrial fibrillation. *International Journal of Clinical and Experimental Pathology*, *13*(8), 1933–1950.

Safaei, S., Tahmasebi-Birgani, M., Bijanzadeh, M., & Seyedian, S. M. (2020). Increased expression level of long noncoding RNA H19 in plasma of patients with myocardial infarction. *International Journal of Molecular and Cellular Medicine*, *9*(2), 122–129. https://doi.org/10.22088/IJMCM.BUMS.9.2.122.

Salamon, I., Serio, S., Bianco, S., Pagiatakis, C., Crasto, S., Chiariello, A. M., et al. (2020). Divergent transcription of the Nkx2-5 locus generates two enhancer RNAs with opposing functions. *iScience*, *23*(9), 101539. https://doi.org/10.1016/j.isci.2020.101539.

Sharma, A. K., Nelson, M. C., Brandt, J. E., Wessman, M., Mahmud, N., Weller, K. P., et al. (2001). Human CD34(+) stem cells express the hiwi gene, a human homologue of the Drosophila gene piwi. *Blood*, *97*(2), 426–434. https://doi.org/10.1182/blood.v97.2.426.

Shemiakova, T., Ivanova, E., Grechko, A. V., Gerasimova, E. V., Sobenin, I. A., & Orekhov, A. N. (2020). Mitochondrial dysfunction and DNA damage in the context of pathogenesis of atherosclerosis. *Biomedicine*, *8*(6), 166. https://doi.org/10.3390/biomedicines8060166.

Shen, S., Ma, L., Shao, F., Jin, L., & Bian, Z. (2020). Long non-coding RNA (lncRNA) NEAT1 aggravates cerebral ischemia-reperfusion injury by suppressing the inhibitory effect of miR-214 on PTEN. *Medical Science Monitor: International Medical Journal of Experimental and Clinical Research, 26*, e924781. https://doi.org/10.12659/MSM.924781.

Si, X., Zheng, H., Wei, G., Li, M., Li, W., Wang, H., et al. (2020). circRNA Hipk3 induces cardiac regeneration after myocardial infarction in mice by binding to Notch1 and miR-133a. *Molecular Therapy—Nucleic Acids, 21*, 636–655. https://doi.org/10.1016/j.omtn.2020.06.024.

Solomon, S. D., Adams, D., Kristen, A., Grogan, M., Gonzalez-Duarte, A., Maurer, M. S., et al. (2019). Effects of patisiran, an RNA interference therapeutic, on cardiac parameters in patients with hereditary transthyretin-mediated amyloidosis. *Circulation, 139*(4), 431–443. https://doi.org/10.1161/CIRCULATIONAHA.118.035831.

Song, J.-J., Yang, M., Liu, Y., Song, J.-W., Wang, J., Chi, H.-J., et al. (2020). MicroRNA-122 aggravates angiotensin II-mediated apoptosis and autophagy imbalance in rat aortic adventitial fibroblasts via the modulation of SIRT6-elabela-ACE2 signaling. *European Journal of Pharmacology, 883*, 173374. https://doi.org/10.1016/j.ejphar.2020.173374.

Su, B., Wang, X., Sun, Y., Long, M., Zheng, J., Wu, W., et al. (2020). miR-30e-3p promotes cardiomyocyte autophagy and inhibits apoptosis via regulating Egr-1 during ischemia/hypoxia. *BioMed Research International, 2020*, 7231243. https://doi.org/10.1155/2020/7231243.

Sun, T., & Han, X. (2019). The disease-related biological functions of PIWI-interacting RNAs (piRNAs) and underlying molecular mechanisms. *ExRNA, 1*(1), 21. https://doi.org/10.1186/s41544-019-0021-1.

Sun, H., Jiang, Q., Sheng, L., & Cui, K. (2020). Downregulation of lncRNA H19 alleviates atherosclerosis through inducing the apoptosis of vascular smooth muscle cells. *Molecular Medicine Reports, 22*(4), 3095–3102. https://doi.org/10.3892/mmr.2020.11394.

Sun, L., Zhu, W., Zhao, P., Wang, Q., Fan, B., Zhu, Y., et al. (2020). Long noncoding RNA UCA1 from hypoxia-conditioned hMSC-derived exosomes: A novel molecular target for cardioprotection through miR-873-5p/XIAP axis. *Cell Death & Disease, 11*(8), 696. https://doi.org/10.1038/s41419-020-02783-5.

Tang, J., Gao, H., Liu, Y., Song, J., Feng, Y., Wang, G., et al. (2020). Network construction of aberrantly expressed miRNAs and their target mRNAs in ventricular myocardium with ischemia-reperfusion arrhythmias. *Journal of Cardiothoracic Surgery, 15*(1), 216. https://doi.org/10.1186/s13019-020-01262-4.

Tang, Y., Wang, Y., Park, K.-M., Hu, Q., Teoh, J.-P., Broskova, Z., et al. (2015). MicroRNA-150 protects the mouse heart from ischaemic injury by regulating cell death. *Cardiovascular Research, 106*(3), 387–397. https://doi.org/10.1093/cvr/cvv121.

Tang, S., Zhong, H., Xiong, T., Yang, X., Mao, Y., & Wang, D. (2020). MiR-489 aggravates H2O2-induced apoptosis of cardiomyocytes via inhibiting IGF1. *Bioscience Reports, 40*(9), BSR20193995. https://doi.org/10.1042/BSR20193995.

Tao, H., Zhang, M., Yang, J. J., & Shi, K. H. (2018). MicroRNA-21 via dysregulation of WW domain-containing protein 1 regulate atrial fibrosis in atrial fibrillation. *Heart, Lung & Circulation, 27*(1), 104–113. https://doi.org/10.1016/j.hlc.2016.01.022.

Thomson, D. W., & Dinger, M. E. (2016). Endogenous microRNA sponges: Evidence and controversy. *Nature Reviews. Genetics, 17*(5), 272–283. https://doi.org/10.1038/nrg.2016.20.

Trembinski, D. J., Bink, D. I., Theodorou, K., Sommer, J., Fischer, A., van Bergen, A., et al. (2020). Aging-regulated anti-apoptotic long non-coding RNA Sarrah augments recovery from acute myocardial infarction. *Nature Communications*, *11*(1), 2039. https://doi.org/10.1038/s41467-020-15995-2.

Ucar, A., Gupta, S. K., Fiedler, J., Erikci, E., Kardasinski, M., Batkai, S., et al. (2012). The miRNA-212/132 family regulates both cardiac hypertrophy and cardiomyocyte autophagy. *Nature Communications*, *3*, 1078. https://doi.org/10.1038/ncomms2090.

van der Kwast, R., T, C., Parma, L., van der Bent, M. L., van Ingen, E., Baganha, F., et al. (2020). Adenosine-to-inosine editing of vasoactive microRNAs alters their targetome and function in ischemia. *Molecular Therapy—Nucleic Acids*, *21*, 932–953. https://doi.org/10.1016/j.omtn.2020.07.020.

van Rooij, E., Sutherland, L. B., Liu, N., Williams, A. H., McAnally, J., Gerard, R. D., et al. (2006). A signature pattern of stress-responsive microRNAs that can evoke cardiac hypertrophy and heart failure. *Proceedings of the National Academy of Sciences of the United States of America*, *103*(48), 18255–18260. https://doi.org/10.1073/pnas.0608791103.

van Rooij, E., Sutherland, L. B., Qi, X., Richardson, J. A., Hill, J., & Olson, E. N. (2007). Control of stress-dependent cardiac growth and gene expression by a microRNA. *Science*, *316*(5824), 575–579. https://doi.org/10.1126/science.1139089.

Vasudevan, S. (2012). Posttranscriptional upregulation by microRNAs. *Wiley Interdisciplinary Reviews: RNA*, *3*(3), 311–330. https://doi.org/10.1002/wrna.121.

Vaze, A., Tran, K.-V., Tanriverdi, K., Sardana, M., Lessard, D., Donahue, J. K., et al. (2020). Relations between plasma microRNAs, echocardiographic markers of atrial remodeling, and atrial fibrillation: Data from the Framingham Offspring study. *PLoS One*, *15*(8), e0236960. https://doi.org/10.1371/journal.pone.0236960.

Vella, S., Gallo, A., Lo Nigro, A., Galvagno, D., Raffa, G. M., Pilato, M., et al. (2016). PIWI-interacting RNA (piRNA) signatures in human cardiac progenitor cells. *The International Journal of Biochemistry & Cell Biology*, *76*, 1–11.

Virani, S. S., Alonso, A., Benjamin, E. J., Bittencourt, M. S., Callaway, C. W., Carson, A. P., et al. (2020). Heart disease and stroke statistics-2020 update: A report from the American heart association. *Circulation*, *141*(9), e139–e596. https://doi.org/10.1161/CIR.0000000000000757.

Wang, K. C., & Chang, H. Y. (2011). Molecular mechanisms of long noncoding RNAs. *Molecular Cell*, *43*(6), 904–914. https://doi.org/10.1016/j.molcel.2011.08.018.

Wang, S., Ge, L., Zhang, D., Wang, L., Liu, H., Ye, X., et al. (2020). MiR-181c-5p promotes inflammatory response during hypoxia/reoxygenation injury by downregulating protein tyrosine phosphatase nonreceptor type 4 in H9C2 cardiomyocytes. *Oxidative Medicine and Cellular Longevity*, *2020*, 7913418. https://doi.org/10.1155/2020/7913418.

Wang, H., Gong, H., Liu, Y., & Feng, L. (2020). Relationship between lncRNA-Ang362 and prognosis of patients with coronary heart disease after percutaneous coronary intervention. *Bioscience Reports*, *40*(7), BSR20201524. https://doi.org/10.1042/BSR20201524.

Wang, X., Guo, S., Hu, Y., Guo, H., Zhang, X., Yan, Y., et al. (2020). Microarray analysis of long non-coding RNA expression profiles in low high-density lipoprotein cholesterol disease. *Lipids in Health and Disease*, *19*(1), 175. https://doi.org/10.1186/s12944-020-01348-x.

Wang, L., Jiang, P., He, Y., Hu, H., Guo, Y., Liu, X., et al. (2019). A novel mechanism of Smads/miR-675/TGFβR1 axis modulating the proliferation and remodeling of mouse cardiac fibroblasts. *Journal of Cellular Physiology*, *234*(11), 20275–20285. https://doi.org/10.1002/jcp.28628.

Wang, M., Li, J., Cai, J., Cheng, L., Wang, X., Xu, P., et al. (2020). Overexpression of microRNA-16 alleviates atherosclerosis by inhibition of inflammatory pathways. *BioMed Research International*, *2020*, 8504238. https://doi.org/10.1155/2020/8504238.

Wang, Z., Li, C., Sun, X., Li, Z., Li, J., Wang, L., et al. (2020). Hypermethylation of miR-181b in monocytes is associated with coronary artery disease and promotes M1 polarized phenotype via PIAS1-KLF4 axis. *Cardiovascular Diagnosis and Therapy*, *10*(4), 738–751. https://doi.org/10.21037/cdt-20-407.

Wang, X., Lu, Y., Zhu, L., Zhang, H., & Feng, L. (2020). Inhibition of miR-27b regulates lipid metabolism in skeletal muscle of obese rats during hypoxic exercise by increasing PPARγ expression. *Frontiers in Physiology*, *11*, 1090. https://doi.org/10.3389/fphys.2020.01090.

Wang, L.-y., Shen, H., Yang, Q., Min, J., Wang, Q., Xi, W., et al. (2019). LncRNA-LINC00472 contributes to the pathogenesis of atrial fibrillation (Af) by reducing expression of JP2 and RyR2 via miR-24. *Biomedicine & Pharmacotherapy*, *120*, 109364.

Wang, D., Xu, H., Wu, B., Jiang, S., Pan, H., Wang, R., et al. (2019). Long noncoding RNA MALAT1 sponges miR1243p.1/KLF5 to promote pulmonary vascular remodeling and cell cycle progression of pulmonary artery hypertension. *International Journal of Molecular Medicine*, *44*(3), 871–884. https://doi.org/10.3892/ijmm.2019.4256.

Wang, Y., Zhao, R., Shen, C., Liu, W., Yuan, J., Li, C., et al. (2020). Exosomal CircHIPK3 released from hypoxia-induced cardiomyocytes regulates cardiac angiogenesis after myocardial infarction. *Oxidative Medicine and Cellular Longevity*, *2020*, 8418407. https://doi.org/10.1155/2020/8418407.

Wei, H., Cao, C., Wei, X., Meng, M., Wu, B., Meng, L., et al. (2020). Circular RNA circVEGFC accelerates high glucose-induced vascular endothelial cells apoptosis through miR-338-3p/HIF-1α/VEGFA axis. *Aging*, *12*(14), 14365–14375. https://doi.org/10.18632/aging.103478.

Wright, F. S., & Okusa, M. D. (1990). Functional role of tubuloglomerular feedback control of glomerular filtration. *Advances in Nephrology from the Necker Hospital*, *19*, 119–133.

Wu, X., Pan, Y., Fang, Y., Zhang, J., Xie, M., Yang, F., et al. (2020). The biogenesis and functions of piRNAs in human diseases. *Molecular Therapy—Nucleic Acids*, *21*, 108–120. https://doi.org/10.1016/j.omtn.2020.05.023.

Xia, M., Jin, Q., Bendahhou, S., He, Y., Larroque, M. M., Chen, Y., et al. (2005). A Kir2.1 gain-of-function mutation underlies familial atrial fibrillation. *Biochemical and Biophysical Research Communications*, *332*(4), 1012–1019. https://doi.org/10.1016/j.bbrc.2005.05.054.

Xiao, W., Li, X., Ji, C., Shi, J., & Pan, Y. (2020). LncRNA Sox2ot modulates the progression of thoracic aortic aneurysm by regulating miR-330-5p/Myh11. *Bioscience Reports*, *40*(7), BSR20194040. https://doi.org/10.1042/BSR20194040.

Xiao, J., Lu, Y., & Yang, X. (2020). THRIL mediates endothelial progenitor cells autophagy via AKT pathway and FUS. *Molecular Medicine*, *26*(1), 86. https://doi.org/10.1186/s10020-020-00201-2.

Xie, M., Hu, C., Li, D., & Li, S. (2020). MicroRNA-377 alleviates myocardial injury induced by hypoxia/reoxygenation via downregulating LILRB2 expression. *Dose-Response: A Publication of International Hormesis Society*, *18*(2), 1559325820936124. https://doi.org/10.1177/1559325820936124.

Xie, X., Ji, Q., Han, X., Zhang, L., & Li, J. (2020). Knockdown of long non-coding RNA TTTY15 protects cardiomyocytes from hypoxia-induced injury by regulating let-7b/MAPK6 axis. *International Journal of Clinical and Experimental Pathology*, *13*(8), 1951–1961.

Xue, Y., Fan, X., Yang, R., Jiao, Y., & Li, Y. (2020). miR-29b-3p inhibits post-infarct cardiac fibrosis by targeting FOS. *Bioscience Reports*, *40*(9), BSR20201227. https://doi.org/10.1042/BSR20201227.

Yang, L., Lu, Y., Ming, J., Pan, Y., Yu, R., Wu, Y., et al. (2020). SNHG16 accelerates the proliferation of primary cardiomyocytes by targeting miRNA-770-5p. *Experimental and Therapeutic Medicine*, *20*(4), 3221–3227. https://doi.org/10.3892/etm.2020.9083.

Yang, S., Wu, M., Li, X., Zhao, R., Zhao, Y., Liu, L., et al. (2020). Role of endoplasmic reticulum stress in atherosclerosis and its potential as a therapeutic target. *Oxidative Medicine and Cellular Longevity*, *2020*, 9270107. https://doi.org/10.1155/2020/9270107.

Yang, T., Yang, P., Roden, D. M., & Darbar, D. (2010). Novel KCNA5 mutation implicates tyrosine kinase signaling in human atrial fibrillation. *Heart Rhythm*, *7*(9), 1246–1252. https://doi.org/10.1016/j.hrthm.2010.05.032.

Yao, X., Xie, L., & Zeng, Y. (2020). MiR-9 promotes angiogenesis via targeting on sphingosine-1-phosphate receptor 1. *Frontiers in Cell and Development Biology*, *8*, 755. https://doi.org/10.3389/fcell.2020.00755.

Yao, L., Zhou, B., You, L., Hu, H., & Xie, R. (2020). LncRNA MIAT/miR-133a-3p axis regulates atrial fibrillation and atrial fibrillation-induced myocardial fibrosis. *Molecular Biology Reports*, *47*(4), 2605–2617. https://doi.org/10.1007/s11033-020-05347-0.

You, L., Chen, H., Xu, L., & Li, X. (2020). Overexpression of miR-29a-3p suppresses proliferation, migration, and invasion of vascular smooth muscle cells in atherosclerosis via targeting TNFRSF1A. *BioMed Research International*, *2020*, 9627974. https://doi.org/10.1155/2020/9627974.

Zahid, K. R., Raza, U., Chen, J., Raj, U. J., & Gou, D. (2020). Pathobiology of pulmonary artery hypertension: Role of long non-coding RNAs. *Cardiovascular Research*, *116*(12), 1937–1947. https://doi.org/10.1093/cvr/cvaa050.

Zhang, Y., & Bian, Y. (2020). Long non-coding RNA SNHG8 plays a key role in myocardial infarction through affecting hypoxia-induced cardiomyocyte injury. *Medical Science Monitor: International Medical Journal of Experimental and Clinical Research*, *26*, e924016. https://doi.org/10.12659/MSM.924016.

Zhang, G., Dou, L., & Chen, Y. (2020). Association of long-chain non-coding RNA MHRT gene single nucleotide polymorphism with risk and prognosis of chronic heart failure. *Medicine*, *99*(29), e19703. https://doi.org/10.1097/MD.0000000000019703.

Zhang, S.-Y., Huang, S.-H., Gao, S.-X., Wang, Y.-B., Jin, P., & Lu, F.-J. (2019). Upregulation of lncRNA RMRP promotes the activation of cardiac fibroblasts by regulating miR-613. *Molecular Medicine Reports*, *20*(4), 3849–3857. https://doi.org/10.3892/mmr.2019.10634.

Zhang, L., Liu, B., Han, J., Wang, T., & Han, L. (2020). Competing endogenous RNA network analysis for screening inflammation-related long non-coding RNAs for acute ischemic stroke. *Molecular Medicine Reports*, *22*(4), 3081–3094. https://doi.org/10.3892/mmr.2020.11415.

Zhang, S.-J., Yun, C.-J., Liu, J., Yao, S.-Y., Li, Y., Wang, M., et al. (2020). MicroRNA-29a attenuates angiotensin-II induced-left ventricular remodeling by inhibiting collagen, TGF-β and SMAD2/3 expression. *Journal of Geriatric Cardiology*, *17*(2), 96–104. https://doi.org/10.11909/j.issn.1671-5411.2020.02.008.

Zhao, J. B., Zhu, N., Lei, Y. H., Zhang, C. J., & Li, Y. H. (2018). Modulative effects of lncRNA TCONS_00202959 on autonomic neural function and myocardial functions in atrial fibrillation rat model. *European Review for Medical and Pharmacological Sciences*, *22*(24), 8891–8897. https://doi.org/10.26355/eurrev_201812_16658.

Zheng, M.-L., Liu, X.-Y., Han, R.-J., Yuan, W., Sun, K., Zhong, J.-C., et al. (2020). Circulating exosomal long non-coding RNAs in patients with acute myocardial infarction. *Journal of Cellular and Molecular Medicine*, *24*(16), 9388–9396. https://doi.org/10.1111/jcmm.15589.

Zheng, J., Peng, B., Zhang, Y., Ai, F., & Hu, X. (2020). FOXD3-AS1 knockdown suppresses hypoxia-induced cardiomyocyte injury by increasing cell survival and inhibiting apoptosis via upregulating cardioprotective molecule miR-150-5p in vitro. *Frontiers in Pharmacology*, *11*, 1284. https://doi.org/10.3389/fphar.2020.01284.

Zheng, S., Zheng, H., Huang, A., Mai, L., Huang, X., Hu, Y., et al. (2020). Piwi-interacting RNAs play a role in vitamin C-mediated effects on endothelial aging. *International Journal of Medical Sciences*, *17*(7), 946–952. https://doi.org/10.7150/ijms.42586 (Accession No. 32308548).

Zhou, G., Chen, T., & Raj, J. U. (2015). MicroRNAs in pulmonary arterial hypertension. *American Journal of Respiratory Cell and Molecular Biology*, *52*(2), 139–151. https://doi.org/10.1165/rcmb.2014-0166TR.

Zhu, B., Liu, J., Zhao, Y., & Yan, J. (2020). lncRNA-SNHG14 promotes atherosclerosis by regulating RORα expression through sponge miR-19a-3p. *Computational and Mathematical Methods in Medicine*, *2020*, 3128053. https://doi.org/10.1155/2020/3128053.

Zhu, X. G., Zhang, T. N., Wen, R., & Liu, C. F. (*2020*). Overexpression of miR-150-5p alleviates apoptosis in sepsis-induced myocardial depression. *BioMed Research International*, *2020*, 3023186. https://doi.org/10.1155/2020/3023186.

Zong, L., & Wang, W. (2020). CircANXA2 promotes myocardial apoptosis in myocardial ischemia-reperfusion injury via inhibiting miRNA-133 expression. *BioMed Research International*, *2020*, 8590861. https://doi.org/10.1155/2020/8590861.

Printed in the United States
by Baker & Taylor Publisher Services